CANADA'S
DEFENCE

NeW
CANADIAN
READINGS
SERIES EDITOR
J. L. GRANATSTEIN

Titles currently available

CANADA'S DEFENCE

PERSPECTIVES ON POLICY IN THE TWENTIETH CENTURY

Edited by

B.D. Hunt

and

R.G. Haycock
Royal Military College

Copp Clark Pitman Ltd.
A Longman Company
Toronto

ISBN: 0-7730-5258-5

editing: Maral Bablanian
design: Susan Hedley, Liz Nyman
cover design: Rob McPhail
typesetting: Marnie Morrissey, Andrea Weiler
printing and binding: Webcom Limited

Canadian Cataloguing in Publication Data

Main entry under title:
Canada's defence: perspectives on policy in the twentieth century
(New Canadian readings)

Includes bibliographic references.
ISBN 0-7730-5258-5

1. Canada—defences.   2. Canada—Military policy.
I. Hunt, Barry D. (Barry Dennis), 1937–1992
II. Haycock, Ronald Graham, 1942–   III. Series.

FC603.C2 1993   335'.033071   C92–095791–9
F1028.C2 1993

Copp Clark Pitman Ltd.
2775 Matheson Blvd. East
Mississauga, Ontario
L4W 4P7

associated companies: *Longman Group Ltd., London* •
Longman Inc., New York • *Longman Cheshire Pty., Melbourne*
• *Longman Paul Pty., Auckland*

Printed and bound in Canada

1 2 3 4 5   5258-5   97 96 95 94 93

FOREWORD

○

Defence has always been the Cinderella of Canadian government policy and, except in wartime when all the pumpkins have been hurriedly converted into golden carriages (or tanks), Cinderella has remained in the scullery. Ordinarily underfunded, ordinarily the target of budget-cutting finance ministers, defence has been a largely neglected sphere of government.

Scholars, however, have studied defence policy, and as the readings in this volume of *New Canadian Readings* demonstrate, they have done so with skill and insight. Should Canada have its own defence industries? our writers ask. How should the Department of National Defence be organized? Must we be dependent on our great power allies or can we become more independent in our defence policy? And what is the impact of changing military technology, not to mention changes in the world situation, on Canada's Arctic? These are all important, even critical, questions, and if they have not often formed the basis of political discourse, then that says much about the nature of that discourse in Canada.

The editors of *Canada's Defence*, Professors Ron Haycock and Barry Hunt of the Royal Military College, have put together a first-rate group of readings that merits the attention of servicemen and women and students of history and political science.

Barry Hunt, the co-editor of this volume, died suddenly the day after he had delivered the final text of this volume. A fine teacher and scholar at the Royal Military College, his research into naval and strategic policy was important and will endure. He will be sorely missed.

J.L. Granatstein

CONTENTS

o

In memory of Barry Hunt,
scholar, soldier, and friend.

INTRODUCTION

o

> It would be better if we all recognised that problems of Canadian
> defence policy are complex. It doesn't help to dramatize them as
> issues between warmongers and peacemongers. It certainly doesn't
> get us very far to say again that we must stand shoulder to shoul-
> der against a threat which is constant, pay for our ride, and gird
> our loins. It is even less helpful to say that the Cold War is over and
> we can make love rather than war—for which we presumably
> ungird our loins.[1]

Military forces have always played an important part in Canada's history,
most obviously in terms of their responsibilities for defending Canadian territory and interests, but also in terms of our economic and political vitality
and, in no small measure, in defining our national sense of self. In 1867, the
new Dominion's first Defence Minister, George Étienne Cartier, called the
militia "the crown of the edifice of nationhood." And yet, since those beginnings, our military establishments have also often seemed to be the
neglected step-child of public policy, a tolerated but imperfectly understood
and expensive presence. Discussions about the armed forces' size, nature,
and use have rarely occupied stage centre in national politics. Except perhaps as a matter of patronage, "defence" has not been an election winner.
Today, in the aftermath of the Cold War, as the Department of National
Defence comes to grips with some of the most severe spending and organi-
sational upheavals since the Second World War, public discussion of official
thinking remains characteristically muted.

Some twenty years ago, the Opposition defence critic—Michael
Forrestal—in anticipation of a new White Paper, commented in the House
of Commons:

> we see the problem as being brought about by the total absence of a
> clear, well-defined national defence policy. Since the government
> came to power we have had ad hoc, irrelevant and inconsistent
> measures taken, and we still have to hear the minister's grand
> design.[2]

That these sentiments still apply—and have throughout this century—says
a good deal about our defence arrangements and how they are made. But
however much experts or critics may lament the seeming absence of design,
the reality remains that defence is an inherently untidy and circumstance-
specific business, especially for a parliamentary democracy. And yet, with
all that, national defence policies of quite remarkable consistency have been
the norm in Canada.

Canadian writers have offered a number of approaches to our under-standing of what constitutes or determines defence policy in Canada. Dan Middlemiss and Joel Sokolsky, for example, suggest that "policy is essentially concerned with the making of choices." And within the range of factors that appear to shape such choices, they have concentrated their studies on the play of the decision-making processes themselves. Colin Gray looked more to the interplay of international (or strategic) and domestic (or structural) spheres, and within that context defined Canadian policy as:

> the capabilities and actions financed through the annual (and, occasionally, supplemental) budget of the Department of National Defence, in conjunction with the authoritative declarations made concerning both Canadian defence and Canadian defence strategy.[3]

Gray's concentration on capabilities, declarations, and actions, and Middlemiss-Sokolsky's on process are complementary approaches. But as the latter also point out, non-decisions or "decisions to do nothing" have also been important in Canadian experience.

> This concept is important, for what often passes as a policy vacuum in the Canadian context—for example the failure to produce a promised review of Canadian defence policy—may be the result of a conscious decision to take no action. Options may have been examined and proposals suggested at the bureaucratic level, but decisions may have been deferred or rejected at the higher political level of government because of their political, military, or economic ramifications. Inaction therefore, is not necessarily synonymous with the absence of decision.[4]

As to environmental forces that have influenced choice making, C.P. Stacey early on underscored the powerful impact of cumulative historical experience and Canada's geopolitical location relative to other countries.[5] R.J. Sutherland, in an important 1962 article, "Canada's Long-Term Strategic Situation," extended Stacey's arguments, adding to his list of "invariants," Canada's economic strength and resources, natural political alignments based on ties of culture and common interests, and the need to balance its European and American links in line with its own objectives and survival as an independent nation. How much latitude Canada's political leaders have enjoyed, or chosen to exercise, in this latter regard especially, can be disputed. Colin Gray, while similarly stressing the role of history in colouring political attitudes, has also noted the effects of previous decisions, such as in the choice of armed forces' equipment, upon future options. Indeed, he asserts that Canada's defence policy at a given point can be easily discerned by looking at existing equipment inventories and capabilities.[6]

The point is that defence policy has been approached from several complementary and even competing perspectives by Canadian experts. Some have focussed on domestic determinants and the interplay of individuals and groups at various levels within the political leadership, the defence bureaucracy, the media, and other public opinion organs. Others have high-lighted the external or international dimensions of the equation. The latter is a natural or perhaps conventional approach. It reflects the university community's preoccupations with such fashionable fields as "strategic studies" that generally have viewed defence choices as being virtually dictated by external factors.

This volume is intended to introduce students to the main features of Canada's defence stances and to some of the explanations of their evolution. The articles that follow give a broad overview of developments mostly since 1904; from the Laurier government's initiatives—by which the militia effectively came of age and Canada assumed responsibility from Britain for its own security—through the World and Cold Wars, to the present. This perspective highlights policy constants over the long haul and gives a balanced sense of the interplay between strategic, structural, and other factors that have been operative at specific points in the story. The intention has not been to advance a specific thesis as to the independence, uniqueness, or otherwise of particular government positions, or the relative importance of domestic versus external considerations. It does suggest that Canada's defence policy has been based on decisions taken mainly by Canada's own leaders; and that, even within the bounds that constrain the ambitions of all twentieth-century states, Canada's latitude for choice has been and remains considerable.

Our desire to make available in a short volume some of the best and more recent writing on Canadian defence policy unfortunately has meant that some important, though essentially internal, administrative, or social policy issues could not be included. References to useful sources in these areas can be found in the notes and in the "Further Reading" section.

The articles that have been included offer several leads as to how and why a nation of such reputedly "unmilitary" instincts has, through the decades of the two World Wars (Section 1) and the latter half-century dominated by the Cold War (Section 2), managed the intricate balancing act of defending itself and its interests in an always changing and frequently hostile world. "A country that cannot be defended and can hardly be attacked may be a delight to its population," Professor Desmond Morton has written, "but it offers its admirals and generals some peculiar problems."[7] How peculiar or characteristically Canadian those problems and their solutions have been is the subject of what follows. Already, some of the assumptions and forecasts contained in the later chapters are being overtaken by recent developments that suggest that the post-Cold War security environment could become even more volatile and uncertain. "Updating" those chapters has not been possible, and perhaps they have become historical documents in their own right. But they nonetheless reinforce the foundations of understanding so crucial to intelligent choice making in a future that threatens to remain decidedly peculiar.

NOTES

1. John W. Holmes, *Canada: A Middle-Aged Power* (Toronto, 1976), 85.

2. Colin S. Gray, *Canadian Defence Priorities: A Question of Relevance* (Toronto, 1972), 2.

3. Ibid., 9–10.

4. D.W. Middlemiss and J.J. Sokolsky, *Canadian Defence: Decisions and Determinants* (Toronto, 1989), 5.

5. C.P. Stacey, *The Military Problems of Canada: A Survey of Present Defence Policies and Strategic Conditions Past and Present* (Toronto, 1940), 1–53.

6. Gray, *Canadian Defence Priorities,* 29–50.

7. Desmond Morton, "The Military Problems of an Unmilitary People," *Revue Internationale d'Histoire Militaire,* no. 51 (1982), 7.

section

1

1904–1945

Over the half-century that included the two World Wars, Canada's defence establishment was utterly transformed from its fledgling militia beginnings into three regular services whose rapid expansion and capabilities contributed vitally to Allied success. The policies that guided this dramatic metamorphosis were driven largely by forces that lay beyond the young Dominion's direct control. Yet even before the turn of the twentieth century—as the American threat had declined, the indefensibility of the border had been admitted, and internal priorities grown larger—British statesmen and advisors could do little to change the Canadian agenda of minimal defence spending. Only in 1904, through the vehicle of a new Militia Act, were Imperial interests that might challenge those of Ottawa placed in realistic balance, and Canada assumed *de facto* control of its defence decisions.

The important reforms of 1904 were the responsibility of Sir Frederick Borden, arguably one of Canada's most energetic and able defence ministers. As Carman Miller demonstrates, Borden's impact was almost immediate in terms of increased funding and training. His longer term objectives included: the creation of a self-contained national army, greater autonomy from Britain in military matters, and self-determined co-operation, rather than integration, in imperial defence arrangements.

Other dimensions of this relationship were laid bare by the controversial decision to create a Canadian Navy. Barry Gough describes the strategic dilemmas of the British who, in meeting Germany's naval challenge, decided upon reform and redistribution policies that included withdrawing the Royal Navy from North American waters. By 1909, with little prior consultation, Canada was left without a standing naval presence and, at the same time, had to deal with calls for contributions to offset Germany's naval building programmes. Prime Minister Laurier's decision in 1910 for a separate Canadian Naval Service satisfied neither strategic nor political needs and contributed to his government's downfall the following year.

Although these developments were overshadowed by the outbreak of the Great War in 1914, they nonetheless continued to affect the ways in which Canada's leaders opted for a fighting role in Flanders, and how their forces were commanded, used, and sustained. By 1917, the dreadful casualty lists and the collapse of voluntary enlistment made conscription a critical political issue. The painful divisions it exposed then (and again in World War II) have lain close to the surface of every subsequent government's defence thinking. The Great War's human and other costs had constitutional, economic, and social ramifications that permanently altered Canadians' views of themselves and their place in the world.

In "Junior but Sovereign Allies," Desmond Morton traces the Canadian Expeditionary Force's transformation from a hastily improvised formation, dependent on British equipment, training, and leadership, into a professionally competent force whose distinguished fighting record directly underpinned Canada's changed status as a national entity and alliance partner. Wartime growth in industrial and administrative efficiency did not, however, extend to all aspects of Canada's war effort. This was especially so with respect to developing supply sources. Ronald Haycock demonstrates that a "munitions dilemma" which had persisted since Confederation, remained unresolved beyond the war.

After 1918, the forces faced new and frustrating struggles for survival. Stephen Harris, concentrating on the Army's part in interwar developments and the creation of a single Ministry of Defence in 1923, examines the difficulties of developing positions that politicians and soldiers alike could accept. Norman Hillmer shows how Commonwealth ties continued to shape those debates in the 1930s and the extent to which, despite continuing governmental wariness of formal imperial commitments, unofficial connections—especially within the military through personal associations, and shared equipment, training, and intelligence—continued to colour the thinking of defence experts through the outbreak of another World War in 1939.

Somewhat surprisingly, perhaps, Canada's massive contributions in World War II did not gain it a powerful voice in the strategic councils of the Allies. In fact, as Adrian Preston argues, Mackenzie King consciously avoided involvement at that level. Even after Dunkirk had made Canada Britain's ranking military partner, the severity of the crisis itself reinforced King's instinct to keep his distance. Thereafter, as the influence of the American military became more pervasive, Canada's leaders accepted relegation to the sidelines in coalition policy making. In the area of munitions production, however, government intervention was decidedly more pronounced. Robert Bothwell examines the successes of government and private business in managing wartime industrial development.

J.L. Granatstein's contribution explores the implications of closer Canada–US defence relations during the War and its aftermath. By 1963, he argues, the two militaries were so closely interconnected that senior Canadian officers sometimes placed their allies' concerns ahead of those of their own political leaders. This suggests interesting parallels with Hillmer's account of Anglo-Canadian connections in the 1930s.

SIR FREDERICK WILLIAM BORDEN AND MILITARY REFORM, 1896-1911 ⬧

CARMAN MILLER

o

. . . Defence is a peculiar portfolio: in peace time it may be relegated to a position of relative insignificance, but during war or the threat of war it becomes a post of responsibility and influence. Twice during Borden's tenure of office, the Canadian government resorted to the use or threat of military force. The Yukon Gold Rush Expedition and the South African Contingents, especially the latter, gave the Militia Department a position of primary importance. When peace returned the department failed to sink into insignificance. Nor indeed could it: for the years between 1896 and 1914 witnessed events which the militia could scarcely ignore. At home, the Venezuela scare and the fear of lawlessness and American intrusion on the Yukon–Alaska border emphasized the need for self-defence. Abroad, the South African War and the growing German menace forced the divisive question of imperial military obligation. Twice departmental conflicts troubled the waters of national politics. The constitutional implications of the heated Hutton affair, for a time, threatened the life of Laurier's first administration. The Dundonald dispute merged into the election of 1904. Both incidents placed Borden and his department at the centre of a national political storm.

In 1871 the British regulars withdrew from Canada leaving Canadian defence to the militia. Canadian defence meant both defence against external aggression and internal defence against rebellion or riots. The militia constituted two units, the volunteers and a small permanent force. The volunteers, a paper force, consisted of consenting civilians who agreed to train sixteen days a year for three years. During sixteen days of annual camp the permanent militia trained the volunteers to form the first line of Canadian

---

⬧ Excerpted from *Canadian Historical Review*, 50, 3 (Sept. 1969), 265–84.

defence. Both the permanent and volunteer militia were divided into the usual three fighting arms of infantry, artillery, and cavalry.

To measure Borden's contribution to military reform, it is important to understand the state of the militia before 1896. Before this date, and for some time after, the militia was a glorified police force. It was unprepared for war: the infantry, artillery, and cavalry, the fighting arms of the militia, were divided into corps of unplanned and unequal strengths; there was no service corps to transport, clothe, and feed a fighting field force; and none of the three fighting arms, at this time, was equipped with adequate weapons. Training was haphazard. During lean years only a third of the volunteer force trained, and in still leaner years, training was dropped altogether. A volunteer could, under this system, complete his three-year contract with no training whatever. This was particularly true of rural corps whose only training opportunity was at annual camp. Moreover, "saturday night" officers, who attained their rank through political or social rewards, commanded many of the local corps. In such a desolate condition the militia was a fertile field for the hand of a reforming minister.

Two decisions, taken in the first years of his administration, demonstrated Borden's determination to fill the role of a military reformer. With all the enthusiasm of a new and untried administrator, Borden asserted his will. He rigidly insisted on an annual drill, and he reformed his headquarters staff. Although no supplies had been voted for an annual drill in 1896, Borden refused the advice of his general officer commanding that drill ought to be dropped for one year.[1] Instead, he rushed his estimates through the House in record time, and had the men in camp before autumn. Both sides of the House supported his decision, and soldiers applauded it. To C. F. Hamilton, an early Canadian military historian, this decision marked the end of the long period of "cool indifference" towards the militia.[2] More profound reforms followed.

Reform of the headquarters staff was more difficult. Here a thorough house-cleaning was long overdue. During his first year and a half in office, Borden superannuated forty-two men, and retired fourteen others on gratuities. One of the retired officers was eighty-one, evidently past his more productive years.[3] Permanent militiamen replaced many of the retired civilians. Borden's purpose was to teach soldiers military administration and to reduce expenditure, since soldiers received lower salaries than civilian employees. He abolished some of the positions or transferred their light tasks to other departments. These retirements were much more than the usual political "changing of the guard"; they were notice of a larger programme of re-organization and reform. . . .

Borden's larger programme of military reform reveals three broad principles of action: to create a self-contained citizen army, to win a larger degree of Canadian military autonomy from Great Britain, and to establish a co-operative policy of imperial defence. Slowly but systematically Borden pushed forward this programme, often in the face of considerable opposition, much of it from within his own party.

To create a self-contained citizen army, Borden followed closely the reports of two of his very capable military advisers, General E. T. H. Hutton and Lord Dundonald, as well as the recommendations of the Canada Defence Committee and the Esher Report. Borden began his campaign for a self-contained army by creating service corps. First on his list of priorities was a Medical Corps, which he authorized in 1898. As a former medical officer Borden took a particular interest in this unit. Soon others followed. In 1903 he established the Army Service Corps, to organize and administer the transportation of food, forage, and military stores, an event which the enthusiastic historian of the corps has described as the beginning of a modern Canadian army.[4] In the same year, the Minister authorized a Corps of Engineers and an Ordnance Corps. In time others were added: a Corps of Signals, a Pay Corps, and a Corps of Guides. The Corps of Guides, part of an intelligence department, collected military information and acted as guides in their military districts. This distinctively Canadian corps, created to meet the needs of a large, often unmapped area, invariably won the high commendation of imperial inspectors general.[5] These service corps provided the Canadian militia with the organizational machinery to maintain an army in the field. Borden deserves much of the credit for the creation of the structure of a self-contained army.

Next, Borden proposed to establish a citizen army; but, his greatest obstacle to this project was the indifference, complacency, and often, the hostility of the Canadian people and government. As the threat of invasion from the south vanished, Canadians, an unmilitary people at best, drifted on a wave of unprecedented prosperity, shielded from any possible aggressor by the wide Atlantic, the Monroe doctrine, and the British navy. The flood-tide of Laurier prosperity carried all before it. Men and resources were in short supply and were desperately needed to open the West. European immigrants had turned their backs on Old World militarism, and willingly sank their interests and energies into working the new land. . . . Few Canadians joined the militia, for the pay was low, and men were scarce.

The South African War had a profound effect on the Canadian militia. After the war, the concept of a nation-in-arms, or a citizen army, replaced the barrack formula for professional military training. In South Africa, Canadians penetrated the proud boasts of former general officers commanding; the British regulars, for all their meticulous drill and display, were not invincible.[6] Canadian soldiers were impressed by the mobility, initiative, and resourcefulness of their enemy, the Boer. Intoxicated with their own success against this enemy, Canadian soldiers returned confident of their own capacity for self-defence. They believed that in the past there had been "a little too much parade, pipe clay, and high collar business."[7] They scoffed at elaborate military organization and preparation. Even men sympathetic to the needs of the militia felt that "providing you have the courage . . . the intelligence . . . the initiative . . . you have . . . the nucleus of the best army in the world."[8] This mentality bred the idea that every man was a potential soldier, and from this idea was born the concept of a citizen army.

Borden determined to use this enthusiasm for a less organized military system to create his citizen army. After a close study of various national military structures, Borden favoured the Swiss. Although he had outlined his citizen army plan before the arrival of Lord Dundonald,[9] the new general officer commanding was of great assistance. Lord Dundonald welcomed the opportunity to introduce a scheme, which he had advocated for Great Britain,[10] into Canada. The plan required a first line of defence composed of 100 000 sharp shooters, two-thirds of whom were to be non-uniformed civilian men, and in some cases women.[11] The remaining one-third were required to attend annual camp for instruction in the organizational, tactical, and administrative problems of warfare. To encourage these hosts of riflemen, the government constructed rifle ranges across the country. Government money grants, rifles, and ammunition were distributed liberally to privately constructed ranges. The Department of Militia believed that during war this plan was capable of expansion into a second line of defence consisting of another 100 000 men. This was Borden's citizen army.

Borden's plan, predicated solely on Canadian defence, met some strong opposition. Imperialists pointed to its failure to prepare an expeditionary force, and its organizational weaknesses. In spite of the plan's weaknesses, it solved a number of difficult problems. It met the charge that Canada was inadequately defended. At the same time, it was inexpensive: fewer uniforms and military camps were needed. Moreover, men were not required to absent themselves from civilian positions at a time when national expansion placed a premium on both. Borden's plan promised the maximum defence at a minimum cost to the country and to individuals.

Borden's second principle of action was Canadian military autonomy. To achieve this end, he revised extensively the Militia Act to sweep away the inequalities between Canadian militiamen and their British equivalents. For example, he altered the qualifications for the appointment of the general officer commanding the Canadian militia, in face of strong opposition from the General Officer Commanding and the Governor-General.[12] Prior to the revised Militia Act of 1904, only officers of "Her Majesty's regular army" were eligible for the highest post in the Canadian militia. Borden, as well as many other Canadians, considered this regulation an affront to the ambitions of competent officers serving the Canadian militia. The revised act abolished discrimination against the appointment of Canadian militia officers to this post. The revised act removed other disabilities. Previous to 1904, British regular army officers serving in Canada, of equal rank to Canadian militia officers but junior in date of appointment, took precedence over Canadian officers for promotion and command. This Sir Frederick Borden regarded as a distinct injustice to Canadians. The revised Militia Act brought this practice to an end.

The new Militia Act solved another irritating problem: the relationship between the officer commanding the British garrison at Halifax and the general officer commanding the Canadian militia. Under the old law, in time of war, the British officer at Halifax, regardless of rank, commanded all British and Canadian troops in Canada. This law had provoked a number of contentious quarrels between the two British officers over their respective peacetime roles. The most recent feud occurred during the Duke of York's visit to Halifax in 1901, in which the Canadian government supported the officer commanding the Canadian militia. The new act simply stated that the general officer commanding the Canadian militia was to command Canadian troops without distinction between peace and war. In this way, Canadian men would be led by a Canadian officer in war as well as in peace.

Sir Frederick extended his policy of autonomy to supply, this time with less conspicuous success. He argued that Canada ought to possess an independent line of supply on the sound hypothesis that Britain might, even for a short duration, lose control of the sea, thus seriously weakening Canadian defence. This was not a new policy; Borden merely extended an old principle. As early as 1884 the Canadian government established the Dominion Cartridge Factory in Quebec City to assure an independent supply of small arms ammunition for the Canadian militia. The government followed a similar policy in ordering military uniforms. Besides its military soundness, this policy had certain political advantages!

Borden's independent rifle policy is of more questionable success. The story of the Ross rifle is long, complex, and too involved to discuss in this paper.[13] . . .

Perhaps a more constructive example of Sir Frederick's approach to Canadian military autonomy was his campaign to take control of the British garrisons at Halifax and Esquimalt, the only two remaining outposts of British troops on Canadian soil. In this instance Borden's actions were motivated more by a desire to relieve British troops of the responsibility of Canadian defence than by a colonial desire for freedom.[14] Consistently opposed by the anti-militarists, and imperialists like Lord Minto, the Canadian Governor-General, who instructed the War Office never to consent "to give up an iota of any little Imperial control we possess,"[15] Borden persisted. After years of prolonged negotiations with the imperial government and his own colleagues, Borden won his point. By 1907 the last British regulars left Canadian soil. Many Conservatives remained skeptical of this innovation; other people greeted this transfer with pride. . . .

The most conspicuous example of Borden's stubborn insistence on Canadian autonomy was his struggle over the nature and role of the proposed Imperial General Staff. Neither the arguments of the War Office nor those of his trusted military adviser, Sir Percy Lake, moved Borden. Firmly he insisted that Canada must create its own General Staff,[16] from which he would select officers to form the Canadian section of a larger Imperial General Staff. At all times Canadian staff officers were responsible to the Canadian government through his Department of Militia.[17] His opposition to imperial integration caused great consternation in both the Colonial and War Offices, but through argument and a refusal to be moved, Borden won his point.[18]

Borden's third rule of political action was imperial co-operation. Opposed to any scheme of imperial integration, his "tender sense of 'amour propre'" never prevented his participation in serviceable plans of imperial co-operation. Neither a self-conscious, resentful colonial, chafing under the baneful influence of the "mother country," nor an unquestioning anglophile,[19]

Borden's imperialism was more pragmatic than ideological. Like many hard-headed businessmen, Borden believed that the empire must justify itself on the base of its utility. For him autonomy and imperialism were complimentary, not contradictory, terms. To Borden the clash of imperial obligation with Canadian interest posed no problem; Canadian interests always claimed his prior allegiance. Co-operation not integration, utility not ideology, were the components of Borden's imperialism.

Borden demonstrated clearly his sense of imperial obligation during the cabinet crisis preceding the first despatch of Canadian troops to South Africa. In the cabinet he led the faction pressing for a full Canadian contingent, equipped, transported, and paid by the Canadian government. He lost this battle and was disappointed, for he realized that if his country was to continue to accept the promise of British protection, or to request a voice in Imperial councils, Canada must pay its way. But his sense of imperial obligation never required an unquestioning submission to imperial judgement. During the same cabinet crisis, Borden fought the War Office decision to divide the Canadian contingents into units of 125 men and to incorporate them into British corps. He wanted to keep the Canadians together in one contingent and to have them commanded, as far as possible, by Canadian officers. His refusal to accept both the War Office's suggestion and the Canadian cabinet's endorsement of this policy ended in triumph when he and Lord Minto, the governor-general, persuaded the War Office to reverse the earlier decision. Thanks largely to Borden's persistence, when the first Canadian contingent sailed from Quebec for South Africa, it sailed as one contingent commanded by Canadian officers.

Often Borden defended Canadian interests against what he considered excessive imperial demands. Nothing illuminates this fact more clearly than his reaction to the special terms of recruitment of the South African Constabulary in Canada. To attract the best cavalrymen and crack-shots to this corps from the Canadian West, the British government offered land grants to men who joined the constabulary and settled in South Africa. In spite of Borden's enthusiasm for Canadian participation in the war in South Africa, he deplored the Canadian cabinet's acceptance of the terms of recruitment. In a critical letter to the prime minister, he informed Sir Wilfrid Laurier that he saw no logic in spending millions of dollars attracting settlers to the Canadian West, if these same men were now encouraged to settle in South Africa.[20] Borden lost this case, yet his protest illustrated his keen understanding of Canadian interests even when in conflict with imperial demands.

Sometimes Borden clashed with his colleagues over Canada's imperial military policy. His participation on the Committee of Imperial Defence is a case in point. In November, 1903, Borden went to London to discuss proposed changes in the Militia Act with the War Office. While in London he accepted a place on the Imperial Defence Committee, the first "colonial" to be so honoured! During the committee's deliberations, Sir Frederick agreed to press the Canadian government to take over the Halifax and Esquimalt garrisons and provide a Canadian regiment for imperial service in India.[21]

Borden thought that a Canadian regiment in India had several advantages. It met the imperial government's perennial complaint that Canada failed to contribute towards imperial defence, and it provided training for Canadian soldiers at no cost to the Canadian government since the Indian government was responsible for the expense of the troops.[22]

Borden's London commitment created a storm of criticism, mostly from members of his own party. His place on the defence committee, which he and many others regarded as an honour, turned into a source of embarrassment. Even Laurier withheld his approval of Sir Frederick's actions. At first Borden tried to explain away the importance of the committee and, finally, he withdrew his membership. Faced with organized hostility led by Senator Dandurand,[23] Sir Frederick instructed Alfred Lyttleton, the colonial secretary, to strike from the minutes any reference to the imperial garrisons in Canada and the proposed regiment for India.[24] Borden's plan for imperial military co-operation found few friends in Laurier's cabinet.

His pragmatic approach to empire is again demonstrated by his introduction into Canada of many of the recommendations of the Esher Report, and his acceptance of an imperial plan for the despatch of an overseas force. The Esher Report stressed decentralization of command and reinforcement of civilian control through the creation of an army council. Borden considered both suggestions even more important to Canada than to Great Britain, and hastened to implement them. Since the Hutton affair, Borden sought a solution to the seemingly perennial conflict of jurisdiction between the government and the British general officer commanding the Canadian militia. The Esher Report's suggestion of a militia council seemed the answer to this troublesome question. In 1904 Borden had just emerged from a public dispute with another meddlesome general, Lord Dundonald. The dismissal of Dundonald provided the opportunity to replace the general with a militia council, composed of four military men and three civilians.[25] The Militia Council had several advantages. It served as a forum for the exchange of military and civilian opinions; it was also a source of information for the minister, who acted as chairman of the council. If an inexperienced minister came to the department, such as Borden in 1896, the council provided continuity and records for the minister's instruction.

Decentralization was the second recommendation of the Esher Report which Borden used to reorganize the Canadian militia. Before 1905 Canada was divided into twelve military districts, commanded by a district officer, with limited powers. Borden consolidated these twelve districts into five larger regional commands with extended powers of initiative and command. To reduce the cumbersome, top-heavy headquarters staff in Ottawa, stores and supplies were delegated to the new regional command centres. In this way Borden intended to train senior officers to assume responsibilities such as would devolve upon them in time of war. With the new system, regional commands could mobilize more efficiently and speedily without awaiting detailed instructions from Ottawa.

Sir Frederick's last major policy decision was the acceptance of a plan for the despatch of an overseas Canadian contingent. This decision has been called one of the most surprising of his whole career, because it contra-dicted many of his previous deeds and actions, particularly Borden's stand against the creation of a special force of colonial reserves for imperial ser-vice.[26] To argue this is to miss the distinction between imperial integration, which Borden opposed and imperial co-operation, of which he approved. Borden's plan was for an overseas contingent with no binding commitment on Canada to participate in any or all imperial wars;[27] it was a necessary emergency blueprint. The precedent for Borden's actions might be found in his authorization of a similar plan some months before the outbreak of the South African War. As an experienced military man Borden was aware of the German menace. As a practical politician he probably understood that Canada would participate in any major imperial conflict. And as minister of militia it was his duty to have a co-ordinated, emergency plan of action. Borden's acceptance of this plan was no death-bed conversion to imperial-ism; it was the result of fifteen years of close association with the problems of autonomy, defence, and empire. It was a practical answer to a very prac-tical question. As such it bears the hallmark of the Minister of Militia.

When Sir Frederick, personally defeated and his party out of office, resigned the department he had held for fifteen years,[28] he could look back upon his career with merited pride. During his administration military education advanced: the Royal Military College was reformed, the skeleton of a general staff was created, schools of instruction were established, senior officers were sent to Britain for advanced training, and examinations were required for promotion to senior posts. Military expenditure more than tripled between 1896 and 1911.[29] Decentralization, rules regulating tenure of command, increased pay, and retirement benefits made the mili-tia more attractive to both officers and men. Borden had successfully insisted on an annual drill; he had equipped the militia with modern weapons, and had acquired a large central training camp at Petawawa.[30] It was from this camp in 1909 that Borden authorized the first military test flight of an airplane in Canada. Above all, he had implemented his three-point programme: to create a self-contained, citizen army; to win a larger degree of Canadian military autonomy; and to establish a co-operative pol-icy of imperial defence. . . .

NOTES

1. N.Y. Gascoigne to F.W. Borden, 14 July 1896, Borden Papers, Public Archives of Canada (hereafter PAC).

2. C.F. Hamilton, "The Canadian Militia: The Beginning of Reform," *Canadian Defence Quarterly* (April 1930), 338.

3. Canada, House of Commons, *Debates*, 8 April 1897, 667.

4. Arnold Warren, *Wait For The Waggon* (Toronto, 1961), 26.

5. Ian Hamilton, *Report On The Military Institutions of Canada* (Ottawa, 1913), 22; J.D.P. French, "Report," *Sessional Papers*, 1911, vol. 45, no. 35a, 16.

6. Criticism of British officers by Canadian troops reached fairly seri-ous proportions, to the point that

Lord Minto felt obliged to publicly caution returned soldiers to restrain "their criticisms of their superior officers." *Mail and Empire*, 6 November 1900. One is reminded of Sam Hughes' colourful, but apt, description of British military tactics on one occasion: ". . . but it was another of those very beautiful incidents in Britain's wars, heroic, but damnably foolish. . . ." *Ottawa Citizen*, 14 April 1900.

7. *Debates*, 10 April 1902, 2534.

8. Ibid., 26 June 1900, 8293.

9. Dundonald arrived in Canada in July 1902. Borden first mentioned the possibility of a Swiss type citizen army for Canada two years earlier. *Debates*, 25 June 1900, 8239. Three months before Dundonald arrived, he outlined plans for the citizen army. *Daily Telegraph*, 12 April 1902.

10. Earl of Dundonald, "Notes on a Citizen Army," *The Fortnightly Review* (October 1905).

11. *Canadian Annual Review* (Toronto, 1909), 284.

12. Minto to Landsdowne, 27 December 1903, Minto Papers, PAC; also, Minto to Broderick, 10 April 1903.

13. For a detailed discussion of the Ross rifle's history, see A. Fortesque Duguid, *History of the Canadian Forces 1914–19* (Ottawa, 1938), appendix 3.

14. F.W. Borden to A.G. Jones, 4 December 1902, Borden Papers, PAC.

15. Minto to Broderick, 10 April 1903, Minto Papers, PAC.

16. Public Record Office (hereafter PRO), WO 32/452/33; see also, A.M. John Hyatt, "Arthur Currie: A Military Biography," (unpublished Ph.D. thesis, Duke University, 1964), chap. 1.

17. Maurice Olliver, *Colonial and Imperial Conferences* (Ottawa, 1954), 1: 244–56. Richard Jebb, *The Imperial Conferences* 2 vols. (London, 1911), 1: 150–51, 2: 141–42. Howard D'Egville, *Imperial Defence and Closer Union* (London, 1913), 150–51.

18. Donald C. Gordon, *The Dominion Partnership in Imperial Defence, 1870–1914* (Baltimore: Johns Hopkins University Press, 1965), 167; PRO, Cd. 5323 (1907), 100.

19. Borden's imperialism could be brutally pragmatic, as evidenced by his reported comment to Hutton during the dark December Days of the South African War: "He (Borden) had come to ask himself . . . in face of the reverses which the British Army had received, if it is worth while for Canada to remain part of the British Empire." Hutton to Minto, 10 January 1902, Hutton Papers, British Museum (hereafter BM), vol. 4 (50,081), 78. Also quoted in John Buchan, *Lord Minto: A Memoir* (London, 1924), 149.

20. F.W. Borden to W. Laurier, 21 January 1901, Borden Papers, PAC. A year and a half later Borden made the same point, that Canada should not be a recruiting ground for the British army. *Globe* (Toronto), 15 October 1902.

21. Arnold Forster Diary, 8 December 1903, Arnold Forster Papers, BM, contains a long, interesting summary of the Borden-Forster pre-committee meeting.

22. PRO, Cab. 38/3, no. 82.

23. F.W. Borden to R. Dandurand, 2 January 1904, Borden Papers, PAC. A corresponding letter would, no doubt, be found in the PAC, Dandurand, Papers, dated 30 December 1903. Henri Bourassa too, joined the campaign against Sir Frederick's actions. *Montreal Star*, 22 February 1904.

24. F.W. Borden to Alfred Lyttelton, 1 March 1904, Borden Papers, PAC.

25. The editor, "The Militia Council," *The Canadian Magazine* (May, 1905).

26. Gordon, *The Dominion Partnership*, 276.

27. The British government was "well aware . . . that no guarantee that contingents of any given strength or composition will be forth coming . . . in

the event of a great war." PRO, Cd., 4475 (1909), 7.

28. Borden's popularity with militia officers lasted until his death. His resignation in 1911 evoked letters of regret from many of the officers: P. Lake to F.W. Borden, 5 October 1911 and F.W. Borden to A.H. Macdonald, 22 December 1911, Borden Papers, PAC. The Militia Council presented him with a gift and a warm letter of thanks for his long service to them and to the militia.

29. Increased military supplies did not come easily from the Laurier government. Twice Borden threatened to resign rather than to accept reductions in his estimates. F.W. Borden to Laurier, 19 November 1908, ibid.; F.W. Borden to W.S. Fielding, 8 October 1902, ibid.; Minto to Broderick, 23 November 1902, Minto Papers, PAC. Once, in the House of Commons, he appealed the government's decision to maintain militia expenditures at par. *Debates,* 10 April 1902, 2513. Borden's strongest support for increased military expenditures came from the Opposition, a fact Borden readily acknowledged. *Montreal Gazette,* 15 February 1901.

30. J.D.P. French considered the establishment of Camp Petawawa "the most important step which has yet been taken towards securing the efficiency of the troops in war." J.D.P. French, "Report", 13.

THE END OF PAX BRITANNICA AND THE ORIGINS OF THE ROYAL CANADIAN NAVY: SHIFTING STRATEGIC DEMANDS OF AN EMPIRE AT SEA ✧

o

In the same critical year that the Royal Canadian Navy came into existence, 1910, its mother service, the Royal Navy still possessed an uncontested preponderance at sea. Great Britain and the Empire held the trident of the seas. That strangely curious interlude in world affairs known as the Pax Britannica was drawing to a close. On nearly every sea and ocean Britain possessed coaling stations and bases, the anchors of empire so to speak. On nearly every coastline she boasted cruisers or gunboats enforcing her mandate of the freedom of the seas. And even if those overseas fleets had been reduced rather savagely in the course of the previous few years, reorganized cruiser squadrons still showed the pirates, the slave traders, and Britain's testy rivals that the United Kingdom and the British Empire intended to keep the peace of the seas. . . .

Why, if Britain remained paramount at sea, did Canada create a separate naval organization? The answer to that is complex and is in keeping with the responsive characteristics of Canadian naval traditions. Here I should like to explore the dimensions of this topic, review the historical studies that have centered on the theme, re-examine the immediate causes in international affairs that induced a Canadian response, and make a distinct appeal that we not only see Canadian naval origins as a search for autonomy in relation to Great Britain and the British Empire but that we also regard them as an acceptance of new obligations in international affairs.

✧ From W.A.B. Douglas, ed. *The R.C.N. in Transition, 1910–1985* (Vancouver: University of British Columbia Press, 1988), 90–102.

The creation of a Canadian Naval Service, the immediate forerunner of the Royal Canadian Navy, constituted a significant step in Canada's search for Dominion status. Yet it was not so much an act of independence as one of co-operation with the Admiralty on terms agreeable to Canada. Put differently, the continuities of Anglo-Canadian naval policy, not the seemingly apparent departures in control from London to Ottawa merit our appreciation. Canada's new authority for a naval service was a logical outgrowth of the imperial-colonial relationship, one which had a parallel in the quest for self-government and for control of the Canadian Militia. The foreign policy of the Empire being indivisible, the question was not whether Canada would fight against an enemy but to what degree and under whose control should that contribution in men, guns, and butter be made. Thus in 1902 when the Rt Hon. Joseph Chamberlain, MP, secretary of state for the colonies, cried "the weary Titan staggers under the too vast orb of its fate" and asked the colonies to shoulder their share of the burden, Canada's answer, given by the prime minister, Sir Wilfrid Laurier, was bluntly "If you want our aid, call us to your councils."[1] And there is some evidence to suggest that Laurier was sufficiently devious to avoid accepting any invitations that might come his way.

INTERNATIONAL NAVAL RIVALRY, 1860–1910

Since 1815 the influence of British naval might had been well nigh ubiquitous on all the major seas and their annexes. In the age of Pax Britannica the Royal Navy had been employed on countless warlike duties, and it can be argued with good reason that the Pax was not as peaceful as we have been led to believe.[2] In keeping "a security for such as pass on the seas upon their lawful occasions," the Royal Navy supplied an enduring contribution to the humanitarian needs of the world.[3] Whether the objects were putting down pirates or freeing slaves, surveying the seas or keeping ocean lanes open for trade, the navy provided the trusted weapon of government, serving especially Foreign Office and Colonial Office needs of an ever-expanding trading network, a burgeoning merchant marine, and a multiplying number of colonial settlements and entrepôts overseas. These obligations meant keeping a gunboat navy on "foreign" stations "showing the flag." Everywhere the white ensign appeared there was armed force to maintain free trade, open trade links, and secure British national advantages. In consequence, the British Empire owed its existence and continued good health to the security of the seas. . . .

With fleets and bases in both home waters and vitally important distant seas, Britain's influence remained profound in 1910.

Nonetheless during that long century new rivals had been presenting themselves. New powers had emerged in the struggle for overseas colonies and markets. Rivals to Britain's empire had acquired bases of repair and supply, beachheads of trade and commerce, railways and entrances to inter-oceanic canals. Britain's once comfortable lead in bases—Viscount Castlereagh's 1816 keys to security of trade and military protection—was shrinking fast.[4] And in guns and gunnery at sea not always did Britain's position rank first. The eighty-ton guns of Her Majesty's newest battleships were "portentous weapons," murmured Prime Minister Gladstone, who added thoughtfully, "I really wonder the human mind can bear such a responsibility!"[5] But bear it the mind did, and this was largely because foreign rivals forced upon an essentially conservative Admiralty Board the necessity of change. Take for instance armament. French gunnery and shells of the 1860s brought their Lordships at the Admiralty to rapt attention. In fighting ships alone, despite continental rivalry, Britain's lead had been maintained from the 1870s onwards. Under terms of the famous Naval Defence Act of 1889, the navy was made theoretically superior to the combination of the fleets of any two European powers.[6] Nonetheless, British naval primacy was facing serious contenders. British power was declining relative to other powers. . . .

But who would be the enemy? As late as the year 1902—cabinet memoranda and the informed periodicals of the day duly attest to this—British strategists were by no means sure which nation would pose the major threat. From time to time France and Russia were mooted as near and present dangers. However, France's weakness at sea relative to Great Britain revealed itself in 1898 during the Fashoda crisis and manifested itself again during the 1900 invasion scare in the United Kingdom. It was even more pronounced by 1902 because France built small battleships and inadequate armed cruisers, while Britain, Germany, Austro-Hungary, and Italy were building large battleships—Dreadnoughts. By 1904 France had entered an *entente cordiale* with Britain and the two ancient rivals settled their overseas differences. Russia, France's ally since 1894, posed a naval threat to Britain in the eastern Mediterranean. But that navy's loss to the Imperial Japanese Navy at the Battle of Tsushima Straits momentarily removed Tsarist Russia as a naval power, and in 1907 Russia and Britain became treaty signatories. In 1902 Japan became a treaty ally; in 1907 that country assumed naval paramouncy in Eastern seas, with British concurrence. Leaving aside Italy and the Austro-Hungarian Empire, that meant that Imperial Germany remained the most likely contestant to British authority at sea.

In fact, not until 1902 did the Admiralty identify Imperial Germany as the most likely opponent.[7] During these early years of the century, British maritime strategy gradually shifted toward preparing for a war against Germany. That involved a concentration of naval strength in what would be the decisive theatre, in northern waters. This modification, Admiral of the Fleet Sir John Fisher liked to point out, was "so unostentatiously carried out that it was only Admiral Mahan's article in *The Scientific American* that drew attention to the fact, when he said that 88 per cent of England's guns were pointed at Germany."[8]

The *Dreadnought* scare of 1909—when Germany's shipbuilding program seemed to threaten British pre-eminence in the number of battleships and other classes of vessels—led to wide-ranging public discussion, heated parliamentary debate, an enhanced construction program, and, of particular interest here, a new need, perhaps never before manifested, with regard to maritime strategic requirements. The key question remained: How could Germany be contained at sea by Britain and the Empire? To this Canadians had their own answer.

CANADA AND THE ROYAL NAVY

At this juxtaposition of world events and new strategic realities a distinct Canadian naval organization came into being. The historian, bewildered by the rhetoric of the Canadian autonomists, can point to a specific act of parliament, which received formal assent on 4 May 1910, establishing a Department of Naval Service and putting in place particulars of organization, training, and regulations. Under this statute, "naval forces" were defined as "those naval forces organized for the defence and protection of the Canadian coasts and trade, or engaged as the Governor in Council may from time to time direct."[9] The historian can also find that the Royal Canadian Navy had its precursors in coastal defence forces, in colonial gunboats, in coast guard vessels, and in fisheries protection vessels. Moreover in the history of provincial marines, naval militias, and naval leagues the historian can discern certain Canadian enthusiasm for a distinctly Canadian naval service. Besides, the country's shipbuilding heritage had given the world one-quarter of its wooden sailing ships of the late nineteenth century. In all these ways and others as well it seems logical that the Dominion came naturally into possession of a naval organization. . . .

In terms of shifting maritime strategic needs, Canada's act of naval independence in 1910 was a logical extension of international realities and imperial needs in the western hemisphere and the Pacific Ocean. Japan had secured dominance in the western Pacific at Tsushima and was an ally of Britain and the Empire. Correspondingly, the United States possessed naval dominance in the eastern Pacific, the Caribbean, and the western Atlantic. Given a state of Anglo-American peace, Canada's defence needs against invasion lay in the Monroe Doctrine and a secure America: that is, a Canada defended by the predominant land power and sea defence capabilities of the United States Navy. Prime Minister Laurier was prepared to acknowledge this Anglo-American Pax as early as 1902, though to this day the thought that at the height of its prowess the British Empire in North America was being sustained by United States military and naval might does not sit easily with the Maple Leaf patriot. And in 1910 that explains why the Canadian naval service took so many pains to be non-American in its organization and to be ready to serve as an integrated imperial force in time of national and imperial emergency.[10] That the Canadian navy had its roots in American security is paradoxical but true.

In the murky background of all this Canadian zeal for self-control lies the distant figure of Sir John Arbuthnot Fisher, later first Baron Fisher of Kilverstone. He unwittingly did the Canadian nationalists a great act of good. Sir John, who became First Sea Lord of the Admiralty on Trafalgar

Day 1904, had been commander-in-chief on the Royal Navy's North American and West Indies station for eighteen months beginning August 1897. In his usual tours of duty in and out of Halifax, Fisher obtained a warm appreciation of American enthusiasm for naval power. . . . A British war against the United States lay beyond Fisher's comprehension. Thus he readily accepted the Americas as the zone of United States' dominance.

By contrast, Fisher tended to see Canadians as colonial laggards, more often than not willing to let the British do the work and pay the bill. . . . What Fisher thought best, and he preached this doctrine when he went to the Admiralty in 1902, was to withdraw the main British fighting units from the North American and West Indies and Pacific Stations and to leave the Canadians to get along more agreeably with their American neighbours.

The Lords Commissioners of the Admiralty did not, on face value at least, see things Fisher's way. Or if they did they could not say so publicly. They continued to preach the unity of the Empire and of the fleet: "One life, one flag, one fleet, one throne, Britons hold your own!" They discussed the idea of the unity of the fleet at colonial, imperial, and naval defence meetings in 1902, 1907, and 1909. They argued for financial contributions to ease the capital costs of new units for the fleet. They argued that this would best benefit all constituent parts of the Empire. Anxious strategic reports emanating from the Admiralty, sometimes on the basis of secret intelligence gathered by plainclothes naval officers who had examined American capabilities, indicated the difficulty if not the impossibility of defending Canada against American invasion.[11] But at the highest political level in Britain such views were unacceptable and not fit for public consumption.

. . . In [Fisher's] view the British fleet was obliged to prepare for a major war, not for another gunboat engagement. He remained worried that some future West Indian or Central American obligation would deflect the Royal Navy away from its main strategical assignment. The Committee of Imperial Defence was of a similar mind but fell just short of Fisher's view when it said in 1904 that Canada could not be defended against a United States attack and thus that only a small garrison ought to be maintained at Halifax.[12] The War Office was seeking economy, too, and in an age of thorough discussion of how to match resources with requirements the British garrison kept at Halifax was one obligation which London sought to pass to the Dominion as expeditiously as possible.

In December 1904, as well, Fisher's scheme to reorganize oversea squadrons was effected. Behind this reform, as others, lay the Fisher motto: "The fighting efficiency of the fleet and its instant readiness for war." To Fisher, "strategical ('and not sentimental') requirements had to be met."[13] It reflected the revolution in Admiralty strategic thought that recognized the most likely danger to be Germany. It announced the end of the gunboat navy. It foreshadowed the end of the Pax Britannica. For all intents and purposes the North American and West Indies Station was disbanded. Halifax became a backwater of Admiralty strategy. The great "Warden of the North" was not included in Fisher's five strategic keys that belonged to England and locked up the world to British naval supremacy: the others were Singapore, the Cape of Good Hope, Alexandria, Gibraltar, and Dover. With the Admiralty having all but abandoned the century-long tradition of keeping a fleet in the western Atlantic for securing British interests, the question of land defences of Canada now assumed a greater importance. Canada's defence needs after 1905 were of continuing concern to British strategists.[14] But in sum what Fisher's reorganization meant was that Britain acknowledged that she could not be supreme everywhere on the seas that she might wish to be. Concentration of force, not dispersal, was what was now needed to secure national ends. If that meant abandoning Canada to American annexation, Fisher and the Admiralty noted in their private documents, so be it.

These views of Fisher and some of his colleagues were not for public consumption. The First Lord of the Admiralty, the Earl of Selborne, who drafted the key memoranda on imperial naval defence for this era, possessed a diplomatic tact not found in Fisher. He was profoundly shocked by Fisher's "violent hostility to Canada."[15] Fisher's view that Britain should not spend one pound for the defence of the Dominion did not strike Selborne or many of his political colleagues as a wholesome means of maintaining and enhancing intragovernmental defence co-operation. In consequence, Selborne anxiously promoted the opinion that Fisher's views were his own, not those of the Admiralty.[16] He also argued that abandoning Canada would be a dishonourable and unnecessary act, one likely to split the Empire without any possibility of reunion.

Thus if Fisher advanced the cause of Britain's hasty strategic withdrawal from the western hemisphere, the Admiralty would be successful in their policies in only two ways: firstly, the withdrawal of the North American and West Indies fleet; and secondly the transfer of Halifax from British to Canadian hands and the closing of the Jamaica dockyard. The War Office, more cautious than the Admiralty, was less anxious for an immediate withdrawal of British troops from Halifax. Fisher acquiesced to the idea. Keeping a few redcoats in Halifax, he wrote (employing the adage of Lord St Vincent), would afford a "comforting spectacle to the old women of both sexes in Halifax."[17] Fisher did not get his way in the case of the West Indian base St Lucia, which was maintained as a place of supply. Nor did he succeed with Bermuda, which was maintained (though on a reduced scale) and used increasingly by the United States Navy. Ascension, too, was reduced. Antigua had already been abandoned as a base.

Fisher's so-called reforms of fleets and bases signalled the end of the Royal Navy's obligations for the specific coastal defence of Canada. They also relinquished Britain's traditional command of the Caribbean to the United States. The withdrawal of forces generated apprehension in Canada. As Governor General Lord Minto complained in 1906, older Canadians bitterly resented these "withdrawals of the flag" for they largely "removed the possibility of the appeal so often made to the flag in the case of some trading schooner that had got into difficulty. . . . In considering imperial defence in an Empire such as ours one must not be guided alone by purely practical considerations but must take human influences into account."[18] But at this time, more frequently than otherwise, sentimentality was not the driving engine of defence planning.

From the Canadian perspective, Fisher's changes marked the need for a national commitment to sea security. They brought into Canadian possession two great bases: one lying athwart the major sea lanes of the North Atlantic, the other situated adjacent to the great circle route of trans-Pacific commerce. As well, Fisher's reforms, endorsed by the War Office, the Committee of Imperial Defence, and the cabinet, led Canada to secure even more than before the garrison and artillery needs of these bases. These bases and garrisons were forerunners to the organization of a naval service which was in itself a preliminary to a Canadian fighting capability at sea.

Fisher alone cannot be blamed or credited with these profound changes. He stands at the centre and around him move the events: the shifts in international relations, the changes of strategy, the revolutions in material. He had his *confrères*. Look, for example, at the activities of Richard Burdon Haldane, MP, Secretary of State for War. . . . By the time Haldane completed his work at home he had remodelled the regular army, restructured the reserve forces into the territorial army, and put in place an expeditionary force to fight a continental war in Europe. In other words, the army had its Fisher too, and one who was acutely aware of how shifting forces of international affairs affect British policy: "The days when splendid isolation was possible were gone," Lord Haldane wrote in reflection. "Our sea-power, even as an instrument of self-defence, was in danger of becoming inadequate in the absence of friendships which would ensure that other navies would remain neutral, if they did not actively co-operate with ours."[19] In short, only in alliances—with Japan, France, Russia, and perhaps even the United States—could ultimate naval preponderance of the British Empire be secured.

BRITAIN AND THE COLONIAL NAVIES

The search for friends and allies actually began at home, within the Empire.[20] No matter how passionately the Colonial Secretary, Chamberlain, might appeal for colonial contributions to imperial defence, no matter how strongly the self-governing colonies might resist London's demands and cry for a voice in councils, the Admiralty was moving toward an increased centralization and organization of naval administration and fighting strength. Colonial naval forces—as authorized by the Colonial Naval Defence Act of 1865—were now being discouraged, eventually to be replaced by the concept of the unitary fleet. In 1902, for instance, the Admiralty policy as presented at the Colonial Conference stated explicitly that any auxiliary squadron, (meaning the New Zealand and Australian subsidized squadron) was to be under Admiralty control in time of peace or war—a departure from the 1865 statute which gave Admiralty control only in an emergency. The intent was to make possible, when necessary, a combination of all units in East Indian, Australian, and Chinese waters.

Even more generally the 1902 Admiralty memorandum strongly criticized the folly of maintaining local navies or of keeping units on a particular coastline. The Admiralty emphasized control of communications. As the memo indicated "the importance which attaches to the command of the sea lies in the control which it gives over sea communications." The advantages of gaining command of the sea were equally illustrated by historical example. To cite one case, "the fall of Quebec and the conquest of French Canada was mainly due to the fact that our superior sea-power closed the Gulf of St. Lawrence to the French and opened it to us. In any similar struggles in the future, this route will be as vital as the past." The Admiralty also laid stress on the importance of the great battle for supremacy, because the great development of French, German, American, and Russian navies (Japan is not mentioned) required a British concentration for the decisive encounter. The enemy would likely threaten detached squadrons and prey on trade, and this would require detachments from the concentrated force. To do this effectively, control of communications remained vital. "The immense importance of the principle of concentration and the facility with which ships and squadrons can be moved from one part of the world to another— it is more easy to move a fleet from Spithead to the Cape or Halifax than it is to move a large army, with its equipment, from Cape Town to Pretoria— points to the necessity of a single navy under one control, by which alone concentrated action between the parts can be assured." The navy, the memorandum concluded, would take the offensive against enemy force, and the navy would have to be prepared to meet the strength and composition of the hostile forces.[21]

As has been suggested, the strategic principles employed in this celebrated document represent the rise of "blue water" thinking.[22] The Admiralty wanted flexibility coupled with control, and that meant possessing sufficient power to conduct a vigorous offensive against outlying hostile squadrons without unduly weakening forces concentrated for the anticipated decisive battle. Professor Gerald Graham, in assessing this stage of Admiralty thinking, indicates that even in 1902 their Lordships were maintaining a traditional role in both hemispheres. "Without aggressively straining for domination, Britain's unique predominance as mistress of the seas remained intact."[23] This was undoubtedly so. Yet as Professor Donald Gordon has stated, there was no universal colonial acceptance of the concept of communications control. Canada and especially Australia fumed about being dictated to by an insensitive mother.[24] "The establishment of special forces," complained an Australian-Canadian memorandum, "set apart for general imperial service, and practically under the absolute control of the imperial government, was objectionable in principle, as derogating from the powers of self-government enjoyed by them, and would be calculated to impede the general improvement in training and organization of their defence forces."[25]

Despite these appeals, one cardinal principle of action was emerging: Admiralty initiative. Concentration of the navy and one fleet by control of communications became strategic bywords. The principle cut right across touchy colonial desires for a share in policy making. Therein was presented a problem which Winston Churchill as First Lord of the Admiralty in 1911 scarcely understood and neglected to his peril.

The Fisher reforms and Canada's own creation of a naval organization made a co-ordinated imperial foreign policy even more vital. Australia had a navy, and New Zealand, Natal, and other possessions were making specific financial contributions to the fleet. Thus, as Secretary of State for Foreign Affairs Sir Edward Grey explained, British sea dominance would have to be effected through imperial organization. Under a separate 1911 conference on strategy held by the Committee of Imperial Defence, the needs of the Admiralty for control and efficiency of an imperial navy had to be reconciled with the requirements of Canada and Australia for financial and administrative autonomy. By agreement, the Royal Canadian and Royal Australian navies were to have training and discipline similar to that of the Royal Navy. In time of peace they were under Dominion control in Dominion waters and under Admiralty control outside those waters. In time of war they were to come under Admiralty control if and when so placed by the Dominions.[26] This was a triumph for dominion autonomy, but it did little to give credence to the idea, now mythical, of "imperial defence."

THE END OF PAX BRITANNICA

On reflection, the end of Pax Britannica and the origins of the Royal Canadian Navy have their roots in the same problem: How could the sea supremacy of the British Empire be maintained in the face of new strategic realities? In 1889, Lord Dunraven, the First Lord of the Admiralty, defended the "two-power standard" and identified the key requirements of naval paramouncy for Britain: "the defence of the Empire, the safeguarding of the trade of the national and imperial commerce, are really the foreign policies of this country."[27] As with Dunraven, so with Castlereagh three generations before. Imperial defence was vital, said the British; indeed it was inseparable from national defence. The same held true in 1910. The Dominions for all their desire for autonomy were part and parcel of the whole reorganization of the naval and military services in the years which closed that era when Britain alone was perceived as being mistress of the seas.

The changing nature of international affairs compelled a Canadian response. The United States, emerging as the predominant power in the Americas and adjacent seas, freed Britain from obligations there and allowed Britain to bequeath to Canada two bases as well as obligations for local sea and shore defence. At the same time, the rise of German naval power initiated another Canadian response. Given the inseparability of Canada from the Empire and the indivisibility of foreign policy, the question emerged, "How could the Canadian naval service best serve the defence of Empire as well as that of the Dominion?" The interminable differences of opinion among various representatives at the imperial and naval conferences, especially in 1902, 1907, 1909, and 1911, were not about loyalty or ultimate purpose. They were really about technique: that is, who would control what in what circumstances. The resulting changes in the world's balance of power, especially the enhanced status of the United States, Japan, and Germany and the reduced influence of France and Russia, and the technological revolution that fueled these rivalries, meant that Britain and in consequence Canada were obliged to face these new realities. Thereby was the Canadian Naval Service born. . . .

As in 1910 so too in future years, in war and in peace, even to the present. The near half-century of conflict which ended in 1945 and the near half-century since 1945 of nuclear peace (that now can be seen as so different from its predecessor) have been characterized, as far as Canadian foreign policy and naval and military commitment are concerned, by four constants: the dictates of geography; the state of international relations; the actions of allies; and the reactive or responsive posture of the nation itself. Canada, by its northern location inseparable from the main currents of international events, has responded to the shifts in the international balance of power and to new strategic realities, has worked in conjunction with its principal allies, most notably Great Britain but also the United States and more recently member states of the North Atlantic Treaty Organization, and has, at bottom, developed policies to fit its own needs.

Only in 1914, by the King's declaration of war on Canada's and the British Empire's enemies, did Canada find itself at war without making an independent decision on the matter. Even then, it must be quickly added, such declaration was a constitutional obligation of fact, the Crown being as of yet indivisible and the technical characteristics of Dominion status not yet defined. All the same, the degree of Canada's commitment in men, material, and finance remained the nation's question alone. Today as yesterday that fundamental premise remains inviolate. The Pax Americana may have replaced the Pax Britannica in some of its particularities; the world's democratic leadership may have shifted westwards across the Atlantic from Westminster to Washington, and Canada's position as principal trading partner to the world's ranking western power may now be even more prominent than it was in the heyday of the British Empire. But, all these matters aside, the choice is Canada's to make as to its material contribution to its own defence; as Canada's history amply demonstrates, the nation's sea security ranks as a pre-eminent need.

NOTES

1. *Colonial Conference of 1902, Summary of Proceedings* (Ottawa: King's Printer, 1902), 4, 31–32; O.D. Skelton, *Life and Letters of Sir Wilfrid Laurier,* vol. 2 (Toronto: Oxford University Press, 1921), 294.

2. Captain Stephen W. Roskill, *The Strategy of Sea Power* (London: Collins, 1962), 83–89.

3. Some duties were not so humanitarian, including putting down slave risings, bombarding villages of innocent native peoples, and coercing Irish peoples who refused to pay their rates.

4. "Our policy," said Castlereagh, "has been to secure the Empire against future attack. In order to do this we had acquired what in former days would have been thought romance— the keys of every great military position" (*Hansard*, first series, vol. 32, 1104).

5. Quoted in James Morris, "A View of the Royal Navy," *Encounter*, 40 (March 1975), 25.

6. Arthur J. Marder, *The Anatomy of British Sea Power: A History of Naval Policy in the Pre-Dreadnought Era, 1880–1905* (New York: Alfred A. Knopf, 1940). Chapters 7 and 8 review the subject of the "two-power standard." For all this book's considerable merits it really does not examine the Admiralty's decision making and decisions. It is anatomy of the body without examining how the brain functioned.

7. The first Admiralty War Plan dates from 1907. See Lieutenant Commander P.K. Kemp ed., *The Papers of Admiral Sir John Fisher* (London: Navy Records Society, 1964), 106: 318–468. Fisher had a hand in this as did Captain G.A. Ballard, Captain Edmond J.W. Slade, and perhaps Captain Maurice Hankey, RMA, but it also bears the arguments of Sir Julian Corbett. For a review of how various British boards and committees reorganized defence planning in this period, see Barry D. Hunt, *Sailor-Scholar: Admiral Sir Herbert Richmond 1871–1946* (Waterloo: Wilfrid Laurier University Press, 1982), chap. 1.

8. Admiral of the Fleet Lord Fisher, *Memories* (London: Hodder and Stoughton, 1919), 5.

9. See especially, Gilbert Norman Tucker, *The Naval Service of Canada: Its Official History*, vol. 1, *Origins and Early Years* (Ottawa: King's Printer, 1952), chaps. 6, 7. The text of *The Naval Service Act 1910*, is in ibid., chap. 43, 377–85.

10. On tradition, Vice-Admiral H.G. DeWolf recalled in 1985: "We grew up with the Royal Navy. As I recall, we were told to go to England and imbibe the spirit of the Royal Navy. We did. When war came along, my experiences convinced me that everything I had learned in my early days with the Royal Navy was sound. I don't think I made any mistakes because of anything I had been taught by the RN. On the other hand, the Royal Navy always regarded us as part of their navy. They were always inclined to look upon our ships or our squadrons as theirs. They didn't seem to grasp the fact that we were developing a service of our own. I think that was simply a lack of knowledge on the part of some individual officers rather than the Admiralty itself. For instance, when the war started they sent over an Admiral to take charge in Halifax. When he arrived, he found we already had somebody in charge. That sort of thing had to be dealt with very tactfully. In the early days, a lot of the officers in the Navy came from the Maritimes or BC and many had British backgrounds. So starting with that, plus the training we received in the UK, there was sometimes the feeling in Canada that we weren't very Canadian, that we were more British than Canadian. We'd spent so much time over there and picked up so much of their way of life that we were perhaps a little bit different than our contemporaries in Canada. I think that was a fair criticism. Now, of course, we had the greatest admiration for the Royal Navy, but we gradually learned that anything they could do we could do as well. On the material side, we learned the hard way during the war that there was too much stuff lost coming across the Atlantic so we gradually shifted to the USN as a source of supply or to our own sources. From this grew closer ties with the USN." *Maritime Warfare Bulletin, Commemorative Edition 1985*, 24.

11. In 1898 Lieutenant-General Sir Percy Lake and Captain Reginald Custance RN examined the defences of Lakes Ontario and Erie. They developed plans so that in the hoped-for expectation that the Americans would cross the border first (and would thus be labelled the aggressors!), Canadian forces could mount a counterattack. Donald M. Schurman, *Julian S. Corbett, 1854-1922: Historian of British Maritime Policy from Drake to Jellicoe* (London: Royal Historical Society Studies in History Series, no. 26, 1981), 110. By 1905, the Admiralty

had concluded that Canada was indefensible (see note 15 below).

12. Samuel F. Wells, jr., "British Strategic Withdrawal from the Western Hemisphere, 1904-1906," *Canadian Historical Review*, 49, 4 (December 1968), 338. However, in the previous year, 1903, the newly created Committee for Imperial Defence had been pressing on Canada to develop a general scheme of defence and suggesting that plans for this ought to have London's expert criticism and advice. For a discussion of this and related points including the Committee for Imperial Defence's agreement with the Canadian Minister of the Militia, Dr. Frederick Borden, that the General Officer Commanding the Canadian Militia should be a Canadian officer, see Donald C. Gordon, *The Dominion Partnership in Imperial Defence 1870-1914* (Baltimore: Johns Hopkins University Press, 1965), 176–77.

13. Quoted in Arthur J. Marder, *From the Dreadnought to Scapa Flow: The Royal Navy in the Fisher Era, 1904–1919*, vol. 1, *The Road to War, 1904–1914* (London: Oxford University Press, 1961), 40.

14. Wells, "Britain's Strategic Withdrawal," 346–47.

15. Wells, "Britain's Strategic Withdrawal." Also, Ottley-Battenburg memo on Defence of Canada 1905, and related documents, Adm. 1/7807, Public Record Office, Kew, England (National Archives of Canada, Reel B3634).

16. Wells, "Britain's Strategic Withdrawal," 349.

17. Ibid., 340.

18. Minto to Lord Morley, 3 April 1906, quoted in Ronald Hyam, *Britain's Imperial Century, 1815–1914* (New York: Barnes and Noble, 1976), 122.

19. Lord Haldane, *Before the War* (London: Cassell and Co., 1920), 7–8;

also Admiral Sir H.W. Richmond, "National Policy and Naval Strength, XVIth to XXth Century," *Raleigh Lecture* (London, 1923), 3.

20. London's war plans included, at the War Office at least, deliberate concealment from the Dominion premiers of its strategic aims on the eve of the 1914 war, that is, that overseas contingents should come to the help of the mother country rather than the other way round. See on this and related matters John Gooch, *The Plans of War: The General Staff and British Military Strategy, c. 1900–1916* (London: Routledge and Kegan Paul, 1974), chaps. 5, 6; John Gooch, *The Prospect of War: Studies in British Defence Policy, 1847–1942* (London: Frank Cass, 1981), especially chap. 3, "Great Britain and the Defence of Canada, 1896–1914."

21. "Admiralty Memorandum . . . 1902," in *Selected Speeches and Documents in British Colonial Policy, 1763–1917* , ed. A.B. Keith (London: Oxford University Press, 1966), Part 2, 230–39.

22. Donald C. Gordon, "The Admiralty and Dominion Navies, 1902–1914," *Journal of Modern History* 33, 4 (December 1961), 409.

23. Gerald S. Graham, *Empire of the North Atlantic: The Maritime Struggle for North America*, 2nd ed., (Toronto: University of Toronto Press, 1958), 291.

24. Gordon, "Admiralty," 409ff.

25. In Skelton, *Laurier*, 2: 297.

26. Brian Tunstall, "Imperial Defence, 1870-1897," in *Cambridge History of the British Empire*, vol. 3 (Cambridge, UK: Cambridge University Press, 1959), 595–97; also *Parliamentary Papers*, 1911, 54, (Cmd. 5746–2), 1–3.

27. *Nineteenth Century*, February 1889, quoted in Admiral Sir Herbert Richmond, *Statesmen and Sea Power* (Oxford: Clarendon Press, 1946), 273.

"JUNIOR BUT SOVEREIGN ALLIES": THE TRANSFORMATION OF THE CANADIAN EXPEDITIONARY FORCE, 1914–1918 ✧

DESMOND MORTON

○

With the expiry of the British ultimatum to Berlin at midnight on August 4th, five self-governing dominions found themselves at war with Germany with the same finality as India, the Falkland Islands, Sierra Leone or Britain herself. The unity was more than formal. At once Canada, Newfoundland, Australia and New Zealand offered troops and economic aid which, by war's end, far exceeded what anyone would have imagined possible in 1914. In the course of four years, almost one Canadian in a hundred would be killed by enemy action; Australian and New Zealand losses were proportionately greater. Even the Union of South Africa, at the cost of a short civil war, contributed major fighting resources for the struggle.[1] Despite British failure to establish a firm system of imperial defence in the pre-war years, Richard Preston concluded, the dominions produced a war effort that could hardly have been bettered by any form of previous imperial defence commitment.[2]

Though imperial solidarity survived the war, it would be among the conflict's casualties. "The challenges and the sacrifices of war," Professor Mansergh wrote of the young dominions, "sharpened their sense of separate identities and strengthened their feelings of nationality."[3] Some have argued that a Canadian identity can be dated from the capture of Vimy Ridge by the Canadian Corps in April 1917.[4] The Australian Imperial Force, claimed J.B.D. Miller, "became the symbol of a new secular religion, in which the best aspects of Australian life were epitomised by the men who fell at Gallipoli and in France."[5] The war found some in the British Empire toying half-heartedly with vague concepts of imperial federation: it put those concepts to a ruthless practical test and left notions of federation in tatters.

✧ *Journal of Imperial & Commonwealth History* 8, 1 (Oct. 1979), 56–67.

Pre-war proposals that Britain should share the burdens of decision making with the white dominions had foundered on the meagre contribution the colonies made to common defence. That argument vanished with the war only to be replaced by intractable problems of consultation, communication and confidence which transformed Canada's Sir Robert Borden, among others, from a devout imperialist to a determined nationalist.[6] Professor Mansergh and others have explored the problems encountered on the high road of prime ministerial consultation. There was also a low road of practical military collaboration along which the military forces of the dominions found their way to greater autonomy. Each experience was different; all arrived at similar goals. The case of the Canadian Expeditionary Force is therefore both unique and illustrative of a wider phenomenon.[7]

In late October 1914, when men of the C.E.F. landed at Plymouth on the first stage of their journey to the Western Front, their legal status was clear. In the terms of Britain's Army Act, they were "Imperial." They were soldiers of the British Army recruited from the Empire.[8] If there was any doubt about the full integration of the Canadians in the British forces, it was laid to rest by the Canadian minister of militia, Colonel Sam Hughes: "We have nothing whatever to say as to the destination of the troops once they cross the water nor have we been informed as to what their destination may be."[9] In London, George Perley, Canada's acting high commissioner, concluded ". . . that as soon as the Canadian troops arrive here they will be entirely under the authority of the War Office and become part of the Imperial army in every sense of the word."[10] In 1914, few presumed otherwise.

Four years later, the presumption was very different. The battle-hardened Canadian Corps was under the operational command of Field Marshal Sir Douglas Haig, the British commander-in-chief, but the Canadians now had their own commanders, their own organization and their own tactical doctrine. They were now considered part of a unified Canadian military organization with a chain of authority which ran past Haig's General Headquarters through a Canadian cabinet minister in London to a prime minister increasingly insistent that Canada have her own say in the issues of war and peace.[11] Like other British officers, Haig grumbled that the Canadians had come to see themselves as junior but sovereign allies. His political superior, the Earl of Derby, counselled resignation in the face of altered circumstances: ". . . we must look upon them in the light in which they wish to be looked upon rather than the light in which we would wish to do so."[12]

Both British and Canadian officials would have been wiser in 1914 if they had taken to heart the precedent for the C.E.F. Fifteen years before Colonel Hughes created his chaotic mobilization camp at Valcartier, a Canadian battalion had been recruited, equipped and despatched to the South African War in only two weeks. When the 2nd (Special Service) Battalion of the Royal Canadian Regiment reached Cape Town, it was immediately incorporated into the British Army and made subject to British military law. Unlike the soldiers of 1914, the volunteers of 1899 had been an embarrassment to the Liberal government of the day. Sir Wilfrid Laurier would cheerfully have explained to French Canadian supporters that his ministry had done no more than transport some perfervid imperialists to an early fate. It had been the governor general, the Earl of Minto, who had insisted that Canada's dignity demanded an organized regiment under Canadian officers.[13] Once in South Africa, the Canadians soon developed some of the tendencies which would reappear in the C.E.F. The commander of the contingent insisted that his men would serve as a unit, not in scattered detachments. In Canada, public opinion demanded a flow of information direct from the contingent commander. Whatever British military regulations said about official channels, senior Canadian officers found themselves servants of two masters. The elderly Lieutenant Colonel William Otter, squatting under a waggon in the pouring rain, dictating despatches to a staff sergeant was inaugurating work performed by massive overseas Canadian military headquarters in both world wars.[14]

Other South African experiences were to be repeated. Otter's men suffered from badly designed and cheaply produced Canadian uniforms and equipment. Canadian soldiers complained about senior officers, Canadian and British, and they discovered a powerful sense of Canadian identity which sometimes supplanted their initial imperial fervour.[15] In the aftermath of the war, Canada's militia shared in the general improvement in efficiency and in the standardization of training, tactics, and staff doctrine. Canadians took great pride in their modest contribution to the war, persuaded themselves of their natural military prowess and acquired a suspicion of British military proficiency.[16] Military nationalism fostered a munitions industry which would soon arm Canadian soldiers with the defective Ross Rifle, MacAdam shield-shovels and boots which disintegrated in the mud of Salisbury Plain and Flanders.[17]

Although British officers like Sir John French and Sir Ian Hamilton continued to advise Canada on [its] traditional defence problem—invasion from the United States—it was apparent after South Africa that the most probable employment for Canadian arms would be in a future imperial war. As early as 1911, a gifted British staff officer, Colonel Willoughby Gwatkin, turned to the task of planning for an overseas contingent of one infantry division and a cavalry brigade. In 1914, at Hughes' insistence, the scheme was officially scrapped to allow the minister a free hand to execute his own eccentric fancies. Even then, the Gwatkin plan was the informal guideline for building both the first and second Canadian contingents.[18] Even Gwatkin's plan, however, overlooked mechanisms for continued Canadian control over the expeditionary force. As a regular victim of his minister's temperament, Gwatkin may well have considered such authority undesirable. What had not been planned would have to be improvised.

Although the South African War furnished valuable precedents for the experience of the C.E.F., there were many differences. In the earlier conflict, the Laurier government had sought no influence over the conduct or resolution of the war. Beyond the cost of delivering contingents to South Africa and meeting the difference between British and Canadian rates of pay, the dominion had assumed no financial responsibility.[19] In 1914, the government of Sir Robert Borden, with full support from the Liberal opposition, proposed that Canada would bear the full cost of her military contribution. Parliament, the watchdog of the treasury, never barked. A Liberal backbencher offered a vague query about the financial consequences of the offer but did not even pause for an answer.[20] In due course, Canada's commitment would add a quarter of a billion dollars to the national debt.[21]

More than the national debt was involved. With thousands of her men engaged in the struggle, Canada had a moral right to a voice in their disposition. Cash flow was the avenue of influence. It was because Canada was helping to pay for the music that the Borden government mustered the will to criticise the orchestra; more important, the British were willing to listen to the complaints. That experience began soon after the first Canadians arrived. Colonel John Wallace Carson, Montreal mining promoter and militia crony of the minister, was despatched to watch over the comfort of the young Canadians. His reports of the appalling conditions on Salisbury Plain and the alleged refusal of the British authorities to remedy the situation led to the first Ottawa misgivings about the management of the war.[22] Armed with additional authority as the minister's "representative," Carson soon denounced the wholesale replacement of Canadian boots, shovels, waggons and motor vehicles with British equipment. A major financial issue was at stake: not only had the equipment of the C.E.F. been costly but its abandonment was a blow to the profitable plans of Canadian manufacturers.[23]

After a few early brushes with Canadian sensitivity, the War Office made no attempt to control Canadian military affairs in England. The British had enough of their own problems. Without more guidance than impatient and sometimes contradictory commands from Sir Sam Hughes, the Canadian base organization grew like Topsy. Organizations were needed to collect records, distribute ordnance stores, train reinforcements and, by early 1915, to care for a growing stream of sick and wounded from the Western Front. One major Canadian camp, Bramshott, opened by mistake when a shipment of reinforcements arrived by mistake.[24] Problems were compounded by Hughes' technique of allowing politicians, businessmen and friends to raise entire new battalions instead of finding recruits for existing C.E.F. units. Dozens of self-confident but untrained regiments reached England only to be stripped of their subalterns and privates. Senior officers, often unqualified but always influential, were left with a burning grievance and an urge to find a suitable position on the staff.[25]

Any military base will be regarded with contempt by fighting soldiers but the Canadian administration in England during 1915 and 1916 was more contemptible than most. Most senior Canadians had experienced no more than a tourist's view of the front. The few genuine veterans had usually been found wanting under the strain of battle.[26] The problem can be traced directly to the minister of militia. Determined to be the only unquestionable authority on matters great and small, he refused to delegate authority to a single officer. The web of command baffled British observers and, by the autumn of 1915, left three senior officers with documentary proof that each was in full command.[27] The most formidable was Carson, now, thanks to his ingenuity, a major-general. Although Hughes refused to substantiate Carson's authority, the Minister's friendship was sufficient to frighten most Canadian officers. The War Office, anxious for some clear channel, grimly resigned itself to deal with Carson.[28]

Hughes's arrangements for liaison with the Canadians in France were equally informal. John Carrick, a real estate speculator, Quaker and Conservative M.P. for Thunder Bay, was sent to General Headquarters with a vague commission as Sam Hughes' agent. Appointed an honorary colonel, Carrick's lack of military knowledge and eagerness to meddle in military appointments humiliated his fellow Canadians.[29] By the autumn of 1915, Carrick had been supplanted by a much more formidable figure, Max Aitken. With the title of Canadian Eyewitness and the rank of colonel, the youthful politician and newspaper proprietor found his Canadian appointment a useful base from which to conduct his merciless vendetta against the British high command.[30] It was a campaign which Sir Sam Hughes cheerfully endorsed. Ever since his own experience in the South African War, Hughes had a low opinion of most British generals. After Second Ypres, he felt free to unburden his tactical expertise on Lord Kitchener, the Secretary of State for War. He made no secret of the views he shared with Aitken: "It is the general opinion that scores of our officers can teach the British Officers for many moons to come."[31]

In the first year of the war, most Canadians regarded their war minister as a marvel. Even the prime minister and cabinet colleagues took months to realize that Hughes' performance was based on brilliant public relations superimposed on needless chaos. Piece by piece, responsibilities were chipped away and handed to more dependable colleagues like Edward Kemp of the War Purchasing Committee. Having granted Hughes a direct channel of communication with the War Office, Borden felt obliged to tug on the reins after learning of his minister's outspoken messages: "You do not represent yourself alone as Minister of Militia but the Government of which you are a member."[32]

Thanks to Perley, who never had been fond of Hughes, the prime minister learned something of the administrative mess in England. After several false starts, he insisted that Hughes go to England in the summer of 1916, prepare a scheme for reorganization and return to win approval from his cabinet colleagues before implementation. Instead the minister spent a joyful few months in Britain and France attending reviews, instructing bayonet fighting and gossiping with cronies. Borden learned from his morning newspaper that Hughes had created an "Acting Sub-Militia Council" headed by Carson, with Hughes's own son-in-law, Major Byron Greene, as secretary and spy.[33]

Years before, Hughes had claimed that Borden was "as gentle-hearted as a girl,"[34] but iron had come into the prime minister's soul. In the same summer, Sir George Perley had returned to Canada and the two friends arrived at a very different solution to the overseas problem. Almost since his arrival in England, Perley had insisted that the high commissionership should be a cabinet portfolio. Only as a cabinet minister could a colonial official carry weight in the councils of the empire.[35] The argument did not persuade Borden but it did suggest a way of separating the minister of militia from his mismanaged overseas responsibilities. When Sir Sam Hughes returned from England, he found his own arrangements overturned and machinery in progress to make Perley the head of a brand new Ministry of Overseas Military Forces of Canada based in London. After a bitter, abusive struggle to save his power, Hughes was dismissed.[36]

Later, Perley confessed that he would never have accepted the position if he had fully understood its problems.[37] Hardly an area of Canadian overseas administration, from the chaplains to the veterinary service, was free from disorder, conflict or outright corruption. The raging battle between the Canadian medical service and its Hughes-appointed critic, Colonel Herbert A. Bruce, would last well into the postwar years.[38] With rare exceptions, Perley's solution was to recall respected, experienced Canadian officers from the Corps in France, give them full backing, and ignore the indignation of the dispossessed.[39] If Canadian soldiers overseas voted massively for the Union government in 1917, Perley's administrative reforms deserve at least part of the credit.[40] Moreover the improvement in Canadian administration was overdue proof that the Borden government could repair a situation which, for two years, had been proof of colonial immaturity. Only by establishing an efficient base could Perley and his successor, Sir Edward Kemp, achieve greater control over the Canadian forces in France.

In the early years of the war, Canadian ministers were very modest about any claim to govern the affairs of Canadian troops in France. Even Hughes, once his own hopes of leading the First Division had been dashed, accepted that only a British officer had the necessary experience. Among those offered by Lord Kitchener, he chose Sir Edwin Alderson who had successfully commanded Canadians in South Africa.[41] Canadian patience snapped, however, when the British named one of their own military amateurs, Brigadier General J.E.B. Seely, to command a cavalry brigade in which two of the three regiments were Canadian. A former Secretary of State for War displaced by the Curragh mutiny, Seely was to prove a successful commander but Canadians complained that they, too, had eligible politicians.[42]

Canada's Second Division was given to the venerable Major General Sam Steele as a sop to western Canadian pride. Steele could be displaced only by promoting Brigadier-General Richard Turner, V.C., a Quebec publisher and loyal Conservative serving in the First Division. When Alderson became commander of the new Canadian Corps, his division passed to Arthur Currie, a Liberal from British Columbia. Both regionalism and bipartisanism were satisfied.[43]

Like earlier British generals with the Canadian forces, Sir Edwin Alderson fought futile battles against the malevolence of political superiors. When incompetent Canadians won promotions and favoured appointments through influence with politicians, Alderson fumed: "The latter really seem as if they wish to destroy the spirit of their officers."[44] Alderson's criticism of the Ross Rifle brought a blistering, widely distributed rebuke from the minister. Hughes had staked his standing as a marksman and rifle expert on the unfortunate weapon.[45] The British general's vulnerability was most apparent after the battle for the St. Eloi craters. Responsibility for the costly fiasco plainly belonged to General Turner and one of his brigadiers.[46] However, Hughes' agent in France, Aitken, made it clear to General Headquarters that it would be highly inexpedient to remove two Canadian generals. Instead, it was Alderson who became the scapegoat.[47]

Alderson's successor, Sir Julian Byng, was promoted in June 1917, after his brilliant capture of Vimy Ridge. By now it was clear that the Corps would pass to a Canadian. In December of 1916, Perley had persuaded Turner that his services were needed in the new Overseas Ministry. For obvious reasons, the British high command preferred that the Corps commander should be Sir Arthur Currie. However, by the spring of 1917, it was now informally but unalterably established that such decisions rested with the Canadian authorities. It was one of Perley's more valuable wartime services that he supported Currie's claims.[48] It was also to the credit of Sir Richard Turner that he demanded no more than a nominal increase in status as a price for surrendering his military ambition.[49]

What was missing, now that a Canadian commanded the Corps, and the Overseas Ministry was equal to greater responsibility, was a satisfactory link between London and the Canadians in the field. Sir Max Aitken's involvement in the frenzied intrigues which led to the Lloyd George government conveniently removed him from his role as Canadian intermediary and fixer in France. His replacement was a former assistant. Colonel R. F. Manly Sims was a former British officer who had worked for Carrick in Port Arthur before the war and again as his assistant in France.[50] However, Sims' liaison role was essentially limited to serving as a travel agent for visiting Canadian celebrities. In the wartime election, Sims served loyally as the Unionist political agent among the troops in France, a necessary and successful service which nonetheless ended his usefulness at General Headquarters.[51]

The order-in-council which created the Overseas Ministry in November 1916, stated for the first time that the men of the C.E.F. were members of the Canadian militia who happened to be serving overseas.[52] This sensible if novel doctrine not only justified the appointment of a cabinet minister; it also furnished a basis for reclaiming full authority over the Canadians in France. "The Canadian Force is an entity," Perley informed the War Office in one of a series of memoranda, "irrespective of where parts of it for the time being may be situated or serving, and it is in the interests of Canada and of the British Empire at large that it should be regarded and respected as such under all circumstances. . . ."[53] Henceforth, he notified the British, all questions relating to appointments in the Corps would pass to and from the ministry through the Canadian representative at General Headquarters.

A Canadian representative would have another even more important role. In addition to the men of the Canadian Corps and a detached cavalry brigade, Canada supplied a growing stream of men for work on the military railways and in the forests first of Scotland and then of France. Since Canada paid the bills, this growing army of uniformed workmen was entitled to Canadian military supervision.[54] While the Overseas Ministry was responsible in the United Kingdom, it slowly became apparent that it could best exercise its authority in France through its representative at Haig's headquarters.

The problem was tackled by Sir George Perley's successor, Sir Edward Kemp. As an M.P. for a Quebec constituency, Sir George had correctly concluded that he was a certain loser in the 1917 election. Persuaded, despite Borden's friendly warnings, that the high commissionership was the most important of his positions, Perley had given up his portfolio.[55] Kemp came to London fresh from a year's experience as Hughes' successor in Ottawa. An experienced businessman and veteran politician, the new minister was determined to impose the full weight of civil authority on his military subordinates and to assert the fullest measure of Canadian autonomy in relations with the War Office and General Headquarters. While at least some of his generals hinted broadly to journalists and subordinates that a politician was no longer needed in London, Kemp resolutely remodelled his organisation on the pattern of the militia department in Ottawa.[56] In a new Overseas Council, Turner and other senior officers had to learn to work under daily political scrutiny.[57] In France, Kemp bullied both British and Canadian senior officers into accepting enhanced authority for a Canadian representative. To fill the post, he appointed Brigadier-General J.F.L. Embury, a capable but elderly brigade commander and a newly appointed member of the Saskatchewan bench.[58]

While Embury's ostensible role was to keep in touch with Canadian railway troops and forestry detachments, his real responsibility was defined in a memorandum of 23 June, 1918. The Canadian representative would be the direct liaison between the Corps commander, the overseas minister and, on important occasions, the British commander-in-chief. With pained reluctance, Sir Douglas Haig and the War Office accepted the arrangement.[59] In the final analysis, the Canadians were no longer colonials nor were they "Imperials": they were allies. More strenuous objection to Embury's new Canadian Section at General Headquarters came from Sir Arthur Currie. The Corps commander wanted no intermediary in communicating with Kemp, particularly on the vexed questions of appointments, promotions and discipline.[60] Fortunately for good relations, Embury possessed Currie's confidence. As a leading Saskatchewan Conservative, he was politically compatible with the minister.

The administrative and institutional changes which gave the Canadian government a voice in all but the operational deployment of the Canadian Corps reflected more subtle changes within the formation itself. By the beginning of 1917, on the testimony of W. B. Kerr, the growing proportion of Canadian-born members of the Corps began to be reflected strongly in the non-commissioned ranks, supplanting the British-born who had predominated in the first two contingents.[61] The British influence never vanished. The senior staff positions in the Corps and the four divisions were filled by excellent British officers until the end of the war. The Corps' trademark of painstaking planning and preparation which had distinguished Vimy Ridge from so many disastrous operations on the Western Front, reflected their work and the thoughtful leadership of Sir Julian Byng.

However, it also depended on the ingenuity and conscientiousness of citizen-soldiers like Brigadier-General A.G.L. McNaughton of the artillery or Major General W.B. Lindsay, the Chief Engineer.

Although there were innovations in the tactical methods of the Canadian Corps—General Brutinel's motorised machine guns, for example, or General McNaughton's counter battery techniques[62]—what characterized the Corps under Currie was a conservative determination to use the maximum possible weight of material in the hope that it would save lives and win objectives. The Canadian Corps used more artillery longer in its barrages.[63] It boasted of the extent of its barbed wire entanglements when, Currie alleged to Borden, British generals were using their labour to build tennis courts.[64] In the spring of 1918, when the British were obliged to cut their infantry divisions from twelve to nine battalions, Currie won a bitter argument with his political superiors to preserve the existing structure in the Canadian Corps rather than building a Canadian army of six small divisions. Instead, the Corps actually increased its infantry manpower.[65] Given the circumstances and tactics of 1918, Currie's insistence probably made his formation a more powerful instrument in the series of offensive battles which filled the last hundred days of the war.[66]

If Currie would now exercise independent judgment, it was because he had secured a direct channel to Canadian authority. If his corps commanded greater resources of manpower and material than neighbouring British formations, it was because the Canadians had accepted, albeit with growing hesitation, full financial responsibility for their soldiers. That gave a practical as well as a sentimental claim to influence in every area where financial considerations mattered: in the final analysis, that was everywhere.

As important as money was competence. As early as May of 1915, Sir George Perley had offered a significant prediction: "Everyone expects that this war will bring great changes in Empire relations and it is especially necessary that we should just now impress on people here that we are both sane and capable in the management of our own affairs...."[67] While Canada's war effort was in the hands of Sir Sam Hughes and his associates, even many Canadians could question their own collective capacity to tackle complex and unfamiliar problems. The Overseas Ministry, with its unified authority, was a necessary solution.

The experiment of establishing a cabinet minister in London was not repeated in the Second World War. It might not have been necessary in 1916 if Borden had determined earlier to dismiss Hughes or if Sir George Perley had not already been a minister without portfolio accidentally stranded in the British capital. Both Kemp and Perley complained of their isolation from Ottawa colleagues and both undoubtedly felt vulnerable to political intrigues and cabinet rivalries. However, a cabinet minister in the Canadian political system possessed the kind of authority which was needed to solve the problems which afflicted overseas Canadian forces at the end of 1916. Real delegation of authority to a mere official, military or civilian, would have been as great an innovation in Canadian public affairs as the creation of the Overseas Ministry itself.

It took more than the Overseas Ministry or the appointment of Canadian-born commanders to transform the C.E.F. from an imperial to a Canadian force. Without the sacrifice and courage of the thousands of men who passed through the Corps, the institutions would have been irrelevant and the pretension of their leaders would have been absurd. It was performance in action which compelled British politicians and soldiers to realize that Canadians must be accepted on their own valuation.

NOTES

1. Nicholas Mansergh, *The Commonwealth Experience* (London, 1969), 167.

2. R.A. Preston, *Canada and "Imperial Defense"* (Durham and Toronto, 1967), 463.

3. Mansergh, *Commonwealth Experience*, 166.

4. C.P. Stacey, "Nationality: The Canadian Experience," *Historical Papers* (1967), 1. See also D.J. Goodspeed, *The Road Past Vimy* (Toronto, 1967); W.B. Kerr, *Shrieks and Crashes* (Toronto, 1930).

5. J.D.B. Miller, *Australia* (London, 1966), 65.

6. Harold A. Wilson, *The Imperial Policy of Sir Robert Borden* (Gainesville, FL, 1966), 34; R.C. Brown, "Whither are we being shoved: Political leadership during World War I" in J.L. Granatstein and R.D. Cuff, *War and Society in North America* (Toronto, 1971).

7. See Mansergh, *Commonwealth Experience*, chap. 6; Preston, "Imperial Defense," chap. 15; G.W.L. Nicholson, *Canadian Expeditionary Force, 1914–1919* (Ottawa, 1962), passim.

8. A.F. Duguid, *Official History of the Canadian Forces in the Great War, 1914–1919, General Series*, vol. 1, Appendices, 24–25; Gwatkin to Christie, 1 October 1914, Gwatkin Papers, Public Archives of Canada (hereafter PAC); Preston, "Imperial Defense," 469; R.L. Borden, *Memoirs*, vol. 1 (Toronto, 1930), 454–55.

9. Canada, House of Commons, *Debates*, 21 August 1914, 56.

10. Perley to Borden, 18 September 1914, Perley Papers, vol. 1, PAC.

11. Nicholson, *Canadian Expeditionary Force*, 381; Preston, "Imperial Defense," 488–92.

12. Derby to Haig, 2 November 1917 in Robert Blake, *The Private Papers of Sir Douglas Haig* (London, 1952), 266.

13. On the Canadian contingent, see Norman Penlington, *Canada and Imperialism, 1896–1899* (Toronto, 1965), chaps. 16–17. See also Desmond Morton, *The Canadian General: Sir William Otter* (Toronto, 1974), 161–65.

14. Ibid., 206–208; *Globe* (Toronto), 4 April 1900; "Lecture on the Paardeberg Campaign," Otter Papers, PAC.

15. See, for example, W.H. McHarg, *From Quebec to Pretoria with the Royal Canadian Regiment* (Toronto, 1902).

16. Preston, "Imperial Defense," 323ff; Desmond Morton, *Ministers and Generals: Politics and the Canadian Militia, 1868–1904* (Toronto, 1970), 186ff.

17. Nicholson, *Canadian Expeditionary Force*, 6–11.

18. Ibid., 14–16; Duguid, *Official History*, vol. 1, Appendices, 18–20; Gwatkin to Christie, 1 October 1914, Gwatkin Papers, PAC.

19. Canada, Department of Militia and Defence, *Supplementary Report: Organization, Equipment, Despatch and Service of the Canadian Contingents during the War in South Africa, 1899–1900* (Ottawa, 1901), 11–13.

20. *Debates*, 21 August 1914.

21. M.G. 24, H.Q. 54-21-23-13, vol. 2, PAC. See also Nicholson, *Canadian Expeditionary Force*, 359–61; Perley to Borden, 16 February 1915, Borden Papers, RLB 997(1), 117471ff., PAC, for arrangements.

22. Borden to Perley, 12 January 1915, Perley Papers, vol. 3.

23. Nicholson, *Canadian Expeditionary Force*, 27–28; Hughes to Borden, 1–2 September, 1916, Borden Papers, OC 308, pp. 33583–88.

24. See Carson to Hughes, 4 October 1915, R.G. 24, Carson File, 8-5-10A.

25. Nicholson, *Canadian Expeditionary Force*, 212–22; Desmond Morton, "French Canada and War, 1868–1917" in Granatstein and Cuff, *War and Society*, 98–99. On surplus officers, see White to Borden, 11 November 1916, Borden Papers, OC 335, p. 39205 and memorandum, p. 39248; Kemp to Borden, 24 March 1917, OC 414.

26. See Alderson to Hutton, 20 September 1915, Hutton Papers, British Museum, Add. 50096, p. 315; Major General G.L. Foster to Deputy Minister, OMFC, 6 April 1919, M.G. 24, O.S. 30-0-0(c); Turner to Carson, 1 December 1915, Carson File, 8-5-10A. To illustrate problems, see Alderson to AMA, 2nd Army, 13 December 1915, Borden Papers, OC 337, p. 39337.

27. See John Swettenham, *To Seize the Victory* (Toronto, 1965), 125–27; Hughes to Carson, 5 December 1915, Carson File, 8-1-26; Carson to Steele, 6 January 1916, 8-5-10B; Carson to MacDougall, 1 February 1916, ibid.; Steele to Sifton, 10 January 1919, A.L. Sifton Papers, PAC, vol. 14.

28. See Carson to G.O.C., Southern Command, 18 February 1915 cited by Nicholson, *Canadian Expeditionary Force*, 202.

29. Perley memorandum, n.d., Perley Papers, file 77; Perley to Borden, 9 May 1915, ibid., file 79; Carson to Hughes, 9 July 1915, Carson File, 6-C-12.

30. Duguid, *Official History*, 1:154, 161; Privy Council order 29, 6 January 1915. In September, Aitken was appointed "General Representative of Canada at the Front." See also Preston, "*Imperial Defense,*" 478–79.

31. Hughes to Aitken, 30 November 1915, Borden Papers, OC 318(1), p. 35582.

32. Borden to Hughes, 8 June 1915, ibid., OC 165(2), p. 12693; see, for example, Hughes to Kitchener, 27 April 1915, ibid.

33. See D.M.A.R. Vince, "The Acting Overseas Sub-Militia Council and the Resignation of Sir Sam Hughes," *Canadian Historical Review*, 31, 1, March 1950.

34. Hughes to McArthur, 23 March 1911, Borden Papers, OCA 36, v. 134.

35. See Perley to Borden, 15 August 1914, Perley Papers, file 16; Borden, *Memoirs*, 2:498; Perley to Borden, 6 December 1914, file 124A; Perley to Borden, 29 May 1916, ibid., file 157A.

36. Summarized by Nicholson, *Canadian Expeditionary Force*, 209–12; the Ministry established by Privy Council order 2656, 31 October 1916. On Hughes's resignation, see Borden Papers, OC 318(2). See also Hughes to Aitken, 8 December 1916, Beaverbrook Papers, F/1/24 letter 65, PAC. Perley's own proposal was for a "small council or committee" see Perley to Borden, 6 September 1916, Borden Papers, OC 318(1), p. 35723.

37. Perley to Borden, 27 January 1917, Borden Papers, OC 176, p. 13638.

38. See Sir Andrew Macphail, *The Official History of the Canadian Forces in the Great War, 1914–1918: Medical Services* (Ottawa, 1925); Colonal H.A. Bruce, *Politics and the Canadian Army Medical Corps* (Toronto, 1919).

39. On Perley's management, see Nicholson, *Canadian Expeditionary Force*, 211–12; Macphail to Sir Thomas White, 5 April 1919, Kemp Papers.

40. Perley to Borden, 10 December 1917, Borden Papers, vol. 79, pp. 41143ff. See Desmond Morton, "Polling the Soldier Vote: The Overseas Campaign in the 1917 General Election," *Journal of Canadian Studies*, 10, 4 November 1975.

41. Preston, *"Imperial Defense,"* 467.

42. Borden to Perley, 6 February 1915, Perley Papers, vol. 3; Nicholson, *Canadian Expeditionary Force*, 39–40.

43. Alderson opposed the promotion of Turner. See Alderson to Hutton, 21 August 1915, Hutton Papers, Add. 50096, p. 310. On Currie, see H.M. Urquhart, *Arthur Currie: The Biography of a Great Canadian* (Toronto, 1950); A.M.J. Hyatt, "The Military Career of Sir Arthur Currie" (Duke University, Ph.D. 1965).

44. Alderson to Hutton, 20 September 1915, Hutton Papers, Add. 50096, p. 315.

45. On the rifle, see Hughes to Alderson, 7 March 1916, Borden Papers, OC 318(1), pp. 35383–85.

46. Nicholson, *Canadian Expeditionary Force*, 136–45; Hyatt, "Currie," 76–77.

47. Nicholson, *Canadian Expeditionary Force*, 146–47; Aitken to Hughes, 26 April 1916, Borden Papers, OC 183(2), pp. 14955ff; Hyatt, "Currie," 86–87; Robert Blake, *The Private Papers of Haig*, 140; A.J.P. Taylor, *Beaverbrook* (London, 1972), 89.

48. Swettenham, *To Seize the Victory*, 170–72; Nicholson, *Canadian Expeditionary Force*, 283–84; Gow to Perley, 29 June 1917, R.G. 24, O.S. 10-8-7; Perley to Lord Derby, 15 June 1917, O.S., file 348. On Canadian appointments, see "Growth and Control of the Overseas Military Forces of Canada," Borden Papers, vol. 298, p. 173819.

49. Overseas Ministry to War Office, 14 June 1917, O.S. 10-8-7; Perley to Kemp, 30 June 1917, ibid.

50. Carson to Hughes, 9 July 1915, Carson file, 6-C-12. On Sims' role see

Perley to War Office, 25 January 1917, O.S. 10-8-1.

51. Perley to Borden, 10 December 1917, Perley Papers, file 307; Colonel Frank Reid to Lt. Col. W.P. Purney, 30 November 1917; 17 December 1917, PAC., Murphy Papers, pp. 16046, 16055.

52. See Perley Papers, file 348; Duguid, *Official History*, vol. 1, Appendices, 5.

53. Overseas Ministry to War Office, 27 November 1917, O.S. 10-8-7.

54. Nicholson, *Canadian Expeditionary Force*, chap. 16, pp. 485–90, 499–500.

55. Borden to Perley, 13 October 1917, Perley Papers, file 281A; Perley to Borden, 16 October 1917, ibid. See Wilson, *Imperial Policy*, 41.

56. Kemp to Borden, 24 February 1918, Kemp Papers, vol. 176; Kemp to Borden, 2 April 1918, Borden Papers, OCA 98, pp. 73957–58.

57. On the Canadian Section, see Privy Council order 885, 11 April 1918.

58. Perley to Milner, 15 June 1918, OS 10-8-7.

59. Lieutenant General A.H. Lawrence to Overseas Minister, 23 June 1918, OS 10-8-7. See also minutes of conference of 2 April 1918, OS 8-52; see Borden Papers, OC 485D, pp. 51362–71.

60. Currie to Kemp, 16 July 1918, O.S. 10-8-7.

61. W.B. Kerr, *Arms and the Maple Leaf* (Seaforth, 1942), 37–40.

62. See, for example, J.A. Swettenham, *McNaughton*, vol. 1, *1887–1939* (Toronto, 1968), 49–165; Larry Worthington, *Amid the Guns Below: The Story of the Canadian Corps, 1914–1918* (Toronto, 1965) on Brutinel; Urquhart, *Currie*, 194–95.

63. Swettenham, *To Seize the Victory*, 204–5.

64. Preston, *"Imperial Defense,"* 490.

65. Ibid., 202; Kemp to Borden, 8 February 1918, Kemp Papers, vol. 182; Currie to Kemp, 7 February

1918, Borden Papers, OC 494, pp. 52798–802.

66. See Nicholson, *Canadian Expeditionary Force*, 232. For a contrary view, see Leslie M. Frost, *Fighting Men* (Toronto, 1967).

67. Perley to Borden, 9 May 1915, Perley Papers, file 79.

"DONE IN OUR OWN COUNTRY": THE POLITICS OF CANADIAN MUNITIONING *

RONALD HAYCOCK

o

The 1987 defence White Paper on Defence Policy promised Canadians, after a long hiatus, a definite strategy for industrial preparedness. Almost imme-diately, the promise collapsed as surely as the Berlin Wall and the Cold War. To industrial mobilization planners at National Defence Headquarters who had been beavering away on these lines since 1984, this turn of events must have been a great disappointment.[1] But it should not have surprised them. From the earliest colonial times, Canadian history is littered with sim-ilar episodes. The processes of Canadian peacetime munitioning have been profoundly affected by changing threat perceptions, considerations of national sovereignty and identity, economic, political, and social concerns, and by resources and skill availability. This paper is concerned, then, with uncovering the factors affecting the creation of Canada's historical muni-tions development from Confederation to the Great War.

In the decades prior to 1867, a series of crises with the United States that had in part made Confederation necessary, alerted Canadians to defence issues. Generally, these crises were resolved by injections of British troops and equipment. In the minds of many British North Americans, since the problems arose from British policy, British taxpayers should therefore foot the bill. During the Crimean War however, many British units were removed from North America. The Canadas reacted by creating the Volunteer Corps. It was a crucial decision, for these soldiers had to be equipped and there were no such resources in the frontier societies of Canada West or East. Consequently, Colonel the Honourable E.P. Taché travelled to London where he purchased all types of war stuffs to accoutre

---

* This paper has not been previously published.

the new Canadian "Volunteers."[2] Interestingly, this act not only hooked Canadian soldiers to the use of British equipment but it pointed out several other ongoing themes relating to policy and procurement.

As Lord Panmure, the British Secretary of State for War cheerily noted in early 1856 to his Commander of British forces in North America, "The fact is we have denuded North America too much, and Jonathan [the USA] is again bumptious."[3] However uncertain or intermittent the American threat, it was obvious that Canadians could not defend, let alone arm themselves. Taché's purchases in far away Britain were expensive, long-delayed, and at best a stopgap. In the end, these events of the mid-1850s "disappeared in smoke," just as Lord Panmure had predicted.[4] Yet, with the five British regiments that the Imperial government had moved into Canada at the end of the Crimean fracas, there was also the arrival of 10 000 rifles and other equipage plus four million rounds of ball ammunition. Canadian volunteers were given access to it on a pay-back basis, calculated as the War Office inventory cost plus shipping and handling. Once again Canada had been saved by imperial troops and imperial factories, and for some Canadians, there seemed to be little need to do more or to keep large munitions reserves.[5]

By the time of Confederation, the American Civil War and the Fenian raids spawned another convulsion to re-arm, together with large injections of the mother country's soldiers and munitions, and the creation of the first Dominion Militia Bill and a new Ministry of Militia and Defence.[6] The latter in turn imposed the question of equipment for the fledgling force. The answer would become a fascinating, and never-ending national munitions saga. On the surface, its history looks like chaotic mismanagement that only reacts to crisis and to impulses beyond Canada's control; below the surface, it is not so erratic or inexplicable. Generally speaking, the period from Confederation to the Great War involves simple munitions with a complex history. Basic items such as boots, uniforms, personal equipment, gunpowder, rifles, and cartridges are at the centre of the story. But behind their production lay the complicated processes of politics, policy, and the role of the state.

In 1867, Canada was protected by the Royal Navy and the tenets of imperial defence, or so Edward Cardwell, the British Secretary of State for War, had promised. Domestic defence was pinned on a militia system mostly of volunteer part-time infantry formed in local battalions hardly larger than companies. They trained about twelve days a year. Equipment needs were simple because the force was simple. Political patronage, according to John English, bridged the gap between the federal government and the local constituencies for the distribution of goods and services.[7] Consequently, like Canadian society generally, the militia was highly political in its districts. During the 1870s at Militia Headquarters in Ottawa, there were few staff officers beyond the British General Officer Commanding and a Canadian Adjutant-General. And there was little perceived need for any more. Military equipment procurement did not even come under the purview of the soldiers; rather, it was controlled by the Deputy Minister of

Militia and Defence and his civilian-staffed Stores Department. Through them, the ever-watchful minister could keep a patronage eye on what, to most politicians, was just another useful means of dispersing favours. There were no inspection services for most war-like stores. Most munitions were kept in district storehouses under little or ineffective supervision.[8] Such then was the immediate equipment environment. There were other considerations best exposed by tracing the individual munitions themselves.

No sooner had British authorities promised that if necessary, all of the resources of the Empire would be used to defend the new Dominion, than the reforming British Secretary of State for War pushed Canada onto the path of "more self-reliance and more self-relying habits" as British Prime Minister Gladstone had flatly put it.[9] The subsequent withdrawal in 1871 of Britain's inland North American garrisons left enough weapons, ammunition, and other kit to keep the small Dominion Militia force supplied for years. Forty thousand .577 inch Snider rifles were consequently added to a bewildering assortment of American small arms that the previous colonial governments had purchased in the United States to shoot back at Americans.[10]

Such benefits made politicians both happy and smug. Prime Minister John A. Macdonald preferred to keep the militia simply equipped. In that state they would not become efficient and demand more; nor would they be attractive to British planners who wanted Canadians to participate officially in the defence of the Empire abroad. With free arms and ammunition and no visible enemy, the defence expenditures were kept at a minimum. The Prime Minister could attend to costly national development in other spheres. Another advantage was the reduced British internal influence once their garrisons had left. For the wily Macdonald, the militia would then be at its useful political best. Macdonald could always point out to imperial planners that Canadians were defended by their own volunteer force, a form of self-reliance that England had been advocating for years.[11]

But these advantages gave the soldiers, particularly the British advisors serving in the embryonic Canadian force little joy. Newly found and existing military equipment was growing obsolete and deteriorating quickly in bad storage conditions. Interestingly, the first serious impulses for new munitions came not as a result of need for modern armament but from the demands for militia uniforms. Put simply, by the mid-1870s the little military wearing apparel that was available was of no particular pattern and in a state of rags. Politically, the militia had to be kept satisfied, and uniforms gave the volunteers identity and local status. There was also a greater likelihood of suitable cloth being manufactured in Canada. Martin van Creveld observed that the best munition is one taken directly from a civilian item currently in commercial production.[12] In the 1870s, Canadian cloth manufacturers were well established and placing government contracts with them was a politically useful option. It was believed that if supply could be obtained, perhaps it would be less expensive than buying clothing through the War Office's government uniform factory at Pimlico in London's garment district.

Unfortunately, when Ottawa first placed orders with the Canadian cloth trade in 1875, the results were unsuccessful. The Canadian product was, according to the Director of Stores, clearly "inferior."[13] When no more success was achieved three years later, the General Officer Commanding (GOC) reported to his minister some higher truths applicable to all Canadian munitions history: such military cloth was not usually produced in Canada; and the militia's demands were so small that manufacturers could not be attracted into the military market. Their production was so costly that, when coupled with the competition of less expensive British products, there was "not much desire by any [Canadian] mill to undertake it."[14] It would be another decade before Canadian factories would produce durable wearing apparel that met specifications. Some mills only became efficient by concentrating on what they did best—such as military great-coats. Expertise and competence built up slowly over time.[15] Until that point, Canadians depended on England for uniforms.

Arms and ammunition however, were much more vital to fighting efficiency. Even when the imperial regiments left in 1871, the Adjutant-General in Ottawa still had to make substantial purchases from British army stocks including artillery, rifles, and ammunition. The following year, Colonel Robertson Ross, the British Officer Commanding the Canadian Militia pressed his minister on strategic grounds for the establishment of some small arms ammunition manufacturing capability for "such warlike stores as are annually required for practice and for the maintenance of a sufficient reserve."[16] His successor, Selby Smyth made similar appeals. But as would frequently happen, larger questions intervened. An intermittent and at times severe economic down-turn over the next twenty years prevented both Macdonald's Conservatives and Alexander Mackenzie's Liberals from spending any more than minimum on the force to keep military enthusiasts politically loyal. Providing more equipment would have meant increased efficiency; in turn, for politicians, that would result in more demands placed on the hard-pressed government by both Canadian soldiers and imperial strategists anxious for Canadian help. Moreover, parsimony in defence spending would not alarm the Americans. Drastically lowered militia budgets, for instance, were translated into severely reduced equipment purchases or cuts in proper inspection or storage of older munitions. Indeed, as the existing supplies deteriorated, the GOC had to cull the actual force numbers to fit what equipment he had to give them.[17]

As would happen again and again, crisis spurred the government into action as was the case with the Russian scare of 1878. At the time, Great Britain and Russia nearly came to conflict over the fate of the "sick man of Europe," Turkey. The British General Officer Commanding the Canadian Militia believed that being a part of the British domains and yet relatively defenceless, Canada could suffer the "Penalty of Empire" by being a target for Russian naval attack. Naval raids would imply two things: coastal artillery and local defence, meaning guns, rifles, and ammunition.

Alarmed by the possibility of war, Selby Smyth wanted Canada's obsolete smooth-bored muzzle loading artillery replaced with modern breach-loading rifled guns. But they were costly and took too long to acquire from the British who were themselves going through the same throes of a gunnery refit. The supply would be even more limited since other parts of the Empire were also badgering British gun makers for heavy artillery. Canada had no choice but to convert the old heavy guns on Sir William Pallisers' principle—a hot, rifled-sleeving system then being adopted in the United Kingdom as a temporary expedient for updating old guns.[18] The urgency of time, distance, and cost forced Canadians to consider having the conversion done at home. And so, in 1878, Selby Smyth urged his minister to react. When Sir Adolphe Caron did not respond, Palliser brought the case directly to the Prime Minister in a bombardment of letters. Selby Smyth also attempted to convince Macdonald by using every argument he could; such as appealing to national pride and national security and, of course, by making the frequently-to-be-heard promise of economic benefits to all. The prime minister found it hard to resist the appeal, especially since he had recently been returned to power on the electoral plank of the National Policy (NP) after a five year stint in the political wilderness. One aspect of NP was to encourage and support struggling domestic industry hurt by the recession. The soldiers had done their homework: they presented the politicians not only with a problem but also a ready-made solution. They had carefully co-ordinated their plans with the Gilbert brothers of the Montreal Engine Works who were the proposed "manufacturers," and with the inventor's brother. Optimistically, Selby Smyth told the prime minister that "Canadian workers using Canadian iron can do the whole work and the cost goes to Canadian citizens." This would make it cheaper than having the job done at Woolwich in the United Kingdom.[19]

The GOC's optimism was short-lived; the Montreal factory could not do the job. The technical process, skill, and facilities were beyond their capabilities. Ottawa had invested over $11 000 of public money to help fund the Gilbert brothers. Six years later in 1884, the effort was still a failure. The embarrassed government threatened a lawsuit to try to recover its investments. Clearly, private enterprise was not the sole solution to munitions production; nor was munitions production going to be as easy as some had thought. And ordnance was to come from England just before World War II.[20]

These same crises and political factors spawned another creation of far more fundamental and lasting results. This was the establishment of the first government-owned and -run ammunition plant. Throughout the 1870s, soldiers lobbied to have the Dominion government build a comprehensive state arsenal system modelled on that of Britain's. They pushed their ambitious plans through the annual Militia Reports and in essays and lectures espousing the benefits of such action, hoping to galvanize opinion in a favourable way. Moreover, the current strategic agency of the British Empire, the Colonial Defence Committee in London, put strong pressure on the senior colonies like Canada to start domestic arms industries as part of accepting larger self-defence responsibilities. Imperial strategists had been aware for some time that the Empire was too large to be munitioned from England. It was also a heavy tax burden. The British arsenal structure was centred on the Royal factories of Woolwich, Enfield, Waltham Abbey, and Pimlico. It was an ancient institution of substantial competence, but it was currently shuddering under technical strain and the subsequent reorganization that the ever-changing face of modern war demanded of munitions manufacture.[21]

In the brief space of thirty years (1867–1897), in the field of small arms alone the British forces witnessed tremendous changes. There was, for instance, the invention of breech-loading magazine rifles, metallic cartridge based on the Boxer system, smokeless gunpowder, full metal jackets for high-speed small bore projectiles, and machine guns that could fire 670 rounds a minute.[22] In addition, since the early nineteenth century, many European states had built the "nation-in-arms" achieved by the *levée en masse*. This meant that the mobilized societies would have to be armed. The British were staggered by all of this and their arsenal system could not keep up with the bewildering changes or with production. With the often feisty Canadian politicians like Macdonald and Mackenzie showing increased independence from imperial fortunes, British-trained Canadian soldiers and the imperial officers who dominated the small Canadian defence organization, saw the promotion of an arsenal as a way of maintaining political influence in Ottawa and promoting Imperial defence. They also knew that the Woolwich connection guaranteed them modern equipment. A British regular soldier could be put in command of the new factory much as the GOC in the Ottawa Headquarters was by law an imperial soldier. They could be certain that Canada would receive sound technical advice, and as such the technical dependence would increase the use of British equipment in Canadian service. These soldiers liked the idea of uniformity of munitions and its sense of familiarity.[23] Another incontestable argument for a public arsenal was demonstrated by the failed Palliser's gun conversion: private industry could not do it. When unable to get artillery gunpowder from British sources during the Russian scare, Ottawa had desperately ordered 29 000 pounds of gunpowder from a new domestic business, the Hamilton Powder Company. The product had not been tested by the usual British authorities and when it was, it failed. Inspite of the risks, Ottawa tried to convince the company to make urgently needed rifle ammunition. The company refused not on the grounds of ability, but on those of profit. Government orders were too intermittent and too small to justify the expensive toolage. And the company's directors also noted that the .577 inch army cartridges had no civilian application.[24]

Interestingly, the shakers and the movers of munitions industries were not politicians but soldiers. The former did little unless required to. For two years after the Russian affair nothing more was done about the arsenal. But it was not all political sloth. Macdonald and his Militia Minister had been bombarded with memos by Selby Smyth about the deteriorating equipment situation. Once the crisis was over, the GOC had to use other, more attractive "peacetime" arguments: the benefits and profits to Canadian workers; a pride in a "national Military enterprise"; and an example to the Empire of Canadian industrial prowess were prominent among the GOC's ideas. But perhaps his creative machinations were also reasons why the politicians stood aloof on the arsenal for some time. Smyth had connected the technical question of national military self-sufficiency with the strategic aim of the creation of a Canadian "Imperial Reserve of troops complete with munitions for War ... for the defence of ... the Empire at large." Macdonald at least, did not want that link. Two later Chiefs of Staff, A.G.L. McNaughton and E.C. Ashton made the same connection in the 1930s, and they received the same reaction from Prime Ministers Bennett and Mackenzie King.[25]

However, by 1880 Macdonald was forced to face facts. Stocks of rifle ammunition were so low that the militiamen had been allowed to fire only twenty rounds each during several of the annual camps. In addition, the numerous rifle associations, themselves considered an inexpensive civilian militia support group, fired off one half the ammunition available each year. English cartridge imports were expensive and took a long time to obtain. In Parliament, the militia "colonels" were complaining bitterly, and, as Desmond Morton shows, in 1878 there were forty of these officers sitting in the Commons.[26] And so, Macdonald acted on the arsenal idea.

What the soldiers got in 1880 was only a shadow of what they had wanted. It was a simple cartridge factory producing nothing but rifle ammunition. It took six years from its inception in 1878 until it produced its first round in 1884. All of the social and political imperatives were considered equally with matters of production and of military efficiency. The first superintendent was Oscar Prévost, a prominent French Canadian with strong connections with the old Tory Quebec lieutenant, Joseph-Adolphe Chapleau. Prévost was a militiaman-lawyer turned regular gunner. Woolwich-trained to prepare himself, he had to secure all his machinery in England and was aided by several English military technicians on loan. Still he had difficulty acquiring the machine tools and the raw materials that Canada had none of. Nevertheless, the new facility was housed in the old "French" barracks of the Citadel in Quebec City. The records reveal a real scramble and much controversy for local patronage employment and what is now called "spinoff."[27] The process was not easy or short, and it was full of political liabilities.

Inspite of this modest beginning, there are several reasons why the cartridge factory was an important step. First, the government took the plunge into the heretofore uncharted waters of public involvement in the private domain, most likely because no one else was willing to. Supported by state funds and responsible to the Militia Minister and the Cabinet, the factory was a military facility run by a regular force officer, and thus relatively safe from direct political patronage in its operation. This small plant was also the first major technical service innovation for the Canadian force. Soldiers, both professional and amateur, were pleased with this step if only because they, the "users," had more ammunition to shoot. But more sophisticated military men knew the broader implications: users were influencing policy, and from this humble start, they hoped expansion could extend to other munitions areas. However, common sense forced the soldiers to admit that a complete arms production facility was simply too expensive at that point. With an increasing anglicization of the militia excluding them at the unit level, French Canadians could take consolation in Prévost's appointment as superintendent at Quebec. If subsequent munitions history is any indicator, Prévost and the little Quebec factory set the stage whereby public munitioning, like the Office of the Deputy Militia Minister itself, became the particular political preserve of French Canada. And it would balance the preponderant British influence on the Headquarters Staff and in the permanent corps.[28]

Despite its slow start, events justified the creation of the factory. During the North West Campaign in 1885, Prévost doubled production in a few months. And the cartridges were of good quality and well liked by the civilian shooters. Even though the plant was nothing more than an assembly facility still dependent on foreign sources for nearly every raw resource, skilled operator, and machine tool, the costs remained reasonable. And quite fortuitously, by the mid-1880s, the Quebec factory was helped by problems in British public and private munitions industries. When poor quality and expensive artillery ammunition began arriving in Canada, Prévost received permission to try making gun rounds at Quebec. Meanwhile, in England, the arsenals were suffering because of apparent strategic policy confusion and financial difficulties. Both the public and "trade" sectors of the British munitions industry were characterized by a series of reform measures, policy debates, and economies until the Great War.[29]

In many ways these events affected the Canadian situation after the 1890s. Given both the English and domestic Canadian situation, the HQ soldiers increasingly saw the solution to their supply problem as one of expanding the product base of the Quebec cartridge facility. In 1883, General Luard, Selby Smyth's successor had lamented that most of the arms and other stores were in frightful decay and in very limited supply. Part of his solution was to recommend the adoption of certain Canadian inventions as substitutes for British items.[30] About the same time, with the Palliser's conversion experiment failing in private hands, one of its promoters tried to convince the Cabinet to add a gun-making plant to the public cartridge factory as a means of saving the enterprise. That suggestion, like the adoption of most other Canadian-invented equipment went nowhere, largely because of cost, too small a Canadian demand, and a faith that the War Office could supply the needed munitions. On top of these limitations, Canadian government officials and military staff simply did not have the organization or the expertise necessary to evaluate inventions. After initial examination most ideas were passed on to the imperial authorities, such as those on the Ordnance Board. Few of these Canadian inventions survived British technical scrutiny. Back in the Dominion, other factors, such as lack of an enemy, time and distance, and rapidly changing technologies made most politicians decide to wait; consequently, they had a policy of no policy.[31] Yet some politicians, like the deputy Militia Minister, Lieutenant-Colonel Charles Panet, indicated that by the late 1880s there was no military equipment that could not be made in this country, "armament excepted." If this was the case, surely then the political reluctance to become more involved in the public sector was due to not wanting to spend precious funds in an area that had much less patronage benefit than that spent in the trade. For example, Sir Adolphe Caron was Minister of Militia for eleven years (1880–1891). He spent an entire career "devoted to the manipulations and negotiations of patronage politics" but he did little for the augmentation of the government factory.[32]

Given this stifling political environment, in the 1890s the supervisor at Quebec's "cartoucherie" started to use a different tactic. He was plagued by poor materials and component delivery delays from Britain that had stopped Quebec's production several times. He was also hit hard by declining budgets. Consequently, Prévost decided to shift from emphasizing only a well-rounded government arsenal system to a combination of private "trade" and public production. The state-owned facility, he now admitted, could not satisfy modern needs, but it should be diversified to produce the more "essential and special" war materials by being given the most modern toolage and research facilities. Furthermore, the state plant was to handle a wide enough assortment of munitions that it would act as a schoolmaster for private industry in major wars. In peacetime, the "arsenal" was to be the first among the equals but not the only one.[33]

No doubt survival was not the only reason Prévost made his proposal: the same was happening in England, from where most technical and policy information came to Canada. However, inspite of Prévost's efforts little happened. Continued depression and, after Macdonald's death in 1891, a collapsing Tory party, left the government in chaos. And, except for the short-lived Venezuelan crisis in 1896, there was little military need. More serious military crises would have to come along before much would shake a government that knew, if possible, that the only realistic policy was to avoid both antagonizing the United States and becoming entangled in British imperial adventures.[34]

In the meantime, quietly and within its existing small budget, Prévost's plant managed to carry out useful research even if it had little production beyond rifle cartridges. It converted machine guns from .45 to .303 calibre, experimented with new bullet gilding metals, and produced solid-drawn brass cartridges as well as a few gun shells. Its staff also informally examined a whole host of local inventions. It even made a prototype rifle in its toolroom. Yet, beyond keeping abreast of some technical knowledge, such efforts indeed gave small satisfaction. Most supplies still came from Great Britain either through the War Office inventory or from the private arms companies such as Vickers, Birmingham Small Arms, London Small Arms, or a host of others. Most were in competition with each other, trying to sell arms to the Canadians whom they knew were holding onto old and decrepit munitions. Most of these "offers" were refused. At home, uniforms and boots were made by a few producers; then at the time of the annual camps there would be a quick flurry of orders, mostly for camp consumables placed with politically acceptable contractors. After that, there was nothing until the next year's annual muster.[35]

These years also pointed out an interesting dichotomy in Britain's attitude to Canadian munitions. For years, strategic bodies like the Colonial Defence Committee or later the Committee for Imperial Defence (CID), had promoted colonial supply self-sufficiency; yet they also demanded imperial uniformity in munitions. Since the 1870s, Britain insisted that the colonies buy supplies through the War Office and not through the private makers. However, that same source often could not meet demands, especially in a crisis; serious delays were not uncommon; prices were sometimes high and occasionally the product was second-rate. Faced with these problems, the Canadians frequently resorted to buying vital stores directly with the British trade through the offices of the Canadian High Commissioner. If there is a relationship between military equipment and strategic policy in an alliance, certainly this practice gave Canada more control. But when Ottawa made such purchases, the British authorities complained bitterly that Canadians were undermining the War Office's control of the "trade." Put another way, colonial purchases gave British trade a new found freedom from the strictures of War Office specification, pricing, and the irregularity of orders. As Clive Trebilcock points out, the British private munitions makers suffered badly from "boom or bust" government impositions. Colonial orders made it less harsh. Most states never resolve the munitions dilemma of being both customers and competitors. No doubt this fact makes governments more cautious of outside control and competition. Indeed governments often wanted to (and did) hold the arms manufacturers in a tight "client" grip. Canada continued to rankle imperial officers by using both British production sectors; but increasingly, shifts in imperial fortunes guaranteed the War Office most of the business.[36] And that would remain so in peacetime as long as Ottawa would do little to diversify the cartridge factory.

As noted before, military threat is always the great catalyst for munitioning. Nevertheless, other conditions must also prevail. So it was in Canada in 1899 at the outbreak of the Anglo-Boer War. The first prerequisite was the presence of new and dynamic political leadership. This was the Liberal Militia Minister, F.W. Borden, who was to serve in that post longer than Caron had for the Conservatives. The second condition was a dual one: an upswing in the economy and a series of aggressive soldiers who wanted to create, "a Canadian national army." Nevertheless, it was perceptions of threat and war itself that were the immediate activators.

In early 1899, E.T.H. Hutton, the strong-willed British Major-General commanding the militia, had "sub rosa" created plans to send a significant Canadian contingent to Africa hoping to coerce his reluctant Ottawa masters into supporting Britain's imperial policy. But he had done little to ensure the forces' equipment for this expected short operation. The general assumed that once out of Canada and safely under imperial command, the troops would be supplied by the British system. Yet, when the call came that October, and for most of the unexpected two and a half year length of the conflict, supplying the 8343 Canadian volunteers was best described as crisis management. The frantic Superintendent of Military Stores had little to give the initial force. Clothing, for instance, "had to be made—the contractors had none on hand." Demands for ammunition caused the government factory to quadruple its production and its staff. Given the circumstances it did a remarkable job.[37] Still the long war had its lessons.

A major lesson was that foreign dependence as a characteristic of peacetime munitioning was in some areas only heightened in war. Moreover, interesting new forms of dependence revealed themselves. The cartridge factory, for instance, could not get semi-manufactured copper-strip or brass cups from the usual British sources that were hard pressed by their own munitions demands. Consequently, the cartoucherie had to install its own rolling mills and smelting facilities. Indeed, after leaving the country, Canadian troops were supplied by the imperial system. Back at home Canadian contractors at first got some very large orders but then little else in direct contracts. As Carman Miller points out, the major ally, Britain, placed some very lucrative contracts for war stuffs in the Dominion. The War Office purchased in total about $7.5 million worth of supplies. And some manufacturers were very buoyant and benefitted substantially because of it. However, these orders did not spawn efforts to create sophisticated munitions products: Canadians were asked for hay, clothing, food, and some small manufactured items. Moreover, these were not orders by the Canadian government. The Canadian force was not large enough, nor did Ottawa's or London's policy allow it to be independent enough to attract new private or public concerns or products into the field. It was only labour and civilian or modified-civilian munitions products that the British were interested in from the Canadians.[38]

There were one or two exceptions. The first was harness equipment for soldiers; the other was a "Canadian-made" rifle. In the former case, the Oliver valise harness, the rigging from which a soldier carried his personal equipment, was actually made in Canada the year before the war started. But it was not a Canadian design; rather it was a British one dating back to the 1870s that the War Office had rejected. Invented by a British army surgeon serving in Halifax, the harness took twenty years of torturous political squirming and the direct interference of a father of Confederation, Sir Leonard Tilley, before Ottawa gave out a contract for its production.[39] Once again, the Canadians had been driven to act by military crises (Venezuela 1896) and by existing obsolete valise equipment of buff leather and pipe clay. And once again they had waited for the British lead hoping to get an imperial pattern. However, the War Office had itself been indecisive in trying to figure out the best harness to match the new magazine rifle, the .303 inch Lee-Enfield. Unable to get the War Office to supply, in 1898 the GOC in Ottawa ordered trials to test several patterns of harness, one of which was a model developed by a Mr. Lewis of the Militia Stores Service and vigorously promoted in political circles by Captain E.F. Wurtele, the manager of a new Canadian concern, the Quebec Military and Supply Company. In the end, Oliver's pattern won and he was given $5000 for his patents—small compensation for twenty years of effort. Certainly Oliver's idea was good: he conceived that in modern war, a soldier's carrying rig must in essence be a complete personal magazine, making him self-sustained in battle for at least twenty-four hours.[40] Made by local manufacturers, this was the soldiers' harness many Canadians took to the Boer War. There, the equipment proved to have flaws: it was not the same as the imperial pattern; it chapped and choked its wearer and, critically, its cartridge pouches held about half as many .303 ball rounds as the English pattern and only one third those of Lewis'. After the war, the government did not change the pattern because of large existing stocks, high costs, and the fact that the available British equipment was also found wanting. Furthermore, the War Office promised Canadians that a new pattern would be forthcoming soon—"soon" took nearly a decade. And Canadians went to the First World War wearing modified versions of the Oliver harness, which by 1914 was all they had, and which by then had entered the risky lists of national pride.[41]

Oliver's harness was a comparatively quiet debate compared to the one that raged over the procurement of the first Canadian-made rifle. Providing an infantry weapon to defence forces made up primarily of militia foot soldiers supported by semi-official civilian rifle associations, was an old issue in this country. Before the turn of the century when the professional soldiers, namely the cartridge factory superintendents, failed to have rifle-making added to their charge, they supported promising local inventors like the young shooting buff Charles Greville Harston. In the 1880s, Harston had come up with a device for converting the venerable single-shot Martini-Henry .45 calibre rifle to a magazine-fed weapon. But, like most Canadian military inventions for years to come, progress was retarded by the absence of any agency for the systematic assessment of ideas. Consequently, inventions usually had to follow the highly political route of the personal lobby to get official attention. Such was the case with Harston. In vain, he tried to get the politicians in Ottawa to attach a plant to the Quebec factory to make 10 000 of his pattern rifle. His argument to Ottawa claimed that it would only cost taxpayers about $5000 and that there were "in Quebec both mechanics and gunsmiths quite capable of doing the job."[42] The soldiers also pressed the militia minister to adopt Harston's design, emphasizing that the rifle and ammunition were the same as the British pattern—which of course it would not be if Harston's conversion was made. Having failed to convince any politician, the inventor then tried the private or "trade" arguments; but when he asked the government for a start-up order of at least 40 000 rifles and import duty concessions on materials and machinery, all he received was stony silence. Frustrated, he then submitted his idea to the British War Office, who rejected it because they wanted a new design, small calibre magazine rifle, not a conversion of an older pattern. Harston achieved nothing and by the early 1890s the British settled on James Paris Lee's .303 rifle amid great controversy and delay. Meanwhile the Canadians waited for the War Office to settle their new weapon's teething problems while keeping their obsolete single shot Snider in service.[43]

To wait may have been a good tactic in peacetime but not in war. So it was for Canada in the Boer War. The militia minister, Frederick Borden, was not willing to tolerate the same British delays or rejections that he had encountered when he requested the latest British pattern small arms for Canada's contingents. Neither private nor public British weapons factories, straining hard to produce for their own armies, would set up in Canada any facility to make "Enfields"; nor would they hand over their designs or toolage.[44] By 1902, all of this made Borden very receptive to the weapons ideas of the wealthy and expansive Scottish entrepreneur, Sir Charles Ross. The Minister's choice of production method was not in a public facility, as the new superintendent of the cartridge factory might have wanted. Instead, Borden chose a private enterprise headed by Ross himself. The plant was to be organized in much the same fashion as the Winchester or Remington Arms Companies in the United States, or Mauser in Germany. And so began a decade and a half of stormy debate over this new weapon.

Laurier's government accepted private manufacture of this particular weapon because the British could not supply one fast enough. More durable considerations included the fact that Ross took all of the financial risks; the decision fitted in with Laurier's "infant industry" policy; and the weapon was advanced in design at a time when rifle-armed infantry were still the "Queen of Battle." Alain Gelly suggests that the decision in 1879 to build the cartridge factory in Quebec City was part of a conscious long-term political direction that helped industrialize the province of Quebec. Similarly, the establishment of the Ross rifle works contributed nicely to both the province's and the city's industrial progress. Certainly having a Canadian-made rifle using Canadian craftsmen and resources had a wider national appeal. This was good for Liberal fortunes in a militia long dominated by the Tory party. There was also a decent prospect that this Canadian-produced rifle could be sold abroad; Ross was already established in the international sporting rifle trade. In the end, the reforming militia minister wanted the best rifle he could get for his troops but he did not have the money to warrant laying down an expensive public plant to make the comparatively small numbers the militia would use.[45]

In pursuit of his rifle policy, Borden had to overcome the natural aversion his British or British-trained professional soldiers had about Canadian equipment differences with that of the British Forces. Adoption of the Ross fractured the old icon of imperial uniformity. Consequently, to get his policy through, Borden simply "packed" the rifle's testing committees in 1902 with militiamen and civilian rifle shooters because he was confident they would vote for the Ross. He was correct. Fiery militiaman and Tory MP, Sam Hughes, and Liberal later Lieutenant Governor of Ontario, J.M. Gibson, were two examples. Both were also on the national executive of the Dominion of Canada Rifle Association (DCRA). If the militia officially adopted the rifle, the DCRA would follow suit. That would mean orders of up to 40 000 Ross rifles from the civilians. Such an idea was not new. Since 1883, the government had allowed the many civilian rifle associations to purchase inexpensive ammunition from the government cartridge factory. At times, it amounted to half the total production. Revenues went back to the government. In short, the procedure was tantamount to finding an "outside" market to help make domestic production viable. It seemed likely that Ross would reap similar advantages. Moreover, his agreements did not limit him from making and selling sporting arms in domestic and foreign markets. Potentially it could mean substantial benefit for all.[46]

Once the militia minister received a sanction for the rifle from as large a political and social spectrum as the subject allowed, he set off in promoting the policy any way he could. To encourage Ross, the government provided a wide variety of indirect subsidies including credit advances before deliveries. There were also duty rebates, and a factory site given for a dollar a year on the Plains of Abraham, near the government factory in Laurier's own political riding. Borden placated the imperial uniformity advocates by agreeing that the rifle's calibre would be the same as the official British .303 cartridge. Unfortunately, in time, this last decision, technically and tactically burdened the Ross with an obsolete British round. Ross' ballistic engineers had envisioned a high-velocity .280 calibre rifle "system" that would have allowed the average soldier with limited training to hit a target as far away as he could see, without complicated sighting changes.[47]

The efficiency and place of weapons must be established before they are truly successful. In the Ross case, Borden and his Tory successor, Sam Hughes, chose the target competition way of doing this. In part the civilians had clearly dominated the users, the soldiers, in choosing this munition, and they naturally gravitated toward the civilian proving ground: the rifle ranges. With the Minister's encouragement, Canadian marksmen won a long series of victories before international audiences with the Ross rifle in the days before the Great War.[48]

Off the rifle ranges, the Ross did not fare well in terms of policy and production. At first, Ross could not meet delivery dates; costs escalated; artisans and components were hard to get; the early products were not well finished; and some users complained and the government seemed unable to put the policy on track. Then, in 1906, in trying to unseat the Laurier juggernaut, the Conservatives seized onto the apparently stalled rifle issue as part of their wider political stratagem of "purity in politics." They drove the Grits to a vote of censure in the Commons in 1908. But the wily Borden completely sidetracked the Tory offensive by striking up a much-needed militia department body, the Standing Small Arms Committee (SSAC). The minister's *coup de grâce* was the appointment of prominent Opposition militia critic and Ross rifle devotee, Sam Hughes, as the SSAC chairman. In the House, Hughes demolished his own party's attack in an emotional defence of a Grit policy. In return, Hughes received a "mentioned in dispatches" for his controversial Boer War record of eight years before.[49]

Munitions need friends in high places. Even more than F.W. Borden, Hughes established the rifle's reputation by competition in target shooting. He interfered with the technical design of the weapon so that it would win matches, no doubt hoping that he could get foreign orders. Indeed, at one time the Australians and New Zealanders flirted with adopting the rifle. But in the end, they decided to stay with the uniform imperial pattern Lee-Enfield. Nevertheless, both British private and public Enfield manufacturers were shaken by the Ross' range victories; some saw the Canadian weapon as a threat to their profitable Empire small arms monopoly. When the War Office technical experts tested one of the early marks of the Ross, they did not find it impressive. Yet as the number of Ross trophies mounted, perhaps they should have tested a later model. To the nationalistic Sam

Hughes, this was just British obstructionism. And he continued to promote the rifle even more vigorously by sending many Canadian MPs and important imperial dignitaries free samples. He also tried to get rifle ammunition orders for the Ross company, a new diversification that could only have challenged the nearby government facility. Whether Hughes realized it or not, by 1914 his constant public harangues of critics had not only politicized the rifle but shunted it into the visceral world of nationalism and partisanship. Whether it was loved or hated, both sides were now blind to the strengths or weakness of the weapon. Moreover, most people had lost sight of what the rifle was meant to be; namely, an efficient service weapon for combat soldiers and not a military target rifle.[50]

The onset of war in 1914 compounded all of the Ross' problems. At first, Hughes did not co-ordinate manpower combat needs with those of industry, resulting in recruiters decreasing the skill level of Sir Charles' work force by enlisting his trained employees. The government allowed "business as usual" practices for Ross so that he continued to supply sporting rifles and to accept huge foreign orders. All of this caused delayed deliveries and the redirection and consumption of vital raw resources into non-operational areas. He could not keep up to his greatly expanded orders. With all of these strains, the quality of the product was threatened. Overseas, technical problems like jamming, caused when the rifle was integrated into an imperial ammunition system that was then supplying bad British ammunition, eroded confidence.[51]

Hughes continued to make ferocious verbal assaults on anyone who pointed to problems. Never once did he give calm counsel or admit that there might be easily rectifiable technical problems. The soldiers, especially Alderson, the British commander of the Canadian Expeditionary Force (CEF), and later, the Canadian Corps, were driven into subterfuge to get the Canadian government to withdraw the rifle. In the spring of 1916, Alderson in France and the Chief of the General Staff, Willoughby Gwatkin, in Ottawa secretly engineered press leaks exposing the rifle's problems. Hughes denied everything and tried to fire one and all who might be involved. While the technical difficulties of the Ross were indeed solved, nothing could mend the soldiers' loss of confidence in it. By the fall of 1916, both the Ross and Sam Hughes were "withdrawn from service." The soldiers and the nation could not afford either, nor did they want them. Shortly after, in early 1917 the factory was expropriated. Ross production stopped because the major allies could now produce enough weapons of their own for all forces including Canada. Most people also realized that the days of the bolt action service rifle were numbered. In turn, Ross sued the Canadian government for breach of contract, arguing that the 1902 agreement committing the government to buy his rifles was open-ended. The Union government settled out of court in 1920 and Ross retired with $2.5 million.[52]

The Ross story all but ended here, save for one brief renaissance in late 1917 when an imaginative French Canadian named Alphonse Huot invented a conversion process to turn the Ross into an efficient light machine gun. The need for the latter was a painful tactical realization coming out of bloody trench warfare. All sides eventually developed such guns as squad weapons to supplement and replace the immobile heavy machine gun. Since late 1915, Canadian and British forces had been using the jam-prone and awkward American-designed Lewis gun. Huot's conversion was easy to make, inexpensive, and to many, like Sir A.W. Currie, better than the Lewis. After a long process of trying to convince the British to test the Canadian gun, they finally did in March 1918. They found it had merit and that the Canadian Corps wanted it. The War Office however, did not see the need to manufacture another arm of this type when British factories were geared to production of the Lewis gun. In short, the dictates of the senior ally determined what was "good or bad" as well as what the Canadians would produce and use in the field. And the Canadian government could see no reason to pay the expenses of setting up a factory in Canada so late in the war.[53]

The Ross rifle policy was a failure, costing Canadian taxpayers about $12 million. Like the similarly unsuccessful Palliser's experiment forty years before, this major defence production effort left soldiers and politicians gun-shy about private munitions manufacture.

Thus the Ross story had substantial future implications on munition-ing, but it and other factors had immediate profound effects on the Government Cartridge Factory after 1902. During the Boer War, the govern-ment plant had acquitted itself well and expanded greatly in meeting the crisis. No doubt the Superintendent, F.M. Gaudet, thought the factory was on the verge of fulfilling the soldiers' hopes of seeing a state-owned produc-tion facility making a complete range of munitions. In 1902, Sir Frederick Borden had even approved an official name change suggesting it might be so. After that date, the cartoucherie was called the Dominion Arsenal. But peacetime reality and the government decision to produce a rifle within the private sector quickly eroded wartime expansion. Government money was going elsewhere. By 1913, limited demands and worn-out machinery were serious problems for the Quebec Arsenal. There were also substantial com-plaints about the quality of its ammunition. Inadequate safety standards had seen some major accidents in the previous decade. Some were deadly explosions. Further inroads were made on the arsenal's mandate when the militia ministry not only promoted its rifle, but, also Ross' manufacture of ammunition. In addition, Sam Hughes announced that he wanted the entire arsenal structure decentralized across the country—a policy he claimed was more in tune with his concept of a broad militia society. His first endeavour would be to build a new factory for Ontario, predictably in his home town of Lindsay. The Quebec factory, he said in explanation, was too susceptible to naval attack.[54]

Another reason the arsenal did not prosper was that there was little evi-dence that ministers of militia, Borden and especially Hughes, understood the basic concepts of industrial strategy concomitant with their ideas of a greatly expanded armed force. As the rifle case shows, both men had con-flicting ideas about maintaining both public and private munitions enter-prises. Both then starved the arsenal; both also laboured under archaic and

inadequate state organization that would have given defence industrial policy a chance. For instance, government inspection services of munitions
were not established until just before the Great War. A mobilization officer
was not created until 1910. Even then, munitions was not part of his charge,
save to say that there was not enough. When Sir John French and Sir Ian
Hamilton, both Imperial Inspectors-General, toured Canada in 1910 and
1913, they pointed out great gaps in organization and in stocks. Hamilton
alone identified shortages of 100 000 rifles and 140 million rounds of ammunition. Most munitions, other than what the Arsenal or Ross produced, continued to come from England, as was the case with the new web harness
and the eighteen-pounder artillery pieces. And both imperial inspectors
warned about the lack of defence policy co-ordination between government
departments and the dangers of improvisation in a national emergency.[55] In
general terms, this latter point was confirmed by Sir George Murray, the
British civil servant brought out to Canada to recommend a much needed
reorganization of faltering government departments. Pointedly, Murray
showed that Canadian ministers co-operated little with each other and that
they spent too much time on patronage and other departmental routine and
not enough on policy development and co-ordination. But Sam Hughes,
like many others in office, guarded his departmental political prerogatives
jealously, lest it undermine the time honoured link between patronage and
personal and party success at the polls.[56] There was little appreciation that
if the size of the militia force was set at 100 000, as ministers and generals
had recommended, then there must be an industrial policy to match the
establishment. Lastly, there must be an understanding that the munitions
needed for the potential mobilized force are different and in addition to
those for the actual one. There is little evidence that politicians thought
along such lines before or after the Great War.[57]

   Despite this Sam Hughes did reform the Arsenal. After many com-
plaints and some serious accidents using Quebec's cartridges, in 1913 he
brought over Woolwich authorities to inspect the Arsenal. They declared
twelve million rounds of SAA too dangerous to shoot and pointed to glar-
ing administrative incompetence and worn out machinery. Whether this
was caused by Gaudet's superintendency or a decade of neglect is not clear.
But Hughes fired Gaudet, claiming that his predecessor Borden had known
of the problems but was afraid to remove the French Canadian because of
Liberal political pressure. That Hughes chose Colonel F.D. Lafferty, a regular force Woolwich-trained gunner with many Tory connections, seemed no
more circumspect than what the Grits had done when they appointed
Gaudet. In public, Hughes made it clear that he would rather "sacrifice an
ornamental superintendent anytime than have all of the heads blown off a
dozen riflemen." Whatever motives the minister may have had in 1913, he
laboured under conditions of recession and disinterest by an unthreatened
Canadian public. When Hughes tried to get $60 000 added to his 1914 bud-
get for the arsenal reforms, one MP scoffed that it would be more useful to
Canada if the Minister turned the munitions plant into a fertilizer factory.[58]

The fate of the Arsenal and the Ross rifle factories was not the only dimension of the munitions history before 1914. In 1910, a reluctant Liberal administration approved the creation of the Royal Canadian Navy (RCN). Laurier assured nationalists that the new force would be under Canadian control. Part of that stand was the very attractive promise that ten modern cruisers and destroyers would be made in Canada, even though it would cost thirty percent more than ships built in British yards. When the Conservative Opposition had pushed the question of imperial naval defence in Parliament the year before, even they had emphasized the benefits to the Canadian ship-building industry. Overseas, the great British armament firms of Vickers and Nobels quickly established Canadian facilities in expectation of getting some military business. The Laurier government approved the merger of the Hamilton Powder Company and one or two others into a Nobel-Dynamite Trust called Canadian Explosives. At the same time the government voted in substantial subsidies that increased Canadian dry dock facilities. They did all of this in the hope that, once the navy became established, the economic "spinoffs" would be in the form of private foreign firms—the only bodies with the technical competence and money—developing advanced Canadian facilities for both naval and civilian work. Unfortunately, the RCN would fall victim to the pitfalls of endless national controversy, French-Canadian nationalism, and political patronage. Indeed, the new Canadian sailors, like the hopeful, mostly British munition firms, could not permeate the deputy minister's office into the cabinet which was stalled in controversy over who should be given contracts. The result was that the navy fizzled; British firms received no contracts and Canadians did not achieve new naval or explosives industrial defence capability. Besides, as the events of the fall of 1914 showed, Winston Churchill, First Lord of the Admiralty since 1911, had steadily opposed allowing any RN or RCN ships to be built in Canadian yards.[59]

However Canadians felt about armament industries, the Great War changed everything. First, no one expected the war to be so long or so bloody. The demand for soldiers and material was beyond pre-war imagination. The response to the conflict was incremental with few leaders, military or civilian, understanding until 1916 that the age of total war was upon them. Trial and error predominated. With respect to munitions, the business community was initially reluctant to redirect scarce pre-war depression-stressed resources into manufacturing risky war stuffs in a conflict likely to be over by Christmas. As in the Boer War, the CEF had simply joined the British armies and with the exception of what equipment Canadians brought with them, were to be supplied by the British system. For their part, the War Office and Admiralty did not place major orders in Canada, preferring for both diplomatic (to involve Americans) and practical reasons (existing plant, resources, and funds) the US market place. This neglect stung Canadians and it was a long-term aggravation against which Robert Borden, the Prime Minister, objected vigorously. Clearly, if Canadians were to pay the alliance price in human life, they deserved some of the economic benefits. For a long time, the usual peacetime patronage expectation persisted among politicians and the party faithful in business, at least until it became too embarrassing and too inefficient. Nevertheless, in the first two years of the conflict, the Borden government purposely used a decentralized munitioning scheme as a lever for bringing the Canadian psyche into the conflict and for redressing economic imbalances across the land.[60]

More than any other politician, Hughes helped Canadians make the transition to war. He initiated what became the massive and successful munitions effort finally achieved by 1918. In 1914, Hughes had thundered at lurking American steel manufacturers anxious for contracts who had scoffed at Canada's likelihood of munitions production that, "By God the work shall be done in our own country; we are not so dependent as you think."[61] He then pushed Canadian business into participation by creating an ad hoc Shell Committee. It was a typically improvised consortium of private Canadian steel makers whom the minister had brought together to do business for the British government in Canada. By the spring of 1915, from almost a dead start, these unpaid "captains" of industry, many of whom Hughes had given honourary military ranks, had shipped 675 000 eighteen-pounder artillery shell bodies and laid the basis for continuing work by placing $170 million worth of orders to over 250 firms across the country.[62]

However, that spring, as R.J.Q. Adams shows, staggering munitions shortages, unclear organization, and much plain incompetence in England produced a serious political crisis there. The result was a major reorganization of Great Britain's war production. The soldiers at the War Office lost their control of war material and the civilian "Wizard," David Lloyd George, became Britain's first Minister of Munitions. Automatically this meant reorganization in Canada simply because Hughes' Shell Committee was the agent not of Canadian, but of British authorities. By the autumn of 1915, rapid expansion, political partisanship, ignorance, and some incompetence in the Canadian committee also demanded change. Consequently, the British ordered Hughes' creation reformed into Sir Joseph Flavelle's Imperial Munitions Board (IMB), the agency that controlled most of Canada's munition production for the rest of the war.[63]

This remarkable body—made purposely apolitical because of the growing public sense of alienation and scandal caused by Hughes' Shell Committee and the Ross rifle, was solely responsible to Lloyd George's British ministry. Yet it was run by Canadians. Significantly, the IMB was the tool of big government. Equally, most of the other Dominions as well as India had similar British bodies established in their countries for the same purpose.[64] And like them, the Canadian government and its soldiers had little real control in policy or production.

As in Britain, Flavelle's IMB held the munitions effort in tight rein, with central control vested in its chair. Almost like a department of government, it had authority and it provided policy direction. As well, it supplied technical and inspection services; it made contracts and purchases and gave other executive and administrative directives to its contractors. To some degree, the IMB broke new ground in social programs. For instance, it employed female labour and it provided services appropriate to them. Where private business could not produce certain war stuffs, Flavelle followed the British model and set up National Factories, such as the fuse plant in Verdun, Quebec and the acetone facility in Toronto. By war's end, Canada had produced 65 million shells, 112 million tons of explosives, 2900 aircraft, 88 ships, and many million rounds of small arms ammunition, to name just some items. Profits from the total munitions production were enough to pay for the cost of Canada's overseas war—a fact later soldiers would not forget. Britain was also in debt to Canada.[65] In addition, it was this phenomenal material effort coupled with the excellent record of Canada's large fighting force that both emboldened self-confidence and won Canadians some voice in post-war imperial and international councils.

Despite the overriding control of a state-sponsored organization, the predominant force of Canadian munitioning in the Great War was private enterprise both in concept and practice; "according to usual business methods" was the phrase most often heard. Yet it was also clear, as it had been for years, that munitions industries, public or private, were the clients of the state. As far as the public Dominion Arsenal was concerned, it could not and did not provide the major effort in wartime production. As Superintendent Prévost had predicted two generations earlier, the Arsenal's function was as the schoolmaster for private industry. Its technical expertise was indispensable, but as far as production was concerned, the government facilities retained two greatly expanded ammunition plants: a new one in Sam Hughes' riding at Lindsay, Ontario established in 1916; and Lafferty's in Quebec City. Indeed, to use Sam Hughes' 1914 words, a great deal was "done in our own country," but there still remained much dependence on outsiders, both in reaction to alliance dictates and for raw resources and machine tools from the United States. Nevertheless, Canada's accomplishments in munitioning during the Great War were nothing short of phenomenal.[66]

The munitioning success of the Great War was by no means the end of the story. However, during the 1920s and 1930s, defence production almost disappeared under a welter of eroding factors, including huge war surpluses and a "merchants of death" syndrome. With almost indecent haste, the IMB organization was quickly dismantled in 1919. The government refused to keep any nucleus munitions capability in spite of exhortations from the industry to do so.[67] Budget cuts, neglect, and strategic confusion within the British alliance plagued the course of munitions until the late 1930s. Several times the Arsenal had to shut down, and, except for its caretaker production of SAA, no munitions were made at all in Canada before 1938. Indeed, in these "years of the locust," Canada was far worse off and more dependent than it had been before the Great War. Only at the last moment and with great political duress were the societal and military imperatives co-ordinated, thus making it possible to achieve the equally phenomenal munitioning effort of World War II. That story is told elsewhere.[68] For our purposes, the factors that influenced munition development after the Great War had all been present in the years since Confederation, proving at least that having it "done in our own country" was never a simple process.

NOTES

1. Department of National Defence (hereafter DND), *Challenge and Commitment: A Defence Policy for Canada* (Ottawa: DND, 1987), 74–76. Also see the *Canadian Defence Industry Guide, 1992* (Toronto: Baxter Publishing, 1992), 5–18.

2. J. Mackay Hitsman, *Military Inspection Services in Canada, 1855–1950* (Ottawa: Department of National Defence, 1962), 11.

3. C.P. Stacey, *Canada and the British Army, 1846–1871: A Study in the Practice of Responsible Government* (Toronto: University of Toronto Press, 1963), 99.

4. Hitsman, *Military Inspection Services,* 11; Stacey, *Canada and the British Army,* 99.

5. Stephen J. Harris, *Canadian Brass: the Making of a Professional Army 1860–1939* (Toronto: University of Toronto Press, 1988), 11–12.

6. R.A. Preston, *Canada and Imperial Defense: A Study in the Origins of the British Commonwealth Defense Organization, 1867–1919* (Durham: Duke University Press, 1967), 54–57.

7. John English, *The Decline of Politics: The Conservatives and the Party System, 1901–1920* (Toronto: University of Toronto Press, 1977), 3–30.

8. Desmond Morton, "Defending the Indefensible: Some Historical Perspectives on Canadian Defence, 1867–1987," *International Journal,* 42 (Autumn 1987), 627–31; Hitsman, *Military Inspection Services,* 11–12.

9. Stacey, *Canada and the British Army,* 204–27.

10. René Chartrand, "American Breech-loading Firearms in the Canadian Service, 1866–1872," *Canadian Journal of Arms Collecting,* 34, 43 (November 1986), 115–21.

11. Preston, *Canada and Imperial Defense,* 77–80; Desmond Morton, *A Military History of Canada* (Edmonton: Hurtig, 1985), 107–108.

12. Martin van Creveld, *Technology and War From 2000 B.C. to the Present* (New York: Free Press, 1989), chap. 15.

13. Canada, Parliament, *Sessional Papers,* "Annual Report of the State of the Militia of the Dominion of Canada for 1876," 2321–23. (Hereafter cited *Militia Report.*)

14. Selby Smyth to the Minister, 1 January 1880, *Militia Report,* 1879.

15. Hitsman, *Military Inspection Services,* 14–15.

16. *Militia Report,* 1872; Hitsman, *Military Inspection Services,* 14.

17. *Militia Report,* 1876, i–xiii; ibid., 17; ibid., 2877, pp. 239 and xxii–xxiii; ibid., 1872; and ibid., 1879, xxi–xxx.

18. Leslie W.C.S. Barnes, *Canada and the Science of Ballistics, 1914–1945* (Ottawa: National Museums of Man, 1985), 3–6; Ian McGibbon, *The Path to Gallipoli: Defending New Zealand, 1840–1915* (G.P. Books, 1991), 27–54, is typical of the other colonies scrambling for artillery from Britain during the "Russian scares."

19. Selby Smyth to Minister, 1 January 1878, *Militia Report,* 1878; Palliser to Macdonald, 8 November 1879, Sir John A. Macdonald Papers, National Archives of Canada (hereafter NAC), MG26A, pp. 167491–94; Palliser to Macdonald, 3 August 1880, ibid., pp. 66028–31; and J.M. Beck, *Pendulum of Power: Canada's Federal Elections* (Scarborough: Prentice Hall, 1969), 30–35.

20. Selby Smyth to Minister, 19 April 1880, Canadian Defence Committee, *Report,* 1886, vol. 487, NAC, RG9; "Correspondence Relating to the Conversion of Smooth Bore into rifled guns, the Palliser's Principle in Canada, 1879–1885," ibid., vol. 486; Palliser to Melgund, October 1884, ibid., vol. 482; Caron to Macdonald, 7 February 1885, Macdonald Papers, NAC, pp. 39696–706.

21. Captain C.E. Montizambert, Captain Oscar Prévost, and Lieutenant-Colonel T.B. Strange, all gunners, all wrote essays advocating aspects of Canadian production: the first was a Dominion Artillery Association Prize essay in 1878 and the last was presented at the Royal United Service Institute, 12 May 1879. See *Militia Report* 1879, Appendix 7, 292. Also see Selby Smyth to Minister, 19 April 1880, CDC dispatch 1878, NAC, RG9, A1, vol. 487.

22. On the technical advances, see Bernard and Fawn Brodie, *From Crossbow to H-Bomb* (Bloomington, IN: Indiana University Press, 1973), 124–50; and on machine guns, Dolf Goldsmith, *The Devil's Paint Brush: Sir Hiram Maxmim's Gun*, ed. R.B. Stevens (Toronto: Collector Grade Publications, 1989), 7–25.

23. Sir Francis Mowat Chairman, "Report of Committee on Reserves of Guns, Stores and etc., for the Army," 31 March 1900, Public Records Office (hereafter PRO), War Office (hereafter WO), 33/163, A617, E001; the Earl of Morley Chairman, "Report of the Committee to Enquire into the Organization and Administration of Manufacturing Departments," 14 July 1887, ibid., 32/6282; and O.F.G. Hogg, *The Royal Arsenal: Its Background, Origin and Subsequent History* (London: Oxford University Press, 1963), vol. 2, chaps. 5, 19–21. Colonel G.F.R. Henderson's classic essay on "War" also covers a contemporary's view of the changes in war and technology. See his *The Science of War: A Collection of Essays and Lectures, 1891–1903*, ed. Colonel Neill Malcolm (London: Longmans, Green, 1919), chap. 1.

24. "Abstracts of the Proceedings of the Directory of Artillery," 31 December 1886, PRO, WO 33/47, 1641/36; Hitsman, *Military Inspection Services*, 14; and *Militia Report*, 1879, 1 January 1880, xxii–xxiii.

25. A.G.L. McNaughton, "The Defence of Canada: A Review of the Present Situation," 28 April 1935, Ian Mackenzie Papers, NAC, MG27, 41

B5, vol. 29, x–4; E.C. Ashton, "An Appreciation of the Defence Problems facing Canada," 16 November 1936, ibid., X–6, Secret; *Militia Report, 1879*, xxxi.

26. On civilian shooters and their role in the defence scheme see William Beahen, "Filling Out the Skeleton: Paramilitary Support Groups, 1904–1914," *Canadian Defence Quarterly*, 13, 4 (Spring 1984), 34–39; and Desmond Morton, *Ministers and Generals: Politics and the Canadian Militia 1868–1904* (Toronto: University of Toronto Press, 1970), 60–64, 201.

27. CDC *Report* 1886, vol. 485, NAC, RG9, A1, pp. 609–12; Dominion Arsenals, *The Dominion Arsenal at Quebec, 1880–1945* (Quebec: 1947), (hereafter cited DAQ); Alain Gelly, *Importance et Incidence de l'Industrie des Munitions sur la structure Industrielle de Quebec, 1879–1956* (Master's thesis, University of Laval, 1989), chap. 1.

28. Bennett to H.W. Brown, 21 December 1931, R.B. Bennett Papers, NAC, MG26 K, M1105, 287435–36; Preston, *Canada and Imperial Defence*, 177–86; and Desmond Morton, "French Canada and War, 1868–1917: The Military Background to the Conscription Crisis of 1917," in *War and Society in North America*, ed. J.L. Granatstein and R. Cuff (Toronto: Nelson, 1971), 88–103.

29. For examples of the confusion and inefficiency in the British armaments industries, see "Report of the Committee on Cost of Ordnance Factory Production," *Interim Report*, 14 April 1909, PRO, WO32, 8984; *Interim Report* 1911, ibid., 8985; Clive Trebilcock, *The Vickers Brothers: Armament and Enterprise, 1854–1914* (London: Europa, 1977), Chaps. 1–3.

30. "Memorandum by the GOC upon the Requirements in Guns, Small Arms, Equipment and etc." 1883, NAC, RG9, A1, 487.

31. See D.W. Middlemiss and Joel Sokolsky, *Canadian Defence: Decisions and Determinants* (Toronto: Harcourt

Brace Jovanovich, 1989), 5, who suggest that a policy of no policy is a normal political tool in Canada.

32. *Militia Report* 1885, xiii; and Morton, *Ministers and Generals*, 57–58.

33. The British had come to this conclusion well before Prévost's conversion. See Sir John Ayde, Director of Artillery, "Government Manufacturing Establishments," WO 24 April 1872, PRO, WO 33/24, N521. The WO usually divided the arms order on a 2:1 ratio between public and private makers. Hogg, *The Royal Arsenal*, 2: 895–96.

34. Morton, "Defending the Indefensible," 629–31.

35. Examples of British, American, and European munitions makers offering to sell Canada nearly any item abound in NAC, RG9 II, A1. Deputy Minister's Office "arms" and "correspondence files," 1867–1903, especially vols. 404 and 405. Typical is London Small Arms' offer to sell Martini-Henry rifles in 1892. By that date the Martini-Henry rifle was obsolete, suggesting LSA wanted to get rid of old inventory: LSA Company to Canadian High Commissioner, 4 May 1892, ibid., vol. 405, 259, A, 11874; E.H. Thurston and Company (boots) to Macdonald, 19 February 1879, Macdonald Papers, NAC, pp. 164032–33.

36. NAC, RG9, A1, vol. 405, 286, 14417, 3 January 1896. Officer-in-Charge for purchase from WO for 40 000 Martini-Melford rifles, 2300 carbines, 5 million rounds of SAA, 50 maxims plus 1.5 million rounds of SAA, and 240 sets of harness—as a result of the Venezuelan scare, most of which was cancelled when the threat went away; WO to Canadian High Commissioner, 27 January 1886, ibid., vol. 490; W.J. Reader, *Imperial Chemical Industries: A History*, vol. 1 (London: Oxford, 1970), Chaps. 1, 9; and Trebilcock, *Vickers*, 1–26.

37. Parks Canada, Historical Research Branch, *The Workers of the Quebec Arsenal, 1879–1964* (Ottawa: Ministry of Supply and Services, 1980), 9–10

(hereafter cited *PC: Arsenal*); *Militia Report*, 1980, Appendix B, 13; and Gelly, Master's thesis, 28–33.

38. Carman Miller, *Painting the Map Red: Canada and the South African War, 1899–1902*. Unpublished manuscript, 744–48.

39. "Report of the Committee on Infantry Equipment," 29 November 1879, PRO, WO 33/36; Memorandum by the GOC upon the Requirements in guns, Small Arms and Equipment, etc., 1883, NAC, RG9, A1, vol. 487; Luard to Caron, 11 April 1883, *CDC Report* 1886, vol. 487, ibid.; ibid., vol. 4896, pp. 639–6891 contains all the letters on Oliver's lobbying; Caron to Macdonald, 12 October 1881, Macdonald Papers, NAC, pp. 141980–84.

40. NAC, RG9, II, A1, vol. 307, F16612; Jack L. Summers, *Tangled Web: Canadian Infantry Accoutrements, 1855–1895* (Bloomfield, ON: Canadian War Museum Historical Publication No. 26 - Museum Restoration Service, 1992), 40–42.

41. Morton, *Military History of Canada*, 137; R.G. Haycock, *Sam Hughes: The Public Career of a Controversial Canadian, 1885–1916* (Waterloo: Wilfred Laurier University Press, 1986), 233.

42. Harston to Caron, 27 March 1885, NAC, RG9, A1, vol. 489; ibid., vol. 489, pp. 701–15, 737.

43. "Abstracts of the Director of Artillery," PRO, WO33/48, No. 3301/168, Min 44077, where the British rejected Harston's conversion as "too complicated and expensive"; Middleton to Caron, 1 January 1886, *Militia Report* 1886, xxii.

44. A.F. Duguid, *Official History of the Canadian Forces in the Great War, 1914–1919* (Ottawa: King's Printer, 1938), appendix 3, "The Ross Rifle. Monograph."

45. Gelly, Master's thesis, 1–38; R. Phillips, et al. *The Ross Rifle Story* (Antigonish, NS: Casket Printing, 1984), 30; and R.G. Haycock, "Early Weapons Acquisition: That Damned

Ross Rifle," in *Canadian Defence Quarterly*, 14 (Winter 1984–1985), 48–59. These contain all of the details of the Ross story.

46. *Militia Report* 1893, 14–16. For instance, in this year the government sold nearly 700 000 rounds to the shooting clubs, the militia fired nearly that at the annual qualification and the government netted about $11 000 revenue.

47. Duguid, appendix 3; Haycock, *Sam Hughes*, 119–129.

48. Phillips, et al., *Ross Rifle Story*, 30.

49. Haycock, *Sam Hughes*, chap. 8, fns. 28, 29, 30.

50. Personal letter. Donald A. Carlyle to the author, 12 February 1991 contains a photograph of the "signed" rifle sent by Hughes to Lieutenant-Colonel F.D. Farquhar, DSO, 14 February 1914. Carlyle still owns this rifle. PRO, WO 140 "School of Musketry Hythe," 140/10, No. 57, "Report of the Commandant of the School of Musketry Hythe 23 January 1907." Ibid., 140/11, No. 58/3308, "Report on the trials of Ross Rifle Mark II, 21 July 1910." Here the British test showed the Ross got so hot in rapid fire that it jammed. They concluded that it had been designed purely as a target rifle; Ross to Haldane, 29 April 1911, ibid., WO32/7072, in which Ross tries to convince the secretary to buy the Canadian weapon.

51. Ross to Hughes, 23 June 1915, Sir Charles Ross Papers, vol. 5, NAC, MG30, A95; Gwatkin to Pease, 2 August 1915 and Gwatkin to Foster, 25 July 1915, Willoughby Gwatkin Papers, F2, ibid.

52. Charles A. Bowman, *Ottawa Editor* (Sydney: Grassy's, 1966), 40–42; House of Lords Record Office (hereafter HLRO), Lord Beaverbrook Papers, E/6, Ross Rifle file; Hughes to Aitken, ibid.; Haycock, *Sam Hughes*, 245–53.

53. Currie to Defensor, 1 October 1918, NAC, RG24, HQC 80-4-63; Phillips et al., *Ross Rifle Story*, 354–71. Currie

wanted the Canadian Corps to get 5000 Ross-Huots.

54. See *PC: Arsenal* for a history of the "explosions" and accidents at the Quebec plant; Hughes to Kemp, 14 June 1912, "Ross Ammunition file," A.E. Kemp Papers, vol. 1, F5, NAC, MG27, II, D9; Canada, *Royal Commission on the Sale of Small Arms Ammunition*, Sir Charles Davidson Commissioner, *Evidence* (Ottawa: King's Printer, 1917), vol. 3, 2118ff., 2572–88 and 2661ff.

55. "Report on the Military Institutions of Canada by General Sir Ian Hamilton, GCB, DSO 1913" (Ottawa: King's Printer, 1913), 7–11, 28; "Report upon the Best Method of Giving Effect to the Recommendations of General Sir John French, GCB, GCVD Regarding the Canadian Militia, by P.N. Lake, Inspector General" (Ottawa: King's Printer, 1910), 14.

56. Canada, Parliament, *Sessional Papers*, 1913, No. 57a, "Report on the Organization of the Public Service in Canada," by Sir George Murray; English, *Decline of Politics*, 80–85.

57. Desmond Morton, *A Peculiar Kind of Politics: Canada's Overseas Ministry in the First World War* (Toronto: University of Toronto Press, 1982), 14–15.

58. Davidson Commission, *Evidence*, 3:2572–88; Hitsman, *Military Inspection Services*, 21; *DAQ*, 43; *Militia Reports* 1906 and 1914, Appendix 4.

59. NAC, RG24, vol. 5604, NS529-6-2; *Canadian Annual Review of Public Affairs for 1909* (hereafter *CAR*), 88–89; Trebilcock, *Vickers*, 133; Reader *ICI*, vol. 1, pp. 1732 and 207–11; and Michael L. Hadley and Roger Sarty, *Tin Pots and Pirate Ships: Canadian Naval Forces and German Sea Raiders, 1880–1918* (Montreal and Kingston: McGill-Queen's University Press, 1991), 53–59, 99–101.

60. The most detailed works on Canada's Great War munitions efforts are David Carnegie, *The History of Munitions Supply in Canada, 1914–1918* (London: Longman's

Green, 1925); Michael Bliss, *A Canadian Millionaire: The Life and Times of Sir Joseph Flavelle, Bart. 1858–1939*; Haycock, *Sam Hughes*, chap. 13, especially sources in fns. 4, 5, 6.

61. Duguid, *Official History*, 112–13; Carnegie, 2, 112; *CAR 1915*, 228–40; C.F. Winter, *Lieutenant-General; the Honourable Sir Sam Hughes, KCB, MP Canada's War Minister 1911–1916* (Toronto: Macmillan, 1931), 94.

62. Shell Committee File, Alex to Henry Bertram, 5 June 1915, Bertram Family Papers, Wentworth County Historical Society Museum, Dundas, Ontario; Great Britain, *Hansard*, 1915, 1205; NAC, RG24, Vol. 413, HQ 54-21-1-20, MGO to DM:DMD, 31 January 1915.

63. PRO, Munitions Ministry *Records*, Mun 5, 173, F1142 contains most of the documents on the Canadian aspect of munitions of both the Shell Committee and the IMB; also see R.J. Q. Adams, *Arms and the Wizard: Lloyd George and the Ministry of Munitions, 1915–1916* (London: Cassells, 1978); Bliss, *A Canadian Millionaire*, chap. 10; Haycock, *Sam Hughes*, chap. 13.

64. Sir T. Holland, "Production of Munitions in India . . . ," 31 January 1918, British Library and India Office Records, L/Mil/7/18987, f10, Military Collection 429; A.T. Ross, "The Arming of Australia: The Politics and Administration of Australia's Self-Containment Strategy for Munitions Supply, 1901–1945"

(Ph.D. thesis, University College, University of New South Wales, 1986) chaps. 1 and 3 contain details of the organization of the Indians and the Australians. On the Canadian scandals of 1916 see Haycock, *Sam Hughes*, chap. 15.

65. Bliss, *A Canadian Millionaire*, 308–11 and 317–18; Carnegie, *Munitions Supply in Canada*, 142–92.

66. Department of National Defence, *Defence Industrial Preparation in Canada, 1914–1945*, by David MacKenzie, a study prepared for the Defence Industrial Task Force, 1986, pp. 2–45; *DAQ*, 43–49; *Militia Report* 1919, Appendix F, 38–46; Lieutenant-Colonel J.R. Allen, "Make Ready Thine Arrows: Ammunition Policy and Practice in the Canadian Forces" (unpublished Master's thesis, Royal Military College, Kingston, Ontario, 1976).

67. Brainerd to MGO, 3 October 1919, NAC, RG24, vol. 1159, HQ 62-2-85; Allan, "Make Ready Thine Arrows," 39–47.

68. See James Eayrs, *In Defence of Canada: From the Great War to the Great Depression* (Toronto: University of Toronto Press, 1964), especially 2: 137–38; Robert Bothwell, "Defence and Industry in Canada, 1935–1970," in *War Business and World Military Industrial Complexes*, ed. B.F. Cooling (Port Washington: Kennikat Press, 1981), 106–19; and W. Johnston, "Our Danger is Small—Defence Preparedness, 1939," in *Forum*, 5, 1 (January 1990), 9–13.

THE CANADIAN GENERAL STAFF AND THE HIGHER ORGANIZATION OF DEFENCE, 1919–1939 ◇

STEPHEN HARRIS

o

> There is a new contentment among us all. We walk with sprightlier
> step . . . clear eyes . . . cleaner cut. . . . The Mad Mullah of Canada
> has been deposed. The Canadian Baron Munchausen will be to less
> effect. . . . The greatest soldier since Napoleon has gone to his gassy
> Elbe, and the greatest block to the successful termination of the war
> has been removed. Joy. Oh Joy!

The obvious relief with which Lt. Col. J.J. Creelman greeted the resignation
of Sir Sam Hughes as Canada's Minister of Militia and Defence in October
1916 reflected much more than simple personal animosity toward the
army's former political master.[1] Instead, Creelman's delight at Hughes' dis-
comfiture attested to the emergence of an entirely new sense of professional
self-awareness within the Canadian Corps in France which decried above
all else the extent of the government's influence over military affairs abroad.

The sharp divisions that appeared between the front-line soldier and
the politicians at home during the Great War were not a uniquely Canadian
phenomenon: the frock coats of Whitehall were widely condemned by the
British Army. Still, the fact that feelings ran as strongly as they did among
Canadian officers was significant. For these were men whose pre-war experience of soldiering had been largely confined to part-time service with a
volunteer and amateurish citizen militia where advancement usually
depended upon an individual's party political affiliation rather than on his
military proficiency.[2] Indeed many of them (Creelman included) had bene-
fitted from Hughes' determination to ignore the advice of the regular staff

◇ *War and Society* 3, 1 (May 1985): 83–98.

and to promote whomever he wished in the units and formations of the Canadian Expeditionary Force (CEF).

But by mid-1916 many of these officers, including the four Canadian divisional commanders—Major Generals Currie, Turner, Mercer, and Watson, militiamen and Hughes appointees all—had come to regard Sir Sam as a menace. Despite their bitter protests he was still appointing political cronies with no experience of battle to senior staff and command positions overseas; he continued to champion arms and equipment already proven useless at the front; and he would not tolerate any reform of the confused and often counter-productive training, reinforcement, and administrative organization he had built up to serve the Corps at home, in England, and in France. Thoroughly dispirited by this interference in what they now regarded as their affairs, Creelman and his colleagues begged that those with recent military experience, not amateur soldier-politicians, be allowed some say in matters affecting the army's fighting efficiency.[3]

Their wishes were by and large fulfilled over the next two years. Sir George Perley, appointed to replace Hughes overseas, proved such a strong ally of the Corps' higher command that his successor, Sir Edward Kemp, was moved to complain that the soldiers actually had too much power and were too free from civilian control. Yet on those occasions when the cabinet in Ottawa attempted to change the administrative, organizational, and personnel policies established by the Corps, Kemp stood by the army every bit as steadfastly as Perley.[4]

Such wholehearted support was as welcome as it was unexpected. It was definitely savoured. Nevertheless, as those officers who chose to join the peacetime permanent force looked to the future they could not forget the Hughes era or most of the preceding fifty years when party patronage and the glorification of part-time soldiering had dominated the management of military affairs in Canada. For Sir Sam had been free to act as his own commander-in-chief not only by virtue of his forceful personality and his powerful political base within the Conservative Party, but also because the traditional balance between the civil power and the Dominion's military establishment so heavily favoured the former. . . .

Looked at in this light, it could be argued that Sam Hughes was guilty of nothing more than adhering to well-established customs and traditions, while the two years when his successors actually listened to the generals in France were a curious and passing aberration. It certainly did not follow that the general staff's advice would be accepted after the war, that its professional independence would be acknowledged, or that a new ethos emphasizing the importance of knowledge, expertise, and experience in the formulation of military policy would be accepted. With Germany defeated and the United States warmly regarded as a friend and ally, why should Canadians worry about national security at all? Why should they want greater military influence over policy making when increased defence expenditures were bound to result? In short, what reasons were there to alter pre-war habits one whit?[5]

Therein lay the problem. While it was essential to guard against the chaos that lurked behind the minister's freedom to act as he saw fit during a crisis, it was also evident that Canadians would not easily by persuaded to criticize such non-expert meddling in military affairs. One solution was to draft mobilization plans (called defence schemes) so comprehensive and so intimidating that the political head of the department would be deterred from taking the initiative.[6] This was not a complete answer, however, for even if the army's representative in cabinet was won over it did not mean that his colleagues or the prime minister would necessarily adopt the staff's professional recommendations. Something more than a portfolio of plans was obviously required, but the general staff's dilemma was how to find a way to convert the government as a whole to the military point of view without issuing a direct challenge to the civil authority which, for historical reasons, the armed forces could never win.

The prospects for changing the nature of the civil-military relationship in Canada were probably at their best after General Sir Arthur Currie was appointed to head the militia in November 1919. The former corps commander agreed wholeheartedly with the need to reform the country's military establishment and despite having come out of the militia he was particularly interested in improving the quality and the image of the permanent force. But when it became clear that the government had other priorities Currie decided that he could not put up with "groping in the dark" until times were better and resigned his appointment, leaving the task of post-war reconstruction to Major General James Howden MacBrien, a tough and uncompromising brigade commander in France.[7]

The story of MacBrien's unhappy experience as militia chief and his stormy relationship with Commodore Walter Hose, head of the Royal Canadian Navy, has already been told, and it is generally conceded that Currie's successor was to blame. . . . [However], MacBrien understood perfectly well what was required to advance the cause of the military profession. More specifically, it was MacBrien who first saw that the army could not put its faith entirely in the mere existence of mobilization plans, but in addition must try to alter the bureaucratic process.

MacBrien's education in these matters had come in Britain. Appointed to represent Canada at an imperial military conference called in 1919 to evaluate the tactical, strategic, and politico-military lessons to be learned from the recent conflict, he came away convinced that the problems surrounding the higher direction of war in Canada could be solved only by fundamental institutional reform. Feeling that ignorance had led to the general staff's not being listened to, and taking the British Committee of Imperial Defence as his model, he held that all government departments should be intimately involved in identifying, analyzing, and resolving the nation's defence problems.

At first sight this approach appears paradoxical, even self-defeating, given the army's desire to escape the traditionally close supervision of its activities by politicians and public servants alike. It certainly would not have been surprising had MacBrien concluded that the role of the civilian in the Department of National Defence should be severely limited. But this would undoubtedly have been taken as a challenge to the civil power. More to the point, MacBrien sincerely believed that the army could win greater sympathy for the general staff and its professional military advice if only the ministers and their deputies would learn that each of them had a stake in maintaining national security—and that all would share the blame for inadequate preparation. It would simply take time—and a carefully orchestrated campaign that had a firm, formal, and institutional base. As broad and hopeful (and ultimately naive) as this strategy was, however, MacBrien realized that he had to begin more modestly. What was the point in aiming for a government-wide consensus if there was sufficient difference of opinion within Militia Headquarters that the politicians could safely do nothing, or get their own way? The key, therefore was to make sure that the army spoke with one voice.[8]

This was no simple matter. Although General Currie had been undisputed head of the militia by virtue of his seniority, his war record, and his status as Inspector General and chief military adviser, MacBrien's powers as Chief of the General Staff were reduced considerably. The Adjutant General and Quartermaster General were both given direct access to the minister, thereby making three voices at headquarters, not one, and all executive decisions had to be made collectively by these officers and the minister sitting together in Militia Council. The politicians, it seems, had also learned a few lessons. With Currie gone and no more wartime heroes to reward with special appointments they had decided to return to a system of command and control in which it would be difficult for any one soldier to dominate and that would once again involve the minister in day-to-day administration as well as policy.

It was not for MacBrien to dispute this decision even though it marked a return to the organization Hughes had manipulated so well. But less than two years later he was presented with an opportunity that promised more than any soldier had ever dreamed possible. After only preliminary study, but at the urging of the prime minister and the naval and militia ministers, cabinet agreed to bring the army, navy, and air force together under one Department of National Defence, both to save money and to improve the co-ordination of policy. MacBrien was understandably concerned about the government's continuing emphasis on the need for economy, but he was even more unhappy about the prospect of discord within the new organization, particularly when he learned that Great Britain and the United States had recently rejected similar proposals for fear that they would actually intensify inter-service rivalries. As he saw it, a prolonged fight between the army and the navy might very well bring all departmental activity to a halt, even that pertaining to the militia alone, while the minister sought to effect a compromise or choose between competing service programs. Anxious to avoid this at any cost, MacBrien hit upon the easy solution of having himself appointed departmental chief of staff with control over all defence planning. He saw, of course, that if this happened he would enjoy all the power Currie had, and more.[9]

Commodore Hose objected strenuously to MacBrien's formula. For although the Director of the Naval Service would continue to sit on the Defence Council (which would replace the old single-service Militia

Council), it was obvious that MacBrien intended to subordinate the navy to the army. After all, had he not argued that the post of chief of staff should be reserved for the head of the militia—as the senior officer in the predominant service—and that Defence Council should not exercise a collective responsibility? Nevertheless, on 24 November 1922, MacBrien's scheme was accepted despite Hose's stout resistance. To add insult to injury the DNS was not informed of this development until 17 January 1923, almost two months later, and even then he was told only after he demanded confirmation of a rumour heard two weeks before. But instead of resigning as he had originally intended, Hose decided to remain in uniform and fight MacBrien from within.[10]

This turned out to be an astute decision as the navy soon had a powerful ally in its struggle to survive. G.J. Desbarats, recently appointed deputy minister in the new department, was already an enthusiastic supporter of Hose and of naval interests in general. He also believed firmly in the paramountcy of the civil authority, and so was more than willing to attack what he regarded as MacBrien's illegitimate seizure of power. The tactics he selected were masterful. Taking full advantage of the minister's complete lack of interest in his portfolio, Desbarats simply refused to implement the directive confirming MacBrien's appointment as Chief of Staff and then appropriated responsibility for co-ordinating defence plans and programs himself.[11]

It was now MacBrien's turn to protest, but he achieved nothing so long as the minister did not permit the chief of staff to exercise the powers vested in his office by the National Defence Act. As a result the army was in precisely that position it had striven so hard to avoid: not only was MacBrien's advice being ignored, but a civilian with no claim to constitutional responsibility was making policy. Equally important, MacBrien was so completely distracted by the departmental in-fighting that he abdicated his responsibility to guide the planning staff. At one point, for example, he refused to settle a dispute among his subordinates over which of two defence schemes should take priority: war with the United States or mobilization of an expeditionary force in the event of another world war.

Although this state of affairs negated everything for which MacBrien was working, the combination of ministerial indifference and the army and navy's absolute refusal to co-operate meant that nothing significant was accomplished for four long years. The government issued no defence policy statements: there was no new equipment; and the planning process remained in limbo. MacBrien finally resigned in 1927, in part for financial reasons but mainly because he had been worn out by what he saw as an unholy alliance of Hose, Desbarats, and the politicians who lacked the moral courage to finish the job of departmental amalgamation by acknowledging his status as Chief of Staff.[12]

MacBrien's successor, Major General H.C. Thacker, was appointed on a purely caretaker basis until the minister's first choice, Brigadier General A.G.L. McNaughton, had served an apprenticeship commanding the military district in British Columbia. In poor health, having no stomach for politics, and not wanting the position in the first place, Thacker did not expect to accomplish much and he did not. Still, he did take one step for which McNaughton never forgave him. Though MacBrien had insisted that his replacement should continue to hold the post of Chief of Staff whether or not he was treated as such, Thacker readily agreed with defence minister J.L. Ralston's suggestion that he serve as militia Chief of the General Staff without authority to co-ordinate departmental policy as a whole. That opened the way for Hose's eventual promotion to Chief of the Naval Staff (CNS) and to his subsequent parity with the CGS. With little fanfare then, and even less protest, the Department of National Defence had been reorganizing along the lines put forward by the navy five years before.[13]

Disheartened by Thacker's surrender, and disillusioned by the government's persistent playing down of defence. McNaughton was tempted to resign within a year of his appointment.[14] It was fortuitous that he did not. In 1930 R.B. Bennett's Conservatives were elected, and with the help of W.D. Herridge, a good friend who also happened to be the prime minister's brother-in-law, the CGS soon found himself numbered among the government's closest advisers. Quick to capitalize on an opportunity when he saw one, McNaughton set out to put all these connections to the best possible use. In 1930, for example, he persuaded Bennett that the CGS alone should speak for the three services at the Imperial Conference and then took advantage of that platform to secure a general agreement that the Dominion's army and air force were more important than her navy. The performance was repeated in 1932, when McNaughton convinced the government that Canada could best meet its obligations to the Geneva disarmament conference by making massive cuts in the naval estimates and only symbolic reductions in the army and air force budgets. Finally, when recommending a return to MacBrien's chief of staff system he stipulated that the top appointment should go only to the army or the air force: no sailors need apply.[15] Although the latter proposal was not accepted, the navy was suffering. Indeed to some it seemed on the way to disappearing altogether.

The deputy ministers of defence between 1930 and 1935 fared little better at McNaughton's hands. Adamant that the doctrine of civilian control of the military did not extend to these men—but only to responsible elected officials—the CGS contended that Desbarats (MacBrien's old nemesis) and Colonel L.R. LaFlèche (appointed in 1933) should be content to manage the civilian side of the Department. Accordingly, when it came to matters affecting policy McNaughton bypassed them and dealt with the minister direct. Sometimes he went even further, ignoring the minister and going straight to the prime minister for a decision.[16] Of course this gave McNaughton more real influence than MacBrien had ever hoped for, and more power than many could accept. As the Liberal Vancouver *Sun* complained, such "brass-hat domination" was disturbing if not unconstitutional.[17] But well-connected as he was, McNaughton had little to fear from such outbursts.

As things turned out the CGS was probably too well protected and too closely identified with the prime minister. As the general election of 1935 drew nearer Bennett began to realize that McNaughton had become a political liability because of his extraordinary influence and unprecedented seven-year term and also on account of his overly enthusiastic and heavy-handed administration of the government's unemployment relief camps. It seemed likely, therefore, that the Liberals would replace him should they win. So, whether it was to serve Tory political interests or to place a trusted friend and colleague in the public service before the Opposition took office, Bennett asked McNaughton to transfer to the National Research Council and to hand over to Major General E.C. Ashton in May 1935.[18]

Mackenzie King's Liberals did win the election, and within a short time they had undone much of what McNaughton had been aiming for. The navy began to expand, albeit slowly, while Colonel LaFlèche success-fully reasserted his right as deputy minister to lend a hand in making pol-icy. In McNaughton's view General Ashton was solely responsible for this debacle because, like Thacker, he had not stood up to the politicians. But McNaughton's criticism was unfair. The former CGS had conveniently lost sight of the fact that he had nearly resigned eight years before following dis-agreements with King's government. He was forgetting, too, that much of his success between 1930 and 1935 had depended entirely on the way in which he had cultivated and manipulated friendships with senior bureau-crats and politicians. Yet even these ties had not convinced Bennett to revive the appointment of Chief of Staff. Having no such friends in the Liberal Party, Ashton could not hope to duplicate his predecessor's achievements.[19]

Still, McNaughton was correct when he identified his successor as com-ing from a different mould. It was not Ashton's style to pick fights; he did not try to cast the navy into oblivion, having accurately appreciated that the prime minister's interest in home defence would inevitably carve out a niche for that service; and, though unsure whether the Royal Canadian Air Force should ever be totally divorced from the militia, he chose not to obstruct the elevation of its senior officer to the post of Chief of the Air Staff (CAS).[20] In short, Ashton's way was compromise, not confrontation.

The new spirit within National Defence Headquarters eventually paid handsome dividends. When it became obvious during the Czechoslovak crisis of September 1938 that the formal requirement to pass all submissions to the defence minister through his deputy was impeding the flow of infor-mation to Ian Mackenzie, Ashton enlisted the support of the CNS and CAS to correct the situation. Together the three service chiefs sought out Mackenzie and, much to their relief, were told to report directly to him. Colonel LaFlèche would be kept abreast of developments, of course, but only as a matter of courtesy.[21]

The role and status of the deputy minister had thus been redefined through the combined efforts of all three service chiefs acting in concert. No doubt it would have been better for MacBrien to have recognized as much before he and Hose began to brawl in 1922–23, but as it was, the intense rivalry between the two did nothing to enhance the reputation of the armed forces. How could anyone take them seriously when the navy refused to participate in Armistice Day ceremonies for no other reason than the administrative order had been issued by the army?

Such childishness was particularly damaging when it came to establishing a good rapport with the Department of External Affairs, the other branch of government involved in interpreting the strategic significance of international events for Canada. This was especially true of the two senior officials in the department, O.D. Skelton and Loring Christie, both of whom had already developed a deep mistrust of men in uniform and who were unlikely to be favourably impressed by the goings-on in National Defence Headquarters.

As External's undersecretary of state and Mackenzie King's close confidant, Skelton was by far the more important of the two and also much less likely to be converted. Eager for Canada to be free of all colonial trappings and suspicious of the militia's close association with the British Army, he feared that the general staff's desire to make plans to help the Mother Country would strengthen the imperialist cause, thereby making it difficult for the Dominion to espouse a foreign policy of its own.[22] Wondering "How many hypotheses make a commitment?" he insisted that the armed forces should not be allowed to assume too much about the nature and extent of Canadian participation in future wars.[23]

Though he also suspected that the army wished to mould public opinion to see the world "through general staff eyes"—so that the only acceptable international treaties would be "agreements to fight,"[24] Christie's opposition to the armed forces was more coldly analytical and more closely focussed on the specifics of military planning. In particular he objected to the general staff's persistence in raising the spectre of another major war when minor crises were so obviously the norm in international affairs. In his view logic dictated that the latter should dominate military planning, especially in the case of a minor power like Canada. By contrast, continuing to predicate all preparations on the remote possibility of an abstract and "extraordinary" event such as a second world war was so demonstrably foolish that it "provoked the query whether a Canadian soldier need bother thinking at all."

Realistic enough to understand that he could not stop the staff from "concocting paper schemes of its own choice," Christie nevertheless thought it wise to pin [the soldiers] down to something that can be intelligently described in public . . . that might incidentally "stimulate the military mind, force it to the initiative, and perhaps even extemporize better." At the very least the government would be guarding against a situation in which the army's expansive plans for commitments abroad became policy by default simply because the politicians had failed to provide leadership in another direction. To Christie, that other direction had to be home defence.[25]

The soldiers were no less eager to be guided by a firm statement of policy—they did not want to plan in a vacuum, always guessing at the government's intentions—but by the late 1920s, it seemed woefully wrong-headed to emphasize preparations for home defence simply to promote national unity or a distinctive Canadian foreign policy.[26] Not only was the Dominion reasonably secure, there being no direct threats to its territory, but the greatest cause for concern was an outbreak of hostilities in Europe or Asia involving the United Kingdom. In that event the staff was convinced that popular support for Britain would compel the Canadian government to act despite its suspected preference for remaining on the sidelines. If the despatch of an expeditionary force was thus inevitable, the staff inquired, did it not make sense to plan for one beforehand for strategic as well as domestic political reasons?[27]

Born of mutual distrust as well as a fundamental disagreement over what was best for the country, the gulf between External Affairs and the general staff was not going to be easily bridged. Frequent association on an interdepartmental committee along the lines put forward by MacBrien might have helped if only by forcing regular communication, but from 1919 to 1935 no such opportunity arose. External Affairs rejected every proposal to establish a Canadian equivalent of the CID in the 1920s, while in the early 1930s Bennett preferred to work with McNaughton on an informal, personal, and non-institutional basis. Instead of trying to co-ordinate their efforts, then, the two departments continued their long-range sniping, the diplomats accusing the soldiers of war-mongering, and the general staff replying that Christie, Skelton, and their kind were isolationists and obstructionists.[28]

The situation changed after the Liberal victory in 1935. For although he has often been criticized for his lack of interest in defence, Mackenzie King had in fact consistently maintained that Canada must at least be capable of making a significant contribution to her own security. Accordingly he was prepared to listen when defence minister Ian Mackenzie repeated General Ashton's warning that the Dominion was now powerless to defend herself.[29] Accepting the need for a measure of rearmament, but not quite sure how far to go or how to proceed, the prime minister delayed for eight months before announcing that he would chair a cabinet-level inter-departmental Canadian Defence Committee (CDC) which would conduct a thorough review of the country's military policy.[30]

The general staff had won a partial victory. However, it remained to set up the many sub-committees MacBrien had always argued were essential for more detailed studies of the CDC's necessarily general strategic appreciations. (It was in these sub-committees that the general staff hoped to win sympathy from the other departments.) Yet despite his willingness to create the CDC in the first place, the prime minister prohibited the Department of External Affairs from participating in the work of these subordinate bodies.[31]

King gave no reason for this decision. General Ashton nevertheless chose not to press for an explanation: too much was at stake to risk alienating the prime minister to the point of dissolving the CDC in a fit of pique. On the other hand, because it was just as important not to create the impression that the CDC sub-committees could function effectively without External Affairs' participation, Ashton also delayed scheduling the first of their planned sessions.[32] But once it seemed that this tactic might play into King's hands the CGS determined that he must do something, and so for the next six months regularly implored External to reconsider its position. Unhappily, when an answer finally arrived it stated bluntly that further cooperation was not "practicable." The task of diplomats, it seemed, was to work for peace, and they would somehow be tainted if they also prepared for war.[33]

For reasons that are not entirely clear, the prime minister and his undersecretary of state reversed their decision within a matter of weeks. Representatives from External Affairs would sit on the inter-departmental sub-committees after all, but only if two preconditions were met. The sub-committees would have to be divorced entirely from the CDC, and all recommendations arising out of their deliberations were to be passed to cabinet above the signature of the defence minister alone.[34] In setting out these terms Skelton had actually fashioned a very crafty compromise, but the manoeuvre did not fool Ashton. The CGS saw that so long as the defence minister was reporting to the government by himself, External Affairs would remain at arm's length from sharing a direct responsibility for military planning, and the diplomats could argue that there was no such thing as collective, inter-departmental advice on national security policy. Perhaps more important, the act of distancing the sub-committees from the CDC (which carried executive authority because of the prime minister's involvement) would rob them of the moral stature enjoyed by their parent organization. Who, therefore, would disagree if External Affairs were to contend that, coming under the aegis of the defence department, these sub-committees were mere puppet organizations designed to mouth only the general staff's point of view?[35]

Desperate to make a start, Ashton nevertheless decided to pay the price and accept Skelton's offer. All that he asked was that External Affairs be represented on all the sub-committees and that it provide staff for the CDC's permanent secretariat. This was still too much for Skelton and Christie, who rejected Ashton's proposal. Convinced, finally, that he had secured the best deal available, Ashton made no further requests. Henceforth the Department of National Defence, and more particularly Colonel Maurice Pope, would supervise all inter-departmental discussions on military preparations and Ian Mackenzie would submit Pope's recommendations to the cabinet.[36]

These sub-committees by and large did useful work. The government War Book, which outlined emergency legislation and the steps each department would take on the outbreak of hostilities, was ready by 10 September 1939, the day Canada declared war on Germany, as were provisions for dealing with enemy aliens, control of shipping, censorship, and the protection of vital ports, industries, and public utilities. Still, these were not the studies of specific military problems that the general staff had been hoping for since the early 1920s and for this reason they were of marginal value in promoting the interests of the armed forces.

The parent Canadian Defence Committee was, if anything, even more of a disappointment. After a good beginning on 26 August 1936, when the way was paved for the limited rearmament program announced the following February, the next five meetings before Canada's entry into the Second World War became mired in budgetary detail and cost-accounting at the expense of a careful review of threat assessments and the army's defence schemes.[37] This failure to discuss contingency plans was the CDC's greatest shortcoming. It meant, for example, that although the defence minister knew of and approved the general staff's mobilization plan to raise an expeditionary force for service abroad, the fact that such work was taking place was never made clear to officials from other departments. As a result Christie and Skelton were both genuinely surprised and annoyed when they learned on 29 August 1939 that the army would be recommending the commitment of at least two divisions overseas to the war expected to break out in Europe. As Christie recalled, this was "something of a hairpin curve from the . . . line of propaganda in recent years about [preparing only for] attacks on Canada." For his part, Skelton fumed at the prospect of active military intervention by the Dominion in a conflict that was "not our war."[38]

Mackenzie King was no happier. Having long argued that Canada must strictly limit her military liabilities abroad, he was taken aback when he discovered that the general staff had been contemplating a major commitment overseas all along, and he blamed Ian Mackenzie for either having connived with the soldiers or for failing to resist them "as he should."[39] Lack of communication had thus laid the groundwork for another round of improvisation and for further undermining of the general staff's plans shortly after the outbreak of a world war.

It did not happen that way. Though King, Skelton, and Christie all favoured restricting Canada's military role, preferring her to be a centre of war production, their view of the Dominion's proper contribution did not prevail. Instead, beginning in September the three service chiefs worked together to ensure that the army's defence scheme calling for an expeditionary force was not put aside, and by August 1945 there were five divisions overseas, just one short of the general staff's pre-war estimate of the country's maximum effort.[40] In addition, tens of thousands of Canadian sailors and airmen were battling the enemy on the Atlantic Ocean and in the skies over Europe. This was hardly war at the cheap.

All this suggests that Currie, MacBrien, and McNaughton had been right about the importance of the services speaking to the politicians with one voice. The three were also correct in thinking that the creation of an inter-departmental national security "bureaucracy" would one day be helpful in overcoming traditional civilian antipathies to the military. Unable to curb their imperiousness, they expected too much too soon. Genuinely close co-operation between the general staff and the Department of External Affairs did not begin until the start of post-hostilities planning in 1943–44, and it did not reach its apex until the 1950s, when both departments agreed that Canada's NATO commitment and her shared responsibility for the air defence of North America required larger peacetime forces than ever before.

It would nevertheless be wrong to attribute Canada's extensive contribution during the Second World War or the convergence between civil and military opinion regarding the Soviet threat solely to inter-service cooperation and the existence of a functioning interdepartmental committee system. Popular support for Britain during the war was so strong that it could not be ignored, while from 1948 Canadians no longer believed that they lived in a "fire-proof" house and so were inclined both to demand insurance and to pay for it. Ultimately, then, as Currie, MacBrien, and McNaughton had always known, public perceptions defined the status of the military profession in Canada and determined the limits of its influence.

NOTES

1. Creelman Diary, 19 November 1916, Creelman Papers, Public Archives of Canada (hereafter PAC). Colonel Creelman added that although he did not like kicking men when they were down, he would gladly "break nine toes to get at Hughes."

2. On the pre-war militia see Desmond Morton, *Ministers and Generals: Politics and the Canadian Militia, 1868–1904* (Toronto, 1970), passim.

3. See Stephen Harris, "Canadian Brass: The Growth of the Canadian Military Profession, 1860–1919" (unpublished Ph.D. diss., Duke University, 1979), chaps. 4, 5.

4. Ibid., 218, 221; Desmond Morton, *Canada and War: A Military and Political History* (Toronto, 1981), 79.

5. Morton, *Ministers and Generals*; James Eayrs, *In Defence of Canada*, vol. 1, *From the Great War to the Great Depression* (Toronto, 1964).

6. Stephen Harris, "Or there would be chaos: the legacy of Sam Hughes and military planning in Canada, 1919–1939," *Military Affairs*, 46:3 (October 1982), 120–27.

7. Currie to Brigadier George Farmar, 5 January 1920, Currie Papers, vol. 11, file 34, PAC.

8. See MacBrien's memoranda, "The military forces of Canada," McNaughton Papers, vol. 109, PAC; "Interim memorandum on the future organization and distribution of the military forces of the Empire," October 1919, PAC, Department of National Defence Records, Record Group 24 (hereafter RG 24), file HQC 3149, microfilm reel C5061. See also MacBrien to Currie, 6 October 1919, MacBrien Papers, PAC; MacBrien to Currie, 18 November 1919 and 1 January 1920, Currie Papers, vol. 11, file 34.

9. See MacBrien memorandum, 19 December 1921, McNaughton Papers, vol. 109; MacBrien to Chief of the Imperial General Staff, 13 October 1922, RG 24, file HQS 3367, microfilm reel C4976.

10. Hose memorandum, 19 January 1923, quoted in Eayrs, *In Defence of Canada*, 1:241ff.

11. See MacBrien to Desbarats, 30 April 1923, and Desbarats to MacBrien, 2 May 1923, McNaughton Papers, vol. 109.

12. Norman Hillmer and William McAndrew, "The cunning of restraint: General J.H. MacBrien and the Problems of Peacetime Soldiering," *Canadian Defence Quarterly*, 8, 4 (Spring 1979), 45–46.

13. Thacker to McNaughton, 23 September 1928, McNaughton Papers, vol. 107; Colonel J.L. Ralston's notes of interview with Thacker, 5 and 8 February 1927, Ralston Papers, vol. 8, PAC.

14. See Thacker to Ralston, 27 December 1928, RG 24, vol. 6541, file HQ 650-77-1; McNaughton to Sir Allan Lascelles (secretary to the Governor General), 8 June 1935, McNaughton Papers, vol. 104. See also John Swettenham, *McNaughton*, vol. 1 (Toronto, 1968), 240.

15. Swettenham, *McNaughton*, 1:251; McNaughton to D.M. Sutherland, 24 October 1932, McNaughton Papers, vol. 16, folder McN(M) 87; McNaughton to Lascelles, 8 June 1935, McNaughton Papers, vol. 104.

16. McNaughton to Hugh Keenleyside, 31 March 1939, McNaughton Papers, vol. 3. See also Eayrs, *In Defence of Canada*, 1:261ff.

17. *Sun* (Vancouver), 30 November 1935.

18. Swettenham, *McNaughton*, 1:314–16.

19. McNaughton to Lascelles, 8 June 1935, McNaughton Papers, vol. 104.

20. See Joint Staff Committee, "The Higher Direction of War," 20 April 1938, RG 24, vol. 2684, file HQS 5199. See also Ashton to Minister of National Defence, 6 December 1935, Ian Mackenzie Papers, vol. 30, file X-13, PAC.

21. Defence Council minutes, 12 September 1938, Ian Mackenzie Papers, vol. 32, file X-52.

22. Hose to Adjutant General, 8 November 1924, Hose Diary, included in Naval File 1440-5 held at Directorate of History, National Defence Headquarters, Ottawa.

23. See Norman Hillmer, "O.D. Skelton: The scholar who set a future pattern," *International Perspectives* (September/October 1973), 46–49; "The Anglo-Canadian neurosis: the case of O.D. Skelton" in P. Lyon (ed.), *Britain and Canada: A survey of a changing relationship* (London, 1976), 61–84; and, most important, "Defence and ideology: the Anglo-Canadian military 'alliance' in the 1930s," *International Journal*, 33 (Summer 1978), 599.

24. Christie to Lord Lothian, 30 May 1935, Christie Papers, vol. 12, file 40, PAC; Christie memorandum on the 1924 Geneva Protocol, ibid., vol. 10; and Christie to King, 20 February 1936, ibid., vol. 26, folder 8.

25. Christie, "Defence policy and organization," 26 February 1937, Christie Papers, vol. 27, folder 9.

26. See Thacker's long warning, 27 December 1928, RG 24, vol. 5623, file HQ 462-18-1; McNaughton to Ralston, 25 December 1928, McNaughton Papers, vol. 3; Lieutenant-Colonel K. Stuart, "The problems and requirements of Canadian defence," Ian Mackenzie Papers, vol. 34, file B-30.

27. See drafts of Defence Scheme No. 3 (Major War) in RG 24, vols. 2643, 2646, and 2648.

28. See for example, Currie to Meighen, 5 August 1920, Currie Papers, vol. 11, file 35; MacBrien to Ralston, 29 January 1927, RG 24, vol. 2683, file HQS 5121; McNaughton to Ralston, 27 January 1929, McNaughton Papers, vol. 111; Ashton, "The Requirements of Canadian Defence,"

12 November 1935, Ian Mackenzie Papers, vol. 29, file X-4.

29. Mackenzie to King, 7 January 1936, King Papers, J1, vol. 220, 189607, PAC.

30. King Diary, 18 August 1936, King Papers, J13, vol. 100.

31. King to Mackenzie, 22 April 1937, RG 24, vol. 2759, file HQS 6615.

32. Ashton to Mackenzie, 14 August 1937, ibid.

33. Escott Reid to Deputy Minister of National Defence, 15 November 1937, and Christie to Skelton, 4 February 1938, Christie Papers, vol. 27, folder 9.

34. Reid to Deputy Minister of National Defence, 15 December 1937, and Deputy Minister to Chiefs of Staff, 21 March 1938, RG 24, vol. 2759, file HQS 6615.

35. Christie to Skelton, 4 February 1938, Christie Papers, vol. 27, folder 9.

36. Deputy Minister to Chiefs of Staff, 21 March 1938, RG 24, vol. 2759, file HQS 6615; and Chief of the General Staff to Colonel Pope, 13 December 1938, ibid.

37. C.P. Stacey, *Arms, Men and Governments: The War Policies of Canada, 1939–1945* (Ottawa, 1970), 69; RG 24, vol. 2084, file HQS 5199-3.

38. Stacey, *Arms, Men and Governments*, 71. See also N. Hillmer, "O.D. Skelton and the Declaration of War," in Robert Bothwell and Norman Hillmer (eds.), *The In-Between Time: Canadian External Policy in the 1930s* (Toronto, 1975), 180.

39. Stacey, *Arms, Men and Governments*, 71.

40. Harris, "Or there would be chaos . . .", 126–27

DEFENCE AND IDEOLOGY:
THE ANGLO-CANADIAN
MILITARY "ALLIANCE" IN THE 1930s ⬦

NORMAN HILLMER

o

The British dominions in the 1930s were a defence planner's nightmare. The empire had become a gouty giant: "disjointed, disconnected and highly vulnerable. It is even open to debate whether it is in reality strategically defensible."[1] Canadians, South Africans, and Irish all expressed an interest in and concern for the welfare and standing of the mother country in a threatening world,[2] but they gave almost nothing in the way of material assistance and no commitments for the future. Though they insisted on separate defence policies, they spent little on their own security, emphasizing instead their pacific nature, geographical situation, and domestic political problems. The Australians and New Zealanders, directly dependent on British sea power, were more positive in their attitude towards the problems of imperial defence. But even they were, or might rapidly become, liabilities rather than assets on the imperial account: "the contribution which they had made at Gallipoli was a debt which London might one day have to repay with heavy interest."[3]

In any imperial enterprise, nevertheless, the dominions were potential backers of proven capacity. Their citizens, for the most part, were of known sympathies. Had not the dominions been a powerful source of strength and support during the First World War? Was not the military relationship between Great Britain and the countries of the emerging Commonwealth so intimate that observers—then and later—described it as nothing less than an alliance?[4] And yet this was an alliance with a difference.[5] As the British government began to rearm, ministers and officials sought, without notable success, assurance of active and moral and material support from the former colonies. The unwillingness of the majority to contribute significantly

⬦ *International Journal*, 33, 3 (Summer 1978), 588–612.

to empire defence and the reluctance of even the most "loyal" to become involved in European complications, agreements, or guarantees can only have added to London's sense of unease and lack of resolution. The attitude of the dominions undoubtedly reinforced an already strong desire on Britain's part to remain aloof from continental commitments: the concept of "limited liability" was in fact embraced by every leading British minister from the mid-1930s onwards and was not seriously challenged until 1939.[6]

Canadians were apt to think that the military alliance with Great Britain was one-sided. The dominion was "off the main strategic highway of the Empire":[7] the defence of Canada played no appreciable role in Britain's imperial defence planning between the wars. At the Imperial Conference of 1923, Prime Minister Mackenzie King took "strong exception" to the New Zealand prime minister's allegation that the dominions had been 'sponging' on Great Britain for their defence. The newspaperman, J.W. Dafoe, who accompanied the Canadian delegation, summarized King's response:

Canada had not given and would not give Empire any problems of defence or obligations; her policies would not involve Empire in trouble. She had made immense contribution, men, money for defence of Great Britain when attacked. Naval policy for Canada must take cognizance of geographical considerations; this was equally true of Great Britain and Australia. If Canada were similarly situated Canadian feelings would be different. Among Canada's many problems one of the most serious was to compete with the U.S. in attracting and holding money and people. Difficulties would be accentuated if impression was created that people of Canada were much more likely than people of U.S. to be involved in war.[8]

British strategists, indeed, accepted that Canada was in the American sphere.[9] An attack by Japan or any other power on Canada, unlikely as that was, would never be accepted by the United States. Canada was vulnerable to attack from the United States itself, but it had been tacitly assumed in Whitehall since early in the century that an Anglo-American war was so improbable that no preparations needed to be made against its occurrence.[10] "It would require a great deal of sentiment indeed," a Canadian study group told the first British Commonwealth Relations Conference in 1933, "to cause the people of Great Britain to fight for the sake of Canada a war with the United States, and in any case . . . they could not fight it successfully."[11] This judgment was surely correct on both counts.

The British government did not stop trying to obtain Canadian contributions to the empire's defences. The dominion's geographical remoteness and proximity to the United States made Canada all the more attractive as a potential arsenal which would pour forth men and munitions in time of war. But these very factors also militated against Canadian participation in imperial defence planning and commitments. Whitehall judged Canada the most "difficult" of the dominions in matters of co-operation on foreign policy and defence.[12] When Sir Maurice Hankey, the secretary of both the

British cabinet and the Committee of Imperial Defence, came to Ottawa on the last leg of a "holiday jaunt"[13] through the self-governing empire in 1934, he encountered a "calculating aloofness" that struck a "chilly note" after the "fervid Imperialism of Australia and New Zealand."[14]

To Hankey, all questions of Anglo-Canadian relations reduced themselves "in the last resort to the brutal question of whether Canada would come to our assistance in another war." The answer seemed to the uninitiated "extremely dubious." Everywhere there were highly vocal minorities of "isolationists," "French Canadians," and "disloyalists" and the politicians, with an election coming up, seemed cautious, timid, helpless.[15]

. . . Hankey was convinced, even so, that Canada would be "sound" when it counted. Only an aggression threatening vital imperial interests "or involving our Treaty obligations (which are really the same thing) could draw us into war. The *casus foederis* would have to be overwhelming." In such circumstances, surely, Canadian neutrality would be impossible. The enemy government would almost certainly be under military domination; Canada would be exposed to a series of incidents and insults (especially if there was a Canadian minister in the enemy country, as there was in Tokyo) which would exasperate public opinion. The intangible bonds of empire would reveal their strength. . . . Canada could no more abstain from a great European or Asian war than it had in 1914. In the meantime, as Hankey had much opportunity to note, the prevailing sentiment in Canada was "definitely opposed to any increased expenditure on military preparedness"; "the wide-spread pacifism, the political situation and especially the need of politicians to bid for the French vote and the votes of the alien communities, the threatened financial embarrassments and the economic difficulties of the times unite to make it impossible."[16]

The Conservative government of R.B. Bennett had in fact begun to increase overall defence expenditure in 1933–34, but the nation's military position did not noticeably improve. Bennett's unemployment relief schemes did give the military "a golden opportunity to get the buildings and establishments they will eventually require,"[17] but, that apart, the depression dealt a severe blow to the Canadian armed forces. The Royal Canadian Navy survived the attempt by the air-minded chief of the general staff (CGS) to scuttle it completely,[18] but its expenditure dropped from $3 597 591 (a figure which admittedly included the costs of the construction of two destroyers for the tiny fleet) in 1930–31 to a low of $2 167 328 in 1932–33 and stayed close to that figure until 1936. No new ships were ordered in the Bennett years. Despite the CGS's support, the Royal Canadian Air Force, which had never been much more than a civilian force—"bush pilots in uniform"[19]— with little or no military training, went from a vote of $7 147 018 in 1930–31 to one of $1 684 562 in 1933–34, although expenditure went up by over $2 million from 1934 to 1936. In May 1935, the RCAF possessed only 25 service aircraft, and not a single machine of a type fit to employ in active operations. Militia expenditure, too, was drastically reduced to a level of under $9 million in each of the years from 1932–33 to 1934–35 and purchases of new arms and equipment virtually ceased.[20]

It is difficult to believe that the military position would have been different had King been prime minister from 1930 to 1935. The Bennett and King policies on national and imperial defence questions were broadly similar. Both leaders placed the emphasis firmly on national defence and since, as one young member of the Foreign Office remarked in 1935, "Canadians have always felt that theirs is about the safest country in the world,"[21] the defence budget was usually a prime candidate for government austerity. Both King and Bennett, furthermore, stressed the political aspect of defence. King wished to assure his supporters, especially those in Quebec, that his government would have no part in building a war machine or conscripting the nation's youth, while Bennett, who had none of King's distaste for the military or his scruples about the use of force, put the armed forces, and particularly the chief of the general staff, to work as a social, economic, and political instrument of the Conservative government.[22]

Nor did King and Bennett, despite the concern and confidence in British leadership exhibited by the latter,[23] differ in their attitude towards visible contributions to imperial defence. Canada would continue to mind its own business at home, "taking the view," as King put it, "that the only Imperial defence we would really countenance, was Canada helping the Empire through looking after herself and maintaining friendly relations on this Continent."[24] Anything more—public undertakings or highly visible involvements which appeared to commit Canada to action in some future hypothetical situation—might be misunderstood and regarded in many circles as a surrender to the centralizing tendencies of Whitehall and the vociferous imperialism of an Anglo-Saxon minority in English-speaking Canada.

For precisely this reason, Mackenzie King had been careful not to make a final decision regarding Canadian participation in the Committee of Imperial Defence (CID). King attended one meeting of the CID during the Imperial Conference of 1926[25] and he authorized his high commissioner to represent Canada at one other gathering of the committee in late 1928 or early 1929.[26] General A.G.L. McNaughton was also allowed to go to four meetings of the Overseas Defence Sub-Committee of the CID in 1927.[27] King's attitude, even so, was clearly grudging and, unlike the Australians, New Zealanders, and South Africans, he was willing to have his high commissioner attend relevant meetings of the CID on a regular basis.[28] Bennett was even less sympathetic towards attempts to involve Canada with the work of the CID. He went to a single meeting of the committee during the Imperial Conference of 1930, but he was disturbed by the mere appearance of his name in the record and he contemplated asking Hankey to remove all trace of the Canadian presence from the final minutes.[29] Bennett's high commissioner never attended a CID meeting during the five Conservative years from 1930 to 1935.

Despite the relative lack of Anglo-Canadian defence planning, or even consultation, at the level of high policy, the military relationship between the two countries was intimate and of long standing. Successive British and dominion governments had established, and colonial and imperial conferences had confirmed, the principle that wherever practicable a system of uniform organization, training, doctrine, and equipment would be adopted in the armed forces of the self-governing empire.[30] The results were predictable. Canada's interwar army, General Maurice Pope wrote in his memoirs, "was indeed British through and through with only minor differences imposed on us by purely local conditions":

> The war establishments of our units and the composition of our formations were precisely those of the British Regular Army. All our manuals were British and so was our tactical training. Practically all our equipment had been obtained in the United Kingdom save perhaps some forms of mechanical transport. True, we ran our own royal and provisional schools for the qualification of N.C.O.'s and officers of the Non-Permanent Active Militia, and of our regular instructors, but to qualify for higher rank our permanent force officers were required to sit for examinations set and marked by the War Office. Thus, while we conducted militia staff courses for the elementary staff training of selected volunteer officers, we were lacking in facilities for the higher command and staff training of our little regular army. Instruction in these matters was sought at the Senior Officers School at Sheerness, the British staff colleges at Camberley and Quetta, and later at the Imperial Defence College in London.[31]

Similar patterns of co-operation and exchange existed between the Royal Air Force and the RCAF and particularly between the Royal Navy and its Canadian counterpart. From 1922, when the Royal Canadian Naval College was closed, most RCN cadets were trained in England, while many Canadian officers served or trained in Royal Navy ships and establishments. The fledgling RCN had contained a sizable proportion of officers and men borrowed from the Royal Navy in the 1920s. By the late 1930s only one RN officer held an appointment in the Royal Canadian Navy, but he was the director of naval intelligence and plans, whose office in Ottawa served as a direct and conscious, if unofficial, link between the Canadian chief of the naval staff and the Admiralty and a reporting centre covering most of the North American continent for Whitehall's worldwide intelligence network. From 1929 through to the outbreak of war, conversations were regularly held between Canadian Naval Headquarters and the Royal Navy's commander-in-chief, America and West Indies station.[32]

These procedures were sanctioned, encouraged, and extended by one Canadian government after another. Co-ordination with the British military on matters of organization, training, and equipment was obviously cheap and convenient, and it was one sense in which, British motives notwithstanding, the "alliance" could be said to have operated to Canada's advantage.[33] It ensured that doctrine, if not arms (there was a limited relationship between the two in the Canada of the 1930s), remained relatively up to date. Such arrangements were supplemented by a wide range of British information and intelligence[34] and a number of specific agreements reached at the political level which outlined Canadian action in support of British forces should a war break out involving both countries. . . .

This series of agreements constituted a formidable limitation on Canada's freedom of action should Great Britain become involved in a major war. "How many hypotheses make a commitment," wondered the under-secretary of state for external affairs, O.D. Skelton, in 1926. "The Department of National Defence was recently requested to prepare for the Prime Minister's consideration a list of tentative naval and military plans and understandings as to co-operation in wartime. The British military 'conversations' with France from 1906 to 1914 are worth bearing in mind."[35] For Skelton, the Anglo-Canadian military "alliance" was a symptom of a larger disease. Canadian opinion was still under the control of an older, jingoistic, British-oriented minority which had the power, public position, and wealth to carry the dominion into a British war whatever the sentiments of the majority. In the final analysis Canada was simply not free to do as it wished in the great questions of peace and war.[36]

[People] of Skelton's frame of mind accused the Canadian armed forces of complicity in the imperialist plot.[37] The military's position in this regard was hardly helped by public statements such as that of the then Major Pope in 1930: in the event of war, Pope wrote in a prize essay which was later published, Canadians should impose upon themselves "the rôle of assisting the other armed forces of the Empire to further the Imperial interest in a measure compatible with our strength and dignity as a nation."[38] Canadian officers endorsed (as did Mackenzie King, at least in theory) the 1923 Imperial Conference resolution which emphasized that the security of the entire self-governing empire's trade and territory was the concern and responsibility of all its governments.[39] Though the direct defence of Canada was deemed their major task, the military made it clear to the government that the nation's "indirect defence" was an important secondary responsibility and, they cautiously (and rightly) added, "possibly one requiring much greater ultimate effort."[40] Thus the army made elaborate plans for the sending of an expeditionary force overseas to participate in a major European war.[41]

There were feelings of empire kinship and sometimes even dependence in the Canadian armed forces of the 1930s. Although Anglo-Canadian military relations were sometimes strained by the patronizing and uninformed attitude of British officers,[42] the regular exchanges, representations, and courses did much to ensure a community of experience, doctrine, and method which facilitated later co-operation. Enduring friendships were formed; common values and common ideals were reaffirmed. The Canadian military, however, based its assumption and policies primarily on a rational assessment of national needs and national realities, not on imperialist impulses or colonialist instincts. . . .

British uncertainty about dominion attitudes gave rise in 1936 to a recommendation by the Defence Policy and Requirements Sub-Committee of the cabinet that the positive co-operation of the self-governing empire should be sought immediately in reducing the load of imperial defence or, alternatively, in increasing the empire's scale of security. The attitude of the reluctant and difficult Canadians was regarded as particularly important.[43] The British government wished, as Ramsay MacDonald had put it on a previous occasion, to emphasize unity where there seemed only disunity and to secure "for Great Britain that equality of status which, in the events which have happened, she no longer possesses in practice."[44] For it had been clear since at least 1926 that Whitehall undertook much of the burden of the empire's negotiations with foreign powers and much of the responsibility for imperial defence, but that it could speak with certainty and act with decision only for the United Kingdom and the dependent empire.

Thus, in March of 1936, the dominion high commissioners were invited to hear Sir Maurice Hankey give an account of recent British defence policy.[45] Hankey pointed out that it was "a cardinal point of our foreign policy to avoid any situation arising in which we were faced with simultaneous complications in the East and the West. It was, however, difficult to arrive at any satisfactory understanding with Japan on the one hand or with Germany on the other if in the latter case we were to carry France with us." These considerations, together with serious deficiencies in the three services which derived from long-standing policies of retrenchment and disarmament,

> made a reconditioning of our defence forces a matter of vital necessity. . . . Had we been stronger Japan might have been more amenable in 1931 and Italy would be more amenable today, while if the theory of collective security survived the present test it could only endure under the leadership of this country and would stand a much better chance of achieving its objects with this country and the Dominions strong in their defences.

The dominions, then, must be contacted and their help and advice requested. This must be done, however, without giving cause for offence. The Australian high commissioner, Stanley Bruce, backed by the Irish and New Zealand representatives, suggested "that it was no use putting the matter in too delicate a way. The problem was a common one, and the United Kingdom should tell the Dominions the limit to which they were able to go and what they would like the Dominions to do. The Dominions could always refuse, and there need be no question of any interference with their full sovereign rights." The Canadian high commissioner remained silent.

The Australian high commissioner's remarks were interpreted as a dominion "initiative." The governments of the Commonwealth must now be given "an opportunity" to assist in concrete ways or "so orientate their own defence policies as to work in with ours."[46] Conventional means—representations by British officials in the dominions—were adopted for the Australians and New Zealanders,[47] but in the Canadian case, notwithstanding Bruce's advice, an approach with "extreme delicacy" and in "deferential terms" was thought necessary. It was decided to put the British case directly to two members of the Canadian cabinet, Charles Dunning, the minister of finance, and Ian Mackenzie, the minister of national defence, who were soon to arrive in London on other government business.[48] Here were two men who would do their best to help. . . .

Dunning and Mackenzie accordingly met in July 1936 with the chiefs of staff and the minister for the co-ordination of defence, Sir Thomas Inskip.

The contributions that the dominion could make to the empire's defences were tactfully spelled out: the purchase (not the loan, as the Canadians wished) of two modern destroyers from the Admiralty; modernization of coast defences; provision for anti-aircraft defences; the improved equipment and training of non-permanent militia and the development of mobile land forces capable of acting with the RCAF to repel raids; the supply of pilots and, in time of war, specialist personnel (wireless telegraphy and medical); and the assurance of munitions delivery to the British government in time of war.[49]

Dunning and Mackenzie were as forthcoming as possible. They sympathized with Whitehall's plight, welcomed technical assistance, and suggested that "although on the issue of peace or war the country would be split . . . unless matters had been very badly handled, Canada would end in being in the war." Nevertheless, there would be no empire commitments or European involvements. As Dunning said, "there was a strong feeling in Canada that they must have the right to decide for themselves."[50] The King government's attitude towards the supply of pilots for the Royal Air Force illustrates the Canadian stand. The government was willing to co-operate in assisting young Canadians who wished to join the Royal Air Force— because the program could hardly be construed as compromising the government's freedom of manoeuvre, because only a small number . . . was involved, and because finding employment for young Canadians, few though the numbers were, could easily be justified in the political arena. But when Mackenzie returned from his talks with the chiefs of staff with a proposal for the establishment of a British government training school for airmen on Canadian soil, the cabinet refused to co-operate, because the plan would be highly visible, give the appearance of commitment, and could not be justified to the pacifists and neutralists in parliament and the country at large on strictly Canadian grounds.[51]

The chairman of the chiefs of staff told Dunning and Mackenzie that indefinite promises of support and tiny contributions were not nearly enough. Rather in the manner of an imperial schoolmaster, Sir Ernle Chatfield combined admonition with instruction, stating

> that from the Naval point of view the uncertain attitude of Canada made it difficult to plan the correct distribution of naval forces. It was difficult to know in what circumstances Canada would look to us for support.
>
> In the event of a war in Europe, Canada would be looking east and the problem of the protection of the lines of sea communication would be comparatively simple. If, however, trouble arose with Japan, the difficulties and distances involved would be much greater. In that event would Canada be anxious for the security of her western seaboard, and would they look to our naval forces for this protection? If such responsibilities were put on the Royal Navy we would want to be assured that there were harbours and facilities on the western coasts where ships could shelter and which would be adequately defended.

    ... If the Admiralty were to prepare for war against Japan in which Canada was not involved, that was one thing, but if Canada was involved and precautions had not been taken in advance, we should not be ready for contingencies which might have to be faced. Admiralty plans were affected by circumstances of this nature and he thought the problem should be tackled.[52]

Chatfield's comment, which surely indicated exasperation rather than real concern, is one of the very few references in interwar British records to the direct defence of Canada. In the late thirties, the chiefs of staff repeatedly pointed to the vulnerability of a far-flung Empire-Commonwealth, and stressed the dangers to the security of the United Kingdom, Eire, Australia, New Zealand, India, and, to some extent, South Africa. There was scarcely a suggestion of a direct threat to Canada from any source.[53] Canada had no defence problem, the deputy chief of the British Air Staff told the RCAF liaison officer in London in early 1939. "All we want of Canada in a war is pilots and aircraft."[54] The British, as we have seen, were moving on these fronts as early as 1936. In 1938–39, after prolonged negotiations, the Canadian prime minister was persuaded to sound the "loyal note" by including British pilots in an expanded RCAF training program. King made it clear, however, that the scheme was a Canadian one, "under the final direction and control of the Minister of National Defence." It was a small concession, and it came to nothing, but the government had been prodded into substantially increasing the defence estimates to create new training schools and, as Colonel Stacey has pointed out, "a precedent had been established which would serve as the foundation for a great structure when war came."[55]

    Whitehall also continued to seek concessions in the area of munitions supply. The British government wished to begin as soon as possible to set up a shadow armaments industry in the dominion with active Canadian government assistance. Enquiries were directed not to the meeting of immediate requirements, an impossibility given the undeveloped state of the Canadian armaments industry and the British policy of concentrating its rearmament program in the United Kingdom, but to "building up ... a potential which could in case of war be relied upon to supplement United Kingdom production."[56] It was assumed that in the early stages of a war, the greater part of the United Kingdom would be subjected to intensive bombing attacks resulting in a serious impairment of industrial capacity. The American government's neutrality legislation further damaged the British position because it seemed unlikely that the United States would again act as an allied arsenal, exporting large amounts of war [material] to the allies as it had done from 1914 to 1917. "The United States policy," wrote Hankey in 1936, "places a premium on the aggressor, who, particularly in the case of a totalitarian State, can prepare for war by a certain date, devoting his whole resources to building up a vast war machine with unlimited reserves and manufacturing capacity—as is happening in Germany and Italy today." Canada was well situated to make up these deficiencies. It was virtually immune from air raids and the prospect of large wartime profits would attract American capital and raw material into Canada on a scale which would largely offset any losses.[57] Canadian peace-time involvement in the supply of munitions to Britain would also, it was hoped, "operate . . . to diminish the tendency in Canada to detachment from concern with imperial defence; and would . . . impress opinion on the Continent."[58]

The Canadian government laid down a policy regarding the manufacture of munitions and war materials in Canada by the British government in September 1936. The King cabinet was agreeable to the placing of British government orders in Canada and to private firms taking the initiative in the establishment of munitions and aircraft plants. Even this approval, however, was subject to the understanding "that Canadian firms are liable to the imposition of any control or regulation whatsoever which the Government may wish to establish at any time with respect to the nationalization of the plants, the complete prohibition of exports of munitions and war materials, or any action which in any future contingency the Government may decide advisable or necessary."[59]

The government made it clear, furthermore, that it was not prepared to act as an intermediary between British service departments and manufacturers in Canada, nor did it wish to enter into negotiations on behalf of either party. Ottawa would simply provide information as required.[60] The Canadian government, however, and particularly the Department of National Defence, had decided interests, views, and biases on supply questions. And these they made no attempt to suppress, as the British air mission found only too quickly when they visited Canada in May 1938 to investigate, among other things, aircraft production potential:

> Throughout their visit to Canada the Mission enjoyed the ready co-operation of the Defence Department and they were supplied by the Canadian Defence authorities with comprehensive data which had been prepared on the capacity of the Canadian firms; but it became clear at the outset . . . that the Canadian government were particularly anxious to prevent the impression that they were themselves directly concerned with the matter. The Canadian Government were, however, anxious that our activities should be in parallel with their own plans for the development of an aircraft industry in Canada.
>
> It became clear in the course of discussion that the establishment of shadow factories would be opposed by the Canadian Defence Department. Moreover, the Mission reached the conclusion that the establishment in Canada of such factories, which would for the most part have to be placed "in vaseline" in peace time, would not afford the necessary guarantee of the availability of labour and materials which would be required to meet the demands which would be made in the event of war. It was also clear that the Defence Department would oppose any scheme which would entrust the development of the necessary potential to one or two of the ten firms which are at present engaged on the manufacture of aircraft in

Canada. The Defence Department feared the hostilities and suspicions latent in such a method and also the detrimental effect of an arbitrary selection of this kind on a young and growing industry.

The Mission, in consequence, reached the conclusion that the only sound method of producing an adequate war potential in Canada was to develop the existing Canadian aircraft industry as a whole on healthy and natural lines by the stimulus of orders from the United Kingdom. This was also the considered view of the Canadian Defence authorities.[61]

The Canadians got what they wanted. In November 1938 the British government placed an "educational order" for eighty Hampden twin-engined bombers with Canadian Associated Aircraft Limited, a central company formed by several Canadian aircraft firms. There would, moreover, be no commitments for the future. "I do not think," the secretary of state for air concluded, "we can expect to obtain from the Canadian Government here and now a specific assurance that the production capacity which the scheme is designed to create will certainly be available to us in war."[62]

This did not mean that the Prime Minister was unsympathetic to the empire's difficulties. King justified the Liberal rearmament program, which began in earnest in 1937, in narrowly national terms—Canadian unity, Canadian responsibility, the exigencies of Canadian politics.[63] He denied the charge of isolationists and neutralists in his party and outside that increases in the defence estimates were for the training and equipment of an expeditionary force to fight alongside the British army in foreign lands. King was willing to make some of the improvements in the "empire's" defences which had been suggested by the chiefs of staff in the previous year, but only if it could be demonstrated that these were measures taken to ensure *Canada's* safety.[64] King did not think, however, there was any major threat to Canadian shores, certainly not in a conflict where more strategic areas were at stake—"Relatively our danger is small." He told the House of Commons he hoped that

> it will not be thought that because we have laid emphasis on the fact that what we are doing we are doing for Canada, we are not thereby making some contribution towards the defence of the British Commonwealth of Nations as a whole, or that we are not making some contribution towards the defence of all English-speaking communities, that we are not making some contribution towards the defence of all democracies, that we are not making some contribution towards the defence of all those countries that may some day necessarily associate themselves together for the purpose of preserving their liberties and freedom against an aggressor, come from wherever he may.[65]

This was as far as King was prepared to go, and understandably so. The strong opposition to increased expenditure voiced during the 1937 defence estimates debate by the socialists and most of the Social Crediters, but also by a significant number of Liberal members of parliament, illustrated the need to move cautiously. Before the debate King reflected on the fight of his Liberal predecessor, Laurier, to preserve national and party unity before 1914. "I am now where Sir Wilfrid was, in a more dangerous time in the world's history—but still between the devil & deep blue sea in having to steer between Imperialism & Nationalism in extreme forms."[66] The debate revealed much insular nationalism: a broadly based disapproval of increased estimates in English and particularly French Canada and little—very little—support of empire defence efforts.[67] The Prime Minister continued to muddy the waters until the end, carefully resisting appeals from those who wished he would do less and demands from those who would have him do much more. But when war broke out, King piloted the government and country to Britain's side, as he had always known (but seldom said) he would.

Ernest Lapointe, the minister of justice and King's closest political associate, told parliament on 9 September 1939 that Canadian involvement in a British war flowed directly and inevitably from "our alliance with Great Britain," an alliance based, *inter alia*, on a common national status and much common legislation. As Lapointe had said in March of that year, neutrality was simply not possible:

> It is contrary to international principles to recognize the possibility of one country being neutral and another a belligerent when they are not separate sovereignties and when one is linked with the other in respect of its own power of legislation. . . .
>
> We have a common national status. A British subject in Canada is a British subject in Britain. We have the use of the diplomatic and consular functions of Britain. Our criminal code would preclude, in many sections of it, any notion of neutrality. Many sections are based on the principle that Canada is engaged in a conflict when Britain is so engaged. The Foreign Enlistment Act of the United Kingdom which was in force in Canada until 1937, made it an offence to enlist Canadians for service in armies of countries at war with the king. We enacted similar legislation in Canada two years ago. . . . If Canada was neutral the entire British merchant marine could shift its registration to Canadian ports, which would be inconsistent with the concept of neutrality. . . . We are bound by contract with Britain to give Britain the full use of the dry docks at Halifax and Esquimalt for British vessels.[68]

All that could be changed, of course. Legislation could be passed, laws repealed, contracts cancelled. Lapointe concluded, however, that Canadians would not wish it, nor did he believe that it would be in their interest to attempt "to protect their neutrality even against British vessels and British sailors and practically wage war against their own king."[69]

The Anglo-Canadian "alliance" was not based on specific commitments to defined ends; it held its vitality through shared allegiances, shared ideals, customs, and institutions, a shared sense of community in a British world. The intimate military relationship between the two countries simply reflected the political facts of life.[70]

NOTES

1. Admiral Sir Ernle Chatfield in August 1936, quoted in Lawrence R. Pratt, *East of Malta, West of Suez: Britain's Mediterranean Crisis, 1936–1939* (London, 1975), 3.

2. See, for example, the speeches made on behalf of the three governments at the Commonwealth prime ministers' meetings of 1935. Minutes, mtgs. 1–4, 3 April–23 May 1935, Public Record Office, London (hereafter PRO), Cabinet Office (hereafter CAB) 32/125.

3. Michael Howard, *The Continental Commitment: The Dilemma of British Defence Policy in the Era of the Two World Wars* (London, 1972), 76.

4. Escott Reid, "An Anglo-Canadian Military Alliance?" *Canadian Forum*, 16 (June 1937), 83–85; H. Duncan Hall, "The British Commonwealth of Nations in War and Peace" in W.Y. Elliot and H.D. Hall, *The British Commonwealth at War* (New York, 1943), 67–68; D.C. Watt, "Imperial Defence Policy and Imperial Foreign Policy, 1911–39: The Substance and the Shadow," in his *Personalities and Policies* (London, 1965), 153–54; R.A. Preston, "The Military Structure of the Old Commonwealth," *International Journal*, 17 (spring 1962), 98–99. See also C.P. Stacey, *Arms, Men and Governments: The War Policies of Canada 1939–1945* (Ottawa, 1970), 94.

5. Watt, "Imperial Defence Policy," 153–54.

6. For a discussion of "limited liability," see Peter Dennis, *Decision by Default: Peacetime Conscription and British Defence 1919–39* (London, 1972) and Howard, *Continental Commitment*, chap. 5.

7. Amery to J.L. Garvin, 25 October 1926, L.S. Amery Papers. I am indebted to the Rt. Hon. Julian Amery for permission to consult his father's papers.

8. Ramsay Cook, ed., "J.W. Dafoe at the Imperial Conference, 1923," *Canadian Historical Review*, 41 (March 1960), 29.

9. See, for example, Stephen Roskill, *Naval Policy between the Wars*, vol. 2, *The Period of Reluctant Rearmament 1930–1939* (London, 1976), 366–68.

10. Memorandum of the Committee of Imperial Defence, 27 November 1928, CAB 24/199, CP 368(28).

11. A.J. Toynbee, ed., *British Commonwealth Relations: Proceedings of the First Unofficial Conference at Toronto, 11–21 September 1933* (London, 1934), 27–28. Toynbee commented in a footnote: "Public opinion in the United Kingdom has never yet envisaged the situation, here imagined, in concrete terms. It is true, no doubt, that if this situation ever did actually arise, it would provide an immediate and most effective touchstone for the traditional assumption in the United Kingdom—an assumption based on sentiment and taken for granted out of long habit—that, as a matter of course, the United Kingdom would come to the assistance of any other member of the Commonwealth with all its strength in any contingency."

12. Minutes of Sir Harry F. Batterbee (assistant under secretary of state, Dominions Office), PRO, Dominions Office (hereafter DO) 35/174/6454C/3; interviews with former Dominions Office officials, Sir Charles Dixon, 30 June 1971, and Sir Stephen Holmes, 15 September 1971.

13. *Citizen* (Ottawa), 24 December 1934. Hankey's real purpose was to inform the dominion governments of Great Britain's defence plans and to gather information on dominion policies and opinions. W.M. Hughes, the former Australian prime minister, had commented in November: "Sir Maurice Hankey is amongst us—taking notes—and God knows what else," Hughes to Lloyd George, 19 November 1934, Lloyd George Papers G/10/8/3, House of Lords Library, London. For a description of the tour, see Stephen Roskill, *Hankey: Man of Secrets*, vol. 3, *1931–1963* (London, 1974), chap. 3; and Ann Trotter, "The

Dominions and Imperial Defence: Hankey's Tour in 1934," *Journal of Imperial and Commonwealth History*, 2 (May 1974), 318–32.

14. Memorandum of Hankey, "Impressions of Canada, December, 1934," CAB 63/81.

15. Ibid.

16. Ibid.

17. Ibid.

18. James Eayrs, *In Defence of Canada*, vol. 1, *From the Great War to the Great Depression* (Toronto, 1964), 266, 274–87.

19. Interview with Air Vice Marshal T.A. Lawrence, 11 October 1973.

20. C.P. Stacey, *The Military Problems of Canada* (Toronto, 1940), 92–95 and appendix C; Memorandum of chief of the general staff, "The Defence of Canada," 28 May 1935, Directorate of History, Department of National Defence (hereafter DHist), 112.3M 2009 (D7).

21. PRO, Foreign Office (hereafter FO) 371/19128/J4117, minutes of I.P. Garran, 27 August 1935.

22. The unemployment relief projects are the best example of this phenomenon. See Eayrs, *In Defence of Canada*, 1:124–48, 260.

23. . . . Meetings of British Commonwealth prime ministers, minutes, 4th mtg., 23 May 1935, CAB 32/125.

24. King Diary, 11 February 1937, Public Archives of Canada (hereafter PAC).

25. Minutes, 217th mtg. of Committee of Imperial Defence, 11 November 1926, CAB 2/4.

26. Memorandum of Hankey, 14 June 1929, CAB 63/41. The instructions arrived too late.

27. DO 114/13/60-5; Minutes, Committee of Imperial Defence, Oversea Defence Committee, 275th, 280th, 284th, and 285th mtgs., CAB 7/9.

28. Minutes, 217th mtg. of Committee of Imperial Defence, 11 November 1926, CAB 2/4.

29. Minutes, 251st mtg. of Committee of Imperial Defence, 28 November 1930, CAB 2/5. Bennett to Hankey, 2 December 1930, Bennett Papers, volume 155, folio 103978, PAC. It is not certain this letter was ever sent. At any rate, Bennett's name and remarks remain in the minutes.

30. Preston, "Military Structure of the Old Commonwealth"; Stacey, *Arms, Men and Governments*, 89.

31. M.A. Pope, *Soldiers and Politicians: The Memoirs of Lt.-Gen. Maurice A. Pope, C.B., M.C.* (Toronto, 1962), 53.

32. Stacey, *Arms, Men and Governments*, 79–80; Preston, "Military Structure of the Old Commonwealth," 116; E.C. Russell, "Interview with Capt. Eric S. Brand, 22 February 1967," DHist 112.3H1.003(D80).

33. Certainly this was the view of the military. "We had no Staff College in Canada at all," recalled one RCAF officer. "That level of thinking, that level of doing, was completely unknown. It was a godsend. Without it, we'd have been absolute neophytes." W.A.B. Douglas, Interview with Air Vice Marshal George Howsam, August 1975, DHist 76/120.

34. For details, see CAB 63/41, memorandum of Hankey, "The Circulation of Documents to the Dominions," 7 May 1929; Eayrs, *In Defence of Canada*, 1:91–93.

35. Papers of Department of External Affairs Historical Division, Conference Report Series, 1-1926/6, "Notes on Imperial Conference, 1926," September ? [sic] 1926.

36. Memorandum of Skelton, "Automatic Belligerency," [1939], Hume Wrong Papers, vol. 3, PAC.

37. Maurice Pope described his participation in a 1937 Canadian Institute of International Affairs round table conference on defence problems thus: "apparently, in the minds of my inquisitors the soldiery were composed of evilly disposed persons. Why were we recommending increased provision for defence now

that the day of collective security had dawned? To what extent were we under the Chief of the Imperial General Staff in London?" Pope, *Memoirs*, 132–34. Cf. A.R.M. Lower, *My First Seventy-Five Years* (Toronto, 1967), 206.

38. "Canadian Defence Quarterly Essay Competition, 1930, Prize Essay," *Canadian Defence Quarterly*, 8 (January 1931), 158.

39. See, for example, memorandum of Joint Staff Committee, 5 September 1936, DHist 74/256. "I hope," King told parliament in 1924. "the day will never come when I will deny that Canada has some responsibility to other parts of the British Empire as well as to herself, and I hope the day will never come when Canada will hesitate to recognize her responsibility . . ." Canada, House of Commons, *Debates*, 20 March 1924, 524.

40. Memorandum of Joint Staff Committee, 5 September 1936, DHist 74/256.

41. See C.P. Stacey, *Six Years of War: The Army in Canada, Britain and the Pacific* (Ottawa, 1955), 30–31.

42. In 1937 liaison letters from the chief of the Imperial General Staff to his Canadian counterpart were addressed to "Chief of the Canadian Section, Imperial General Staff, Ottawa, Canada." See generally PRO, War Office 32/4124/0585119, and Stacey, *Arms, Men and Governments*, 72–76.

43. The committee decision is referred to in DO 35/174/6454C/3, dominions secretary to Admiralty, War Office, and Air Ministry, 13 April 1936.

44. CAB 32/70, Imperial Conference and Economic Conference, 1930, Policy Committee, minutes, 2nd mtg., 7 May 1930.

45. Minutes, mtg. held at Dominions Office, 5 March 1936, DO 35/174/6454C/1.

46. Dominions secretary to Admiralty, War Office, and Air Ministry, 13 April 1936 and passim, DO 35/174/6454C/3.

47. See note by the Joint Planning Sub-Committee of the Chiefs of Staff Committee, 1 May 1936, CAB 53/28; Minutes of C.W. Dixon, 28 April 1936, DO 35/174/6454C/8.

48. CAB 53/28; Report by the Joint Planning Sub-Committee, [July 1936], COS 489 (J.P.).

49. Minutes, mtg. held at 2 Whitehall Gardens, 21 July 1936, CAB 53/6. CAB 53/28, COS 489 (J.P.), enclosure 1.

50. Minutes, mtg. held at 2 Whitehall Gardens, 21 July 1936, CAB 53/6.

51. See Stacey, *Arms, Men and Governments*, 81–83.

52. Minutes, mtg. held at 2 Whitehall Gardens, 21 July 1936, CAB 53/6.

53. See, for example, Review of Imperial Defence by the Chiefs of Staff Sub-Committee, 22 February 1937, Imperial Conference, 1937, CAB 32/127.

54. Squadron Leader F.V. Heakes, "Liaison Notes," 8 February 1939, DHist 77/51.

55. Stacey, *Arms, Men and Governments*, 89. See also H. Blair Neatby, *William Lyon Mackenzie King*, vol. 3, *1932–1939; The Prism of Unity* (Toronto, 1976), 282–83.

56. "Air Mission to United States and Canada: Creation of a War Potential for Aircraft Production in Canada," 16 June 1938, CAB 21/671, CP 143(38).

57. Memorandum of Hankey, 22 October 1936, CAB 21/267; CAB 53/6, COS 489 (J.P.), enclosure 1. Cf. War Office 32/4122/0585106.

58. Secretary of state for air, "Creation of a War Potential for Aircraft Production in Canada," 13 October 1938, CAB 21/671, CP 224(38).

59. Memorandum of 11 September 1936, King Papers J4/148/F1226/C107967, PAC.

60. Norman Hillmer, "The Pursuit of Peace: Mackenzie King and the 1937 Imperial Conference" in John English and J.O. Stubbs, eds.,

Mackenzie King: Widening the Debate (Toronto, 1978), 158; James Eayrs, *In Defence of Canada*, vol. 2, *Appeasement and Rearmament* (Toronto, 1965), 119.

61. "Air Mission to United States and Canada: Creation of a War Potential for Aircraft Production in Canada," 16 June 1938, CAB 21/671, CP 143(38). On the Canadian government's involvement in the 1937–38 negotiation between a private Canadian firm and the British government, leading to a major contract for the manufacture of Bren light machine guns, see Eayrs, *In Defence of Canada*, 2:119–21.

62. Secretary of state for air, "Creation of a War Potential for Aircraft Production in Canada," 13 October 1938, CAB 21/671, CP 224(38).

63. The Prime Minister, and not the Minister of National Defence, put the case for an increased defence budget to the Liberal caucus. For the notes of his speech to the Liberal members, see Ian Mackenzie Papers, file X-6, PAC (partly quoted in Stacey, *Six Years of War*, 14).

64. The 1937 defence estimates, for example, made provision for the refurbishing of coastal defences, particularly on the Pacific.

65. *Debates*, 19 February 1937, 1058. The Minister of National Defence, a less skilful politician, laid more emphasis on "the protection of Canadian shores, the protection of Canadian homes, the protection of Canadian shipping terminals and harbours" (ibid., 15 February 1937, 895). King himself later came to fear that the outbreak of a world war would lead to an effort to seize Canada as a war prize, "the one land capable of colonization by large numbers of persons of other lands": King Diary, 18 January 1938. I owe this quotation to Anne M. Trowell.

66. King Diary, 11 February 1937.

67. *Debates*, 15–19 February 1937, 876–965, 992–1073 passim.

68. Ibid., 31 March 1939, 2466–67.

69. Ibid., 2467.

70. See Lieutenant-Colonel H.D.G. Crerar, "The Development of Closer Relations between the Military Forces of the Empire," *Journal of the Royal United Service Institution*, 71 (August 1926), 441–42.

CANADA AND THE HIGHER DIRECTION OF THE SECOND WORLD WAR 1939–1945 ⬦

ADRIAN W. PRESTON

○

Our standing as a military nation was not as important as many of our people had led themselves to believe. We did not dispose of enough "battalions." No opportunity was ever given to proffer advice as to how the war should be directed, and if it had been I wonder if our knowledge of the general situation and our limited experience in matters of this kind would have made us competent to give it effectively. As a consequence, we remained in the second rank. We contributed to the Allied war effort, and this handsomely, but the United Kingdom and, later, the United States did not [call us to their councils]. Thus many of our people, while conscious of the fact that we were going all out in the production of the means of making war, felt themselves denied . . . any sense of achievement.

Thus, in retrospect, wrote Major-General Maurice A. Pope, late Chairman of the Canadian Joint Staff Mission to Washington during the latter part of the Second World War.

The machinery for the higher direction of the war developed in three stages. Initially, during what is commonly referred to as the twilight period, it was invested in a modified version of the 1917–18 Anglo-French Supreme War Council; this functioned until the fall of France in June 1940. The second phase, which lasted until Pearl Harbour and American entry into the war, was characterised by the dominance of Churchill's position and his extraordinary grip upon the reins of strategy. In the third and final phase, this supreme authority was transferred to the uniquely combined

⬦ *Journal of the Royal United Services Institute* 110, 637 (Feb. 1965), 28–44.

Anglo-American agencies where, answering only to President and Prime Minister, it remained until the war's end.

The Second World War was the first international conflict in which Canada was engaged as a fully autonomous State. In this paper I have in turn looked at each of the above-mentioned stages from the point of view of Canada's attitude to the question of who should run the war. This attitude changed as the war progressed; as France, with which a large and influential section of Canada's population had strong if remote ties, surrendered; as the Russians, Japanese, and Americans came in; as . . . belief in [its] geographical inviolability changed to dread of transoceanic invasion; as the strategic control of the war began increasingly to focus in American hands and the strategic pendulum to swing from Europe to the Far East; and as [Canada's] own initially meagre contributions in men and materials grew to proportions large enough to justify . . . demands for an increased share of strategic responsibility. Throughout this whole process [Canada] was continually obsessed with the idea of status, with the fear of a national split through having to introduce conscription, and with the frustration, especially during the latter stages of the war, that because . . . pre-war preparations had been so small she had forfeited irretrievably any real say in how the war was to be won. [Canada's] position was at any single moment so novel, peculiar, and complex that it defied satisfactory solution, and at the end of the war [the] only consolation was that she had established herself as a middle or auxiliary power, high in prestige but without any real responsibility in world affairs.

The Canadian approach to the Anglo-French conduct of the twilight war had its tap-root in a long and at times turbulent tradition of "no commitments"—political or military—which would act as precedents to cramp its freedom of action; a tradition which aimed at stripping every trace of British influence from Canadian military affairs. . . .

In the "low dishonest decade" preceding the outbreak of war, this anxiety to disentangle her foreign policy from Britain's and reduce it to the lowest common denominator of public agreement—subordinating it to the excuse of a unity dangerously strained "by economic depression and other consequence of the last war and its aftermath"—had the not unnatural result as war approached of creating the impression in British minds that Canadians were no more interested in the management of war than they had been in the practical maintenance of peace by force. . . . The League of Nations and Locarno had made war remote and unthinkable, and encouraged Canadians to concentrate on questions of status and underestimate the urgency of problems of security and foreign policy which as an autonomous people they would have to face in the future. Isolationism dwelt in geography, and accentuated a distaste for foreign dealings. The revulsion against war and European entanglements—and the power politics, secret treaties, and the old diplomacy which had made them possible—was periodically expressed throughout the inter-war years in stern lectures from Geneva on Europe's need for moral reconstruction, a Europe now virtually an armed camp, full of intrigue and recriminations.

If this priggish attitude was meant to have a wholesome, corrective influence, it soon disappeared when the formidable totalitarian menace of the late 1930s made Canada fumble her own policy-making, with the ironic result that she came to endorse and rely on that British policy of appeasement which has since been labelled so disreputable. Canadians as a whole, in a sensible realisation of the limitations of their own power and knowledge, could not and did not favour resolute resistance to aggressors. They did not aspire to fill the role of "nature's last crusaders," and eschewed the thought of crusades for any cause which were not led by Britain or the United States. Several British statesmen have since gone so far as to state that Dominion opposition to war, especially by Canada and South Africa, was used by Mr. Chamberlain to "discourage those within the British Cabinet who would urge a less cowardly posture in the face of the German threats." There can be little doubt that Canadian insistence upon this accentuated form of "no entanglements" contributed to the breakdown of a strong British policy.

British attempts to initiate measures of military and industrial co-operation in advance of war were consistently rebuffed....

...What few plans the Canadian Government had in mind were enough to confirm the suspicion of the British High Command that Canadians intended so far as possible to remain outside a European war....

...To Mackenzie King, it was inherently dangerous to accept political, military, or industrial commitments, or otherwise indulge in pre-war planning with Great Britain which might predispose the nature and scope of Canada's contribution. The paradoxical result was that, in spite of her new-found sovereignty, the colonial tradition of leaving to Britain the supreme direction of any war, particularly continental, was allowed to persist, arousing in British minds the sensation that as a whole Canadians wished no truck or trade with war and its direction.

If, in principle, dissociation was well-advised, Canadian slowness to equip herself for an independent foreign policy was less so. For the resultant condition of her intelligence system, armed forces, and armaments industry rendered her incapable, once war had broken out, of exercising a modicum of influence upon the higher direction of war then and for some time in the future, even had she wished to do so. In 1935, a General Staff "Appreciation of the Defence Problems confronting Canada, with Recommendations for the Development of the Armed Forces" disclosed the ludicrous state of Canadian defence. "There was not a single modern anti-tank gun of any sort in Canada.... The coastal defence armament was obsolescent and ... defective.... Obsolescent field gun ammunition inherited from the Great War represented 90 minutes' fire.... Not a single aircraft is ... fit to be employed in active operations.... No provision has been made for the supply of mechanised transport for war purposes.... We possess no tanks or service armoured cars.... No facilities whatsoever exist for the production of rifles, machine-guns, and artillery weapons in Canada."

Over the next four years there was little improvement, and if Mackenzie King's rearmament program was "comprehensive and workmanlike," it was not notable for its urgency or realism. His ideas were wrapped up in that theory of limited liability which Liddell Hart was shortly to prescribe—blockade, economic warfare, and but a token expeditionary force—as the foundation of Britain's war policy. The British Government's momentary adoption of Liddell Hart's views encouraged Mackenzie King to place an exaggerated value on Canada as the arsenal of raw materials and supply, and to keep his military preparations at the lowest level. In spite of the spectacular demonstration of German arms in Poland—with the implication that blockade might prove less telling in mobile than in static warfare—this emphasis upon the economic weapon in war continued to dominate Canadian war policy long after its rejection by Britain in February 1939. The direct result was that "the preparations made by Canada before the outbreak of war were so small that she was unable to make any really large contribution to the sum of the Allied military effort for years after war broke out."

Indefensible was the Government's rejection, over three successive administrations, of their military adviser's recommendations to create the nucleus of a native armament industry. . . .

The state of her intelligence system, a system vital to the formulation of independent and balanced viewpoints, strategic or diplomatic, was no less incredible. It had its genesis in the Chanak affair and the fresh accession to power of Mackenzie King. Prior to Churchill's trumpet call to the Dominions, he had received no cabled despatches from any department of the United Kingdom Government, in spite of the promise made by the Foreign Secretary the year before to send "once a month, or, if necessary, once a week, a selection of those extremely confidential Foreign Office Papers which are now seen only by the Cabinet and circulated to our representatives in our Embassies and Legations abroad." This rather mishandled affair set up in Mackenzie King's mind the germ of that anti-Downing Street complex which was to grow and colour much of his business with Whitehall over the next 20 years. Assuming a genuine anxiety on the part of the Canadian Government to develop a solid foreign policy, however, "the only logical reaction to the Chanak *débâcle*," writes Professor Eayrs, "would have been to have insisted on improved communications and to exercise the right to Legation granted to Canada in 1920.

"No effort was made to move in either of these directions. An active and independent role in world affairs was about the last thing the Canadian people wanted at this time. . . ."

Of the three world capitals where aggressions leading to the Second World War were spawned and hatched, only one in 1939 was the site of a Canadian diplomatic mission." There were no military attachés at any foreign post. The pre-war American–Canadian military conversations were conducted in such a clandestine fashion that the stranglehold of security probably choked them meaningless. This state of affairs precluded the Canadian Government from sounding out the trends and pitfalls of international politics, developing an effective foreign policy of her own, and offering, had she wished to do so, another authoritative and perhaps decisive opinion to the crucible of British policy.

In one respect at least—that of General Staff liaison and training—the policy of dissociation was not carried so far, though ultimately it was in its own way as much to blame for the passive attitude of the Canadian High Command towards the running of the war. . . .

. . . By 1939, a significant proportion of the Canadian officer corps had attended the British Staff and Imperial Defence Colleges, or had served at the War Office. Imperceptibly over the years the British approach to strategy had ingrained itself in the Canadian military mind. In the absence of any political measures or machinery for the formulation of an independent strategic doctrine, and in view of the inexperience of her generals in the handling of large formations or the planning of campaigns, it insensibly predisposed the Canadian High Command to fight according to British strategy and, at least in the early stages of a war, to an inferior or even negligible role in its direction.

At the outbreak of war, therefore, and for the most part throughout its first phase, the Canadian Government not only felt unwilling to cast its weight into the scales of a strategy into which through its own political diffidence it had been incautiously drawn, but was quite unable, militarily and industrially, to do so. Its necessarily blinkered notions of the coming struggle envisaged a small-bore World War I in which Germany would knock itself out upon the Maginot Line while being steadily strangled by the Royal Navy. The moral revulsion against war (especially its total implications) still lingered in spite of the destruction of Poland. And as there yet seemed no necessity for the full-scale mobilisation of men and industry which was later demanded, governmental policy simply confined itself to the provision of raw materials, facilities for training aircrews, and a few reinforcements without taking the slightest interest in how the Anglo-French Supreme War Council was going to win the war.

PHASE I

The official discarding of limited liability in early 1939 had suggested the need for some Anglo-French agreement on co-operative machinery for the inter-Allied conduct of the war. During the spring the two Prime Ministers had agreed upon a modified version of the 1917–18 model:

"There was to be a Supreme War Council on which France and the United Kingdom should each be represented by the Prime Minister and one other minister, and other Allied Powers, perhaps, by their ambassadors. The Council was to have no executive authority, final decisions being reserved to the governments. Each of the two Powers would appoint Permanent Military representatives to advise on technical matters, working as a joint staff but subordinate to their own Service chiefs; the British representatives would also be collectively subordinate to the Chiefs of Staff Committee. The French and British secretariat was set up, and branches of the new organisation functioned on both sides of the Channel."

By September, the Anglo-French staff discussions had hatched "the broad strategic policy for the conduct of the war." It applied primarily to Europe, although it was also agreed upon naval strategy, operations in the Mediterranean, Middle East, and North Africa, and the effects of possible Japanese intervention. Its essential features were (1) an initial defensive strategy based on naval and economic warfare; (2) the defeat of Italy before Germany; and (3) securing the benevolent neutrality or active assistance of the United States.

But having decided the strategy, the key element in making it work—the fusion of interest and sentiment, and complete harmony of thought and action—was never allowed to develop in an atmosphere free of suspicion and resentment. As a result there was no "permanent Anglo-French Secretariat to maintain records, preserve continuity, and ensure systematic procedure," and the lack of delegated power and proper liaison ensured complete breakdown when the crisis came to a head in May 1940. But none of these shortcomings was regarded as critical, if they were even noticed, before the crisis arose. What we may now justly describe as the sluggish and gangling nature of Chamberlain's war machinery was not then considered so extraordinary. It provided the means for inter-governmental consultation sufficiently satisfactory to the Canadian Government to be kept apprised of Anglo-French decisions without arousing any desire further to participate in them or in those of a strictly Imperial War Cabinet.

Mackenzie King's attitude to the strategic direction of the war at this time is in distinct contrast to that of Sir Robert Borden in 1915, and the motives for the difference can be measured in terms of contemporaneous developments in status and technology. In Borden's day the haphazard machinery for transmitting continuous and reliable information from the front to Canada, the slow system of telegraphic or sea communications so open to rupture by the enemy, the danger that Canada's increasing military contribution might be squandered in a senseless strategy, and the need to establish yet another precedent in the struggle for autonomy, all combined to force Borden to demand and receive a clearer idea as to how the war was being run. Thereafter, he was sent War Council Circulars, General Staff appreciations, and situation reports, and in December 1916, was invited to attend "a series of special and continuous meetings of the War Cabinet in order to consider urgent questions affecting the prosecution of the war," an invitation which he exploited to the full for the remainder of the war.

In 1940, conditions were quite different. Communications were multiple, expeditious, and dependable; and brought governmental scrutiny and influence directly to bear on the very field of battle. Mackenzie King's version of limited liability decreed that the extensive slaughter of large Canadian armies would never take place. Public opinion in this way was not likely to become divided over manpower issues, or incensed over the exceptional amount of blood spilt by the incompetence of foreign generals. The basic Allied strategy, necessarily limited and defensive, coincided with these views and in the twilight period there seemed no reason to disagree with it. Even after the German spring offensive in Scandinavia, the idea that Canada might be invaded never seriously entered people's heads. And the little Canadian army squatted in England much like an aborigine in a strange wilderness, self-conscious but safe; fulfilment of that promise to be "at Britain's side." But of prime importance and the bed-rock of Mackenzie King's attitude was that full autonomy had been so recently accomplished that it was best emphasized by demanding independent consultation while at the same time remaining far from the seat of war. The paramount danger of always discussing things in London was that it invited the suspicion that the Canadian representative was either too readily acceding to the British arguments or was being brow-beaten into doing so. But if this impression of subordination was likely to arouse angry comment in some sensitive areas, the idea of blunt opposition to any form of Imperial Council was calculated in other areas to be construed as evidence of a faint-hearted American-like war effort. The crux of the matter depended upon the promptitude and quality of intergovernmental information and consultation; only in this way could a safe balance be struck between the necessity of at least some appearance of strategic responsibility and the hazard of commitment to politically suicidal projects.

Better to juggle these matters, Mackenzie King was prepared to divest some of his anti-Imperial affectations and encourage closer liaison. Chamberlain (and later Churchill) sent secret daily (after the invasion of the Low Countries and France, twice daily) telegrams recording the progress of operations, while copies of the Chiefs of Staff Weekly Résumé followed by sea. A Canadian Military Headquarters was set up in London as a forward echelon of N.D.H.Q. to maintain close liaison with the War Office. . . .

The Canadian High Commissioner, with his Dominion colleagues, met the Foreign or Dominions Secretary daily for a general discussion of the war situation and problems of mutual interest. "We tell them everything we can," wrote the Dominions Secretary, "on all and every subject . . . foreign affairs, economic development, military co-operation, even domestic issues here which are likely to interest our partners," though Churchill was constrained to chide him "not to scatter so much deadly and secret information over this very large circle," or to "get into the habit of running a kind of newspaper full of deadly secrets."

Though this elaborate communications system made Mackenzie King very happy, there were oddly enough some doubts in Whitehall as to whether even this was enough to bring the Dominions to the centre of discussions. Those early apprehensions, however, were quickly snuffed out. Chamberlain did not feel that it was time for Dominion membership on the Supreme War Council or that a new Imperial War Cabinet would be very practicable. . . ." The closest contact is maintained," he assured the House of Commons in September 1939, "both directly between governments, through the High Commissioners in London and in the Dominion capitals, and through an immense variety of technical and subordinate arrangements covering every field of relations."

Nor, it seems, was Canada poorly served or seriously misled even at times of crisis, on the actual or probable course of events in Europe. The spring blitzkrieg left little to transmit except a catalogue of military disasters sent secretly in special Cabinet papers. This steady outflow of information and opinion did much to check panic and wild imaginings in foreign capitals as to Britain's attitude towards the continuation of the war. On 5 June, for example, Churchill warned Mackenzie King that France might collapse, and on the 16th composed a message to all the Dominion Premiers "showing them that our resolve to continue the struggle although alone was not based upon mere obstinacy or desperation but to convince them by practical and technical reasons, of which they might well be unaware, of the real strength of our position." This confirmed Mackenzie King's determination "that Canada should stay in the fight so long as England and France, or England itself, continued to oppose Nazi aggression," a fact reported by Churchill to the British House of Commons on 18 June.

Lapses were inevitable; action often preceded consultation. The noble but abortive offer of union with France is in this context understandable. In spite of obvious implications, the Canadian Government was not consulted. More serious omissions raised the dread of disunity. The bombardment of the French Fleet at Mers-el-Kebir provoked misgivings in Ottawa. While Mackenzie King understood that for Britain this had been a painful decision, he nevertheless remained anxious that some foolish impetuosity might push Vichy actively into the Axis camp. "The possibility of France becoming directly involved in war with Britain," he wrote, "was unthinkable." He "could not imagine . . . calling Parliament for the purpose of bringing Canada into war against France," and was thankful when the failure of the Dakar operation produced no further complications.

In spite of the occasional breakdown, therefore, Mackenzie King was at this time highly satisfied with the dissemination and quality of politico-strategic information. It enabled him to keep sufficiently apprised of what was going on and what was contemplated to satisfy the noisier Imperialistic elements in Parliament while at the same time it gave the impression of great distance and independence from London, most gratifying to the more extreme nationalists and status-watchers. Throughout the Second World War, as in the First, Mackenzie King was hounded by the spectre of a crippling disunity—a disunity largely between English and French, most easily brought about through the introduction of conscription. While invasion remained remote, and the war was confined to Europe, the sea, and the air, the less he insisted in sharing in its direction, the less likely would he be expected to vitiate his strategic voice by producing such armies as would make conscription necessary. Some representation was obviously necessary, as was machinery for direct communication in case of emergency. This he had on every official level. But for the present, largely for domestic and financial reasons, he was content to require no more of it than that it should keep him reasonably up-to-date on events.

PHASE II

With the collapse of France, Canada became Britain's most active and valu-able ally; her importance as a secure arsenal for food and raw material was enormously increased. Clearly, had she demanded it, there was a strong case for a larger share of strategic control; but paradoxically the fall of France, in exposing Britain to the possibility of attack, and even collapse, produced two circumstances which prevented this occurring. Churchill began rapidly to concentrate an extraordinary degree of personal power and control, and while attempting to enlist American support brooked little outside interference in the management of the war; while Mackenzie King, suddenly confronted with the unexpected possibility of invasion across the high seas, adopted the unusual expedient of forming a defensive agreement with the United States, against whom all Canadian defence plans had been traditionally aimed. This reversal of past policy was accompanied by a determination to act as interpreter or lynch-pin between Great Britain and the United States (or, more specifically, between Churchill and Roosevelt), a determination which soon grew into an obsession that he was almost indis-pensable to a harmoniously conceived and integrated Anglo-American strategy.

Churchill's assumption of power after the Norwegian *débâcle* brought swift alterations in the machinery of war direction. The key change was "the supervision and direction of the Chiefs of Staff Committee by a Minister of Defence with undefined powers," viz., the Prime Minister. . . . Thereafter Churchill's massive though uneven genius, helped along by the polished competence of the Chiefs of Staff—"a super-Chief of a War Staff in commission"—dominated the military scene. . . .

Churchill's position, at first unchallenged, in time became unassail-able. . . . He was the very core of power, assured by his unique grasp of strategy and the techniques of war, his vigilance, drive, eloquence, and courage. But unfortunately for Mackenzie King there were times when this genius skipped the rails, causing Churchill to take criticism for timidity or inertia and making him intolerant of those whose temperament was unlike his own.

Enjoying this position and its acute responsibilities, Churchill did not encourage, "on both constitutional and practical grounds," the creation of formal machinery for Dominion participation in the conduct of war. Indeed, no devisable system at the time could have ensured multiple control with-out loss of efficiency, for it is an axiomatic military principle that every addition to a War Council must tend against speed of decision and preser-vation of secrecy. But this state of affairs was not destined to remain undis-turbed for long. By November 1940, it had become apparent to many Canadians that the successful outcome of the Battle of Britain had consider-ably reduced, if not entirely eliminated, the danger of an imminent trans-oceanic invasion. Moreover, peering out from behind the rather unexciting cloak of hemispheric defence, they were attracted by the glitter of Wavell's victories in the Western Desert and the potentially rapid collapse of Italian military power in North Africa. They reasoned, furthermore, that if

Canadian troops could join their Dominion colleagues in the fighting, there would be a better case for resurrecting the old Imperial War Cabinet and obtaining a greater share of strategic responsibility. These views, expressed in Press and Parliament, were supported by the generals' assertion that battle experience, especially for commanders, would prove invaluable as the war progressed. But Mackenzie King was in no hurry to undermine his personal position by revising the cautious policy of limited commitments upon which he had launched Canada into war, and earmarking Canadian forces for operations which would bring no immediate political advantages either at home or abroad. . . .

Before Parliament on 17 February 1941, in an authoritative examination of the machinery for intergovernmental consultation, Mackenzie King rejected . . . demands for a modified form of Imperial War Cabinet. Scorning those who mistake appearance for reality, he suggested that "they had in existence today, in actual practice, although in no compact physical form, the most perfect continuous conference of Cabinets that any group of nations could possibly have." It was made possible by a system of constant and instantaneous communication. It possessed, moreover, one indispensable feature denied any imperial gathering; the opportunity of immediate discussion with Cabinet colleagues on every issue raised. . . .

An Imperial Cabinet, confronted with the need for a prompt decision on a delicate issue, would flounder; members could either act on their own, and thereby incur the possibility of eventual disownment, or postpone decision while consultation was arranged. Little, if anything, was gained. The present system was simpler and safer; it enabled a Prime Minister to consult his colleagues and reach a decision in the sure knowledge that it represented authoritatively and finally the view of the government as a whole. Mackenzie King did not dispute the desirability of short periodic visits for special purposes, but added the important rider that influence would tend to diminish if ministers were continually travelling back and forth. More important still, no political leader could contemplate divorcing himself for any lengthy period, especially in times of crisis, from the electorate to whom he was responsible without the most careful consideration of possible consequences; and to select any other minister than the Prime Minister would simply duplicate the position of High Commissioner. His general conclusion to this carefully worded declaration of policy was that he should "limit the visits of ministers to Britain" for it was "inadvisable to have it appear too much to the public that all matters were being settled [by Cabinet sitting in Downing Street] in London rather than that we were managing our own war effort satisfactorily here, and were meeting, by the agencies of communication, all that was most essential for conference." But there were further considerations in this repudiation of the Imperial War Cabinet suggestion.

Unquestionably, the least acceptable of these was the idea that a single resident minister might represent the corporate Dominion viewpoint vis-à-vis the British. "Most certainly not"; he objected to Churchill, "it was a mistake to be lumping all the Dominions together. . . . Each Dominion had their own problems and had to make their decisions in the light of them."

As to the more technical aspects of strategy, Mackenzie King refused to offer advice in London without the assistance of his Chiefs of Staff, whom, he added, could not be spared from Canada. But there was another aspect to this question which becomes more important when it is remembered that Mackenzie King, when confronted with a man of Churchill's stature as a student and practitioner of war, was acutely conscious of his own shortcomings and ignorance of military affairs. The potential dangers implicit in this situation had been graphically described to Mackenzie King by Robert Menzies during a visit to Canada in August 1940. "There was no British Cabinet, no War Cabinet," Menzies told him. "Churchill was the whole show, and that those who were around him were 'yes-men' and nothing else . . . what happened was, when all were together, Churchill reviewed the whole war situation, doing so in eloquent manner. He did the talking, and no one dared to say anything. . . . He himself was practically the only one who had ventured to question anything that was done." "One could not but feel," wrote Mackenzie King, after having been subjected to a similar experience; "how unfair the position was with matters of foreign policy and the rest of it carefully studied, as Churchill said, by a special Committee selected in advance on which none of the Dominions had been represented. That we should be there without any advisers. That there should be six British ministers to the four other self-governing parts, all of them with their colleagues and officials close at hand. Also British ministers were prepared to lead off in discussion knowing what they were to say, but expecting Premiers to deal with matters in reply. Personally, I felt the hopelessness of the situation. . . ."

Unstressed in his Ottawa and London speeches but perhaps of greatest importance in his own mind, was the advantage, especially at times of crises, of immediate personal contact with the President of the United States. The indispensability of American assistance had been recognised by the Supreme War Council at the outbreak of war as a strategic goal fundamental to eventual victory. And when the imminent collapse of England in mid-1940 raised the whole vexatious question of the future disposition of the British fleet, Mackenzie King was quick to appreciate that the promotion of "close and friendly understanding between the Allied Powers and the Roosevelt administration" would have much political value and tend to magnify Canada's position in world affairs.

In spite of the legend which has subsequently grown around this particular episode and which now forms an ineraseable part of the Canadian tradition, it is fairly certain that his interpretations on the two vital points at issue were rejected. Roosevelt refused to accept his explanation of Churchill's constitutional inability to bind a future (perhaps Quisling) administration to any decision concerning the fate of the fleet; while Churchill mildly scolded him for suggesting that he might be holding up the fleet "to make a bargain with the United States about their entry into the war." At the end of these discussions, which culminated in the "destroyer-bases" deal, it is plain that Churchill had not budged an inch; he had never contemplated the surrender nor the scuttling of the British fleet—and never would. Nor had he considered using this contingency as a bargaining lever to get the United States into the war. On Roosevelt's part, mainly due to the continuing British resistance and the Battle of Britain, there is no doubt at all that he was more willing to accept Churchill's attitude about the disposition of the fleet than he had been earlier. It is almost certain, therefore, that Mackenzie King was not the catalyst in forging the Churchill-Roosevelt accord he later would have others believe.

It should be remembered that since the outbreak of war, as "Naval Person" and "Former Naval Person," Churchill had developed an informal confidential correspondence with Roosevelt, and from 19 May onwards had privately sent him copies of those daily telegrams about military operations which had been compiled for the Dominion Premiers. Unaware of the real sympathy that Roosevelt and Churchill by this time had developed for each other's position, therefore, Mackenzie King was inclined to exaggerate the part he had played in the negotiations—and the fact that at times both Roosevelt and Churchill in their bargaining had appealed to Mackenzie King to put pressure on the other tended to encourage this idea. Undoubtedly, this was the origin of the myth, so carefully cultivated by Mackenzie King as fodder for the electorate, that Canadian mediation was necessary for the formulation of a combined Anglo-American strategy; for hereafter, every attempt by either Britain or the United States to encourage closer liaison strictly between themselves was greeted with increasing animosity by the Canadian Government.

The first occasion which was to indicate how Canada would react to exclusively Anglo-American dealings presented itself in the exploratory staff talks between those two countries in August 1940. At this time, the possibility of Britain's collapse was still a very real and imminent danger and had raised the whole question of the future direction of the war. For Canada, the implications were large and serious, for should the British Government and the Royal Navy eventually be compelled to move to and conduct the war from North America, Canada would become the next likely object in any Nazi plan for overseas aggression. On [Canada] would fall the full weight of the war effort. In these circumstances, therefore, Mackenzie King could be expected to claim at least an equal share in the direction of the war, and probably the commanding voice on all questions of North America defence. The Anglo-American staff talks, however, secretly conducted in London at a most critical stage of Britain's existence, threatened to disturb these prospects by promising to exclude Canada from all future consultations.

It was undoubtedly as a direct result of his being excluded from these staff talks that Mackenzie King began thinking about forming a defensive alliance with the United States so that in the event of Britain collapsing, he would be at least party to the higher direction of the war so far as it concerned the defence of North America. This in part accounts for the seriousness with which Mackenzie King was inclined to treat the Ogdensburg Agreement and the Permanent Joint Board on Defence. Not only was it included in the Canadian Treaty Series, but, in his eyes, it far surpassed in its ultimate importance the formation of the triple axis, a fact that suggests he was already thinking of Canada as an active third partner. But these aspirations of more strategic responsibility were based on the contingency of Britain's collapse and, outwardly at least, confined at the moment solely to the defence of North America. When the immediate danger to Britain had passed, therefore, and Japanese action at Pearl Harbour brought America into the war at once as an offensive-minded belligerent without passing through an initial defensive stage, the Permanent Joint Board on Defence overnight lost much of its importance, and the hopes attached to it as a potential stepping-stone into the councils of war were rudely disappointed. As a result of the rapidly growing intimacy in Anglo-American cooperation which followed the exploratory staff talks of August 1940, the likelihood of Canadian participation in the deliberations of a Grand Alliance—even as interpreter—became increasingly remote; and Mackenzie King, deeply irritated that circumstances should have cut off his chances at such a promising point and sensitive to the apparent coldness of this exclusion from Anglo-American dealings, "became increasingly occupied with the vexing task of trying to keep himself informed of what the strategists of the Grand Alliance were up to."

The tentative machinery and broad strategical policies for the Anglo-American conduct of the war had been hammered out in Washington at a series of staff talks, from 29 January to 29 March 1941, and later incorporated into a joint Report commonly known as A.B.C.-1. No Commonwealth country had been separately represented, and the Report simply provided that "the High Command of the United States and United Kingdom will collaborate continuously in the formulation and execution of strategical policies and plans which shall govern the conduct of the war." While it was agreed that America might exchange liaison officers with Dominion forces, it was the establishment in early April of a permanent British military mission—then, because of American neutrality, referred to as "Advisers to the British Supply Council in North America"—which instantly provoked from Canada a series of demands for a similar full-scale mission.

In one respect, at least, there were grounds for these demands; for the Service components of the Permanent Joint Board on Defence had at one stage in their planning agreed in principle on the exchange of military missions. As we have seen, however, this Canadian–American board now played little, if any, part in the formulation of strategy; and to have used it as an instrument for insisting on a mission would simply have led to a dead end. On 1 July 1941, therefore, the Canadian Government formally requested permission to establish a mission in Washington, but this was opposed by the American War and Navy Departments "on the grounds that representation through the Permanent Joint Board on Defence and the British Joint Staff Mission met all the Canadian needs for liaison, and that an undesirable precedent would be established for similar requests by other Dominions and the American republics."

The Argentia Conference, held the following month tantalisingly close to the Canadian shore-line, added more coals to the fires of . . . discontent. "I feel it is taking a gambler's risk," wrote Mackenzie King with great bitterness, with large stakes, appalling losses, even to that of an Empire, should some disaster overtake the gamble. To me, it is the apotheosis of a craze for publicity and show. At the bottom, it is a matter of vanity. There is no need for any meeting of the kind . . . the public in Canada and certainly some of my colleagues and my own officials will think it extraordinary that Churchill should have brought his own staff to negotiate with the United States staff, and ignored Canada altogether. While I had expected a personal visitation between Churchill and Roosevelt I had never thought of their bringing their representatives on foreign affairs, and Chiefs of Staff, etc., for conference on war plans, leaving Canada completely to one side—simply saying we would be told what had been done, though having no voice in the arrangements.

Though the Conference drafted the Atlantic Charter, and came to "important decisions concerning the Canadian war role" (namely, fleet convoy duties), the Canadian Government, as a co-belligerent almost within hailing distance, was not asked for its opinions or suggestions. Not unnaturally, the Churchill–Roosevelt press release of 12 August provoked a fresh outburst from Mackenzie King. "Here again Canada was being ignored. Here, at once, the very opposite [of consultation] was taking place and over a matter in which we, of all parts of the Empire, were very immediately concerned, as it relates to both military and economic co-operation." Within a week, on 18 August, he was telling the American Minister, J. Pierrepont Moffat, that "the prolonged refusal of Washington to approve a military mission was the only aspect of United States–Canadian relationships that seriously troubled him . . ." and urged that "the proposal be reconsidered in the light of its political implications and of the greater confidence that would be engendered in the Canadian public mind by direct military representation at Washington." Though sympathetic and resourceful, the only reply that Moffat could squeeze from the American Chiefs of Staff and Secretary for War was "that foreign political considerations inimical to our military interests should not be allowed to determine the attitude of the War Department."

This cavalier attitude on the part of the United States was again displayed in November when the State Department failed to consult Canada (as [it] did Britain, China, the Netherlands, and Australia) on the approaching Japanese crisis, in spite of her obvious interest in the political and security problems of the Pacific. But this was merely the beginning of a long, vexatious tradition. Her dreams of playing a larger part in the running of the war were continually cut short in this way. And the fact that she had by this time adopted a more realistic outlook in her war effort, and continued to be excluded, only increased her frustration at having made so few pre-war preparations. It is apparent from the odd statement to Parliament that Mackenzie King was prepared to accept this situation as implicit in Canada's role as a small Power; but he never relented in the pursuit for greater representation at the seat of war; and in this reversal of his earlier policy he was driven equally by his domestic need for some striking international feat and by a growing awareness of the regrettable treatment which Canada's position had suffered, and would continue to suffer, at American hands.

PHASE III

Immediately following the Japanese attack on Pearl Harbour, Churchill rushed to Washington where, at the Arcadia Conference, he helped to put into effect the machinery and policies for the combined conduct of the war which had been tentatively agreed upon almost exactly a year before. Four civilian Boards—raw materials, shipping adjustment, production resources, and food—were established "to muster the economic strength of the United Nations"; along with a special mixed (i.e. civilian and military) board for munitions assignment. These were subordinate to the Combined Chiefs of Staff, who, "under the direction of the heads of the United Nations," were charged with collaborating "in the formulation and execution of policies and plans concerning: (a), the strategic conduct of the war; (b), the broad program of war requirements based on approved strategic policy; (c), the allocation of munition resources based on strategic needs and the availability of means of transportation; and (d), the broad program of war requirements based on approved strategic priority." It was almost certain from the start that as American strength grew to its huge proportions, their "big-punch" strategy should take precedence over that peripheral strategy of the British which had served so well in the past, and that the responsibility for the conduct of the war should fall almost entirely into American hands.

None of this boded well for Canada and [its] hitherto thwarted ambitions, and when Churchill visited Ottawa on his return from Washington in late December 1941, Mackenzie King again took up the question of creating a Canadian joint military mission at Washington: "Said quite openly to him the problem we faced was that while we had been in the war during two and a quarter years, things would be so arranged that the United States and Britain would settle everything between themselves, and that our Services, Chiefs of Staff, etc., would not have a say in what was to be done. That in the last war there had been a Military Mission at Washington. People thought, in Canada, there should be a Military Mission there now, watching Canada's interests. That he would understand our political problem in that regard."

Churchill was to go over the same ground again with the Canadian Chiefs of Staff, but his only recorded opinion was that while he felt Canada "should have representation there," he hoped [it] "would take a large view of the relationships of the large countries [and] to avoid anything in the way of antagonisms." This brought little comfort to Mackenzie King, but it was Churchill's view, supported by much experience, that the "most sure way to lose the war," was to put every Power contributing forces "on all the councils and organisations which have to be set up and to require that everybody is consulted before anything is done." The United States were inclined to adopt a much less conciliatory attitude, openly referring to Canada as "a nuisance," and "as a part of Britain," remarks which Mackenzie King not unnaturally resented. . . .

Undoubtedly, one of the chief reasons for these repeated demands for a mission in Washington was the growing Parliamentary dissatisfaction with Canada's role in the making of strategy. Questions were tabled weekly calling for some explanation of the Government's attitude towards a seat among the Combined Chiefs of Staff, upon an Imperial War Cabinet, or the recently created Pacific War Council. It was pointed out that not to insist upon greater representation was to accept colonial status, that "the Government must get out of its mind that Canada is merely contributing something towards the waging of someone else's war," and that "this manifestation of an inferiority complex was putting Canada in a false light." Having for so long publicly stressed the part he had played in the destroyer-bases negotiations, it was not easy to dismiss or explain away these charges, which, as Mackenzie King well knew, possessed a great deal of substance. As a check, therefore, to the increasing likelihood of political embarrassment in this matter, a temporary expedient was found in appointing the military representative on the Permanent Joint Board on Defence, Major-General Maurice Pope, who was presumed to have already some idea of American defence planning methods, to act as "representative of the War Committee in Washington for the purpose of maintaining continuous contact with the United Kingdom–United States Combined Staffs and the Combined Planning Committees, and to represent the War Committee before the Combined Staffs when questions affecting Canada were under consideration." When this appointment was almost immediately followed by the establishment in Washington of a full-scale Canadian Joint Staff Mission designed "to keep Ottawa informed as to trends in the overall direction to the Allied war effort," Mackenzie King was able to tell Parliament with a discernible measure of satisfaction that "we are entitled to follow discussions in Washington and to be heard respecting any aspect of the war situation we may wish to present. Canada's interests in these matters are placed fully before the boards that are continuously sitting and dealing with these matters."

In fact, however, this remark was an exaggeration of the truth. The Combined Chiefs of Staff were not disposed to share their secrets; and a more accurate representation of Canada's position has been recorded by General Pope. "When an item of business particularly affecting one or other of the lesser Powers was under consideration," he wrote, "its military representatives were sometimes permitted to join in the discussions. But this was largely a formality, and it consisted of little else than affording the representatives of the country in question an opportunity of expressing their assent to the conclusion, or recommendation, that had previously been worked out in committee. Unhappily, even this general rule was honoured more in the breach than the observance. Apart from this occasional nod to us, the Combined Chiefs of Staff jealously kept information relating to the higher direction of the war almost entirely to themselves."

In spite of these obstacles, General Pope had the knack of getting and interpreting such information as to give Mackenzie King the impression that Canada's interests were being amply taken care of. "If we were precluded from asking direct questions," he writes, "what with a word here and a phrase there, added to an eloquent reticence on another occasion, not only could a fair picture of the situation at the moment be assembled, but also an intelligent forecast could be made of things that were to come." These reports might have given the Cabinet a clear idea of how the war was being prosecuted, but there was little they could do if they were dissatisfied.

Moreover, there were small but significant indications that the Canadian Government was still not physically prepared to play a larger part in the running of the war. McNaughton's proposed mission to Moscow was quickly rejected on grounds of risk and security; at the Quebec Conference there were "hardly excusable" lapses in Canadian staff-preparations; and even Pope's reports from Washington were inclined to get bogged down in the inter-governmental machinery, leaving the Prime Minister at times ignorant of current developments in strategy. Nevertheless, Parliamentary pressure upon Mackenzie King "to be asserting more strongly Canada's position" continued to increase until the invasion of Europe in June 1944, when the possibility of a drawn-out war due to an unexpected degree of enemy resistance both in Europe and Asia once again held out the hope that Canada's increasing economic and material contribution would at last earn her a place at the council table of the Grand Alliance.

In particular did this apply to the Pacific War in which Mackenzie King, because of Canada's Pacific seaboard, felt he had a real legitimate interest, in its consequences considerably more promising than the European war had ever been. Moreover, since the Pacific War Council had turned out to be powerless in the making of strategy, its military representatives in consequence not having met for a year, a more effective substitute or alternative to its membership appeared to be necessary. At the Commonwealth Prime Ministers' Conference in May 1944, therefore, Mackenzie King refused to commit himself to British plans and policies for the Far Eastern war and began to toy with the idea of playing a more independent role, by bargaining more shrewdly with his raw materials and resources which he knew would soon be called for in a protracted war; a complete reversal of his attitude when war began. Whether he sensed that Britain was nearing the end of her productive capacity, and that her voice in the war's direction would proportionately diminish is not known; but henceforth it was to become a plain principle of his policy, as enunciated to Parliament, that "representation especially in economic matters should be determined on a functional basis which will admit to full membership those countries, large or small, which have the greatest contribution to make."

By this time Canada had been made a member of the Combined Production Resources and Combined Food Boards; but as these had nothing to do with the making of strategy, they brought little consolation. More irritating still was the refusal of the Combined Chiefs to take Canada on the Munitions Assignment Board, in spite of the personal protests of Mackenzie King and her position, below the United States and Great Britain, as the largest munitions producer among the western allies. Undoubtedly the Combined Chiefs appreciated Canada's contribution, but hesitated to recognise it by virtue of the pseudo-military and influential nature of munitions assignment, the responsibility for which could not be shared further without implications for the processes of strategy-making. It was this situation which most probably spurred Mackenzie King to adopt his "theory of functional representation" in case the war should turn out to be lengthier and more exhausting than expected.

As it happened, of course, events falsified these hopes. The scale of American war production by the war's sudden end gave no indication that it would have to maintain itself on Canadian raw materials if the war had continued. Moreover, the American "broad-front" strategy in Europe so wholly absorbed the Canadian war effort that when the fighting there was brought to an end in May, and a quick turnaround and re-application of force, involving a complete re-orientation of her war effort and policies to the Far Eastern theatre became necessary, it was found impossible to accomplish it sufficiently expeditiously to obtain some voice in a war which was rapidly terminated with the bombings of Hiroshima and Nagasaki.

What readjustment in strategic responsibility might have occurred in a longer war, no one can say; though Canada's repeated failure to obtain more in the way of war direction was testimony enough that neither Churchill nor Roosevelt subscribed fully to Mackenzie King's narrowly economic theory of functional representation, and that unless she had men and equipment to offer, she must remain a muzzled partner. Setting the domestic necessity of the linch-pin legend aside, moreover, the conclusion is almost inescapable that, in spite of his constant wrangling and carping, Mackenzie King never really expected to carry much weight in the making of strategy. What did displease him, however, and rightly so, was not to be consulted, however superficially, especially on matters of clear Canadian interest.

In this, the Americans undoubtedly grew to be the greater offenders, though Britons were not wholly free of blame. Had Great Britain and the United States been more conscientious in this respect, even to only paying it lip-service, Mackenzie King would have been a happier man. As it was, he resented the continued British indifference as a colonial hangover, and yet could not bring himself to face the fact that the Ogdensburg Agreement, in embalming the myth of the unguarded frontier, was a watershed in Canadian military and external policy, that geography and the overwhelming disparity of industrial and military might had swept Canada into the whirlpool of American affairs, and that the old policy of filing away at British shackles would have to begin all over again on tougher American ones.

Several conclusions seem clear. Though the Statute of Westminster, passed barely eight years before war began, had technically completed the trend towards autonomy, Canadians regretted to learn that there still existed both in Whitehall and Ottawa some traces of a colonial complex and "Downing Street domination" in their relations with Britain. In their determination to remove these, in the pursuit of status and a national identity, defences and armaments were neglected. No independent foreign or strategic policy was developed, so that the very positivism of "no-entanglements" committed them, paradoxically, to the delusion and consequences of Munich. General Staff thinking, in the absence of any official doctrine, automatically assumed the integration of the Canadian with the British war effort without, however, retaining much say in how it was to be used. Nor, in the beginning, did the Government much care. But as it became painfully evident as the war progressed that he had unwittingly forfeited all say in the grand strategic design to which, for domestic reasons, it became increasingly prudent to commit his forces, Mackenzie King's obsession with status took another form; he assumed the common but dangerous role of an indispensable partner and began to forge the legend of the linch-pin as a sop to the electorate. Undoubtedly, much of the resultant frustration and acrimony in Anglo–Canadian and later Canadian–American relations in these matters sprang from the failure of many Canadians to make or appreciate this distinction; that the Statute of Westminster, in dispensing with colonial status, had conferred the function of a *minor*, not a *world* Power; and that while the idea of indispensability was a useful political device in domestic affairs, it became simply a psychological obstacle on the world stage when it was proved to be false. Equality of status in wartime *could* not, by virtue of a surplus of raw materials, automatically extend to that of function; and the scale of war production, *per se*, is no guaranteed determinant, much less a yardstick, of strategic responsibility.

The closest Canadians ever came to taking an active part in the making of strategy was with the Ogdensburg Agreement and the Permanent Joint Board of Defence. But even this had an air of unreality about it. It was made in a moment of panic with a neutral nation on the contingency of Britain's collapse. Its responsibilities were confined solely to the defence of the northern half of the western hemisphere; its composition was strongly civilian in character; and its functions were merely advisory and consultative. Its policies and plans were all speculative; and the overall strategic direction in North America was conceded, in spite of her neutral status, to America. As the war progressed, the scope of its interests was narrowed to mean just Canada; Canadians remained ignorant of American defensive plans, and the use of Canadian forces in the United States was not seriously considered or planned; while "mutual assistance" quickly evolved into unilateral American reinforcement of Canadian Pacific and Maritime regions.

But if Canada was excluded from all strategic responsibility, she kept a close watch on the disposition of her armed forces and insisted that her generals reject plans which appeared operationally impracticable, and refer all dubious cases to the home Government. In the case of operations in which Canada was to be the predominant partner (such as Dieppe or the abortive Norwegian expedition), this reservation was understandable and justifiable as a check to hastily committing Canadian forces to hazardous schemes, simply because, by 1942, the public were clamorous for blood and no longer content to see their armies sitting in apparent idleness in England waiting for an invasion that would never come. Dieppe, however, was but a single thread of the grand strategic tapestry, and pointed to the unedifying position the Canadian High Command had by this time worked itself into; that it felt compelled by public pressure to take a more active part in the war, and yet felt equally compelled, in the interests of status, to insist that her generals exercise their right of withdrawal from these same operations, which had, incidentally, been drawn up and approved by the Combined Chiefs of Staffs, and upon which the Canadian abstention would have no material effect.

Finally, it is an historical commonplace that in coalition war the sheer eminence of a statesman, commander, or *confidante* may exercise an influence upon grand strategy out of all proportion to the national or physical resources for which he is primarily and particularly responsible. Smuts, Rasputin, and Harry Hopkins are cases in point, examples of the huge but indefinable influence which personality sometimes exerts in the waging of war and the formulation of strategy. Could Mackenzie King, therefore, to all outward appearances ideally fitted by temperament, education, and geography to assume an important conciliatory role in Anglo-American negotiations, have played a more decisive and self-gratifying part in the higher direction of the war?

Of course, all answers must be speculative. As a politician, the Canadian Prime Minister was a shrewd counter-puncher, ruthless in his pursuit of power, whose obsessive nationalism, when focused on the world stage, appeared to many a paranoic form of parochialism. Trained in the colourless school of industrial arbitration, he lacked the grand manner and speech of the public hero. His interest in military affairs and other outside activities was non-existent. Since 1923, he had remained in power, much as Stalin did, on the basis of past associations with former leaders (e.g. Sir Wilfrid Laurier); an uncommon grasp of the machinery and potentialities of party politics; and a very real sense of power manipulation. Politics for him was less the art of the possible than the science of preventing the probable. He was by nature cautious, suspicious, and self-centred.

These personal characteristics, together with his somewhat negative conception of politics, did not predispose him for the role of a Chatham or a Churchill. He lacked the quality to arouse mass devotion or the personal magnetism to inspire intense loyalty. To the soldiers, above all, he remained throughout the war a remote and slightly contemptible figure. Among the allies he moved like a gnome among giants, the echo of untapped power rather than the eloquent strategist called to every council table. Yet this latter role might not have been so improbable of fulfilment when one considers the more than peripheral nature of Canada's other economic contributions. A dominant personality with a grasp of the techniques of modern warfare might have bridged the gap. To press the issue further would be mere speculation; but one thing is clear, that whatever influence personality exerted in the shaping of strategy, Mackenzie King's contribution was negligible.

In a war dominated by personalities, this might seem an unduly severe charge, but within its own limitations it is fair. Unlike those of Smuts, Mackenzie King's attempts to negotiate and influence strategy were primarily motivated by prospects of increased personal power and national prestige. The catalogue of failures to impose his personality upon the Allied chiefs, however, points to the fundamental and overriding dilemma of his position which even a close personal friendship with Roosevelt and

Churchill, a personal disinterestedness, and a grasp of strategy would have found both delicate and difficult of solution, namely, that strategic control, to be effective, must be concentrated in as few hands as possible, regardless of national feeling. Canada's misfortune was that she was compelled to change horses in the midstream of war so that neither before the fall of France nor after Pearl Harbour was she able to fasten a grasp on the reins of strategy. Her peculiar and special position vis-à-vis Britain and the United States made her failure to do so seem all the more like a calculated exclusion on their part. But there is no evidence to suggest that had Canada played a larger part in the higher direction of the Second World War, it would have been brought to a swifter, more satisfactory conclusion.

"WHO'S PAYING FOR ANYTHING THESE DAYS?" WAR PRODUCTION IN CANADA, 1939–1945 ⬦

ROBERT BOTHWELL

○

It is an understatement to observe that the Second World War turned out unexpectedly for Canada. A rational social scientist, surveying the Canadian scene in the late 1930s, might have been tempted to conclude that the Dominion of the North could barely eke out enough political stability and economic strength to preserve a precarious and unsatisfactory existence in North America, without indulging in any European adventures. O.D. Skelton, that eminently rational man, in fact did utter just such a prediction at the beginning of September 1939, when the Canadian government confirmed its fateful intention to follow Britain into war with Hitler's Germany.[1]

For Skelton, entrance into the war meant subordination and control of Canada by Great Britain. It threatened political and racial division, and it promised economic disharmony and financial bankruptcy. After all, as Skelton frequently observed, Canada was still having a hard time paying for the First World War; how could it even begin to pay for a Second? The Canadian government did not know, nor did the British. The war promised to become the most costly and ruinous proposition to which Mackenzie King's government had ever subscribed. His Liberal government approached the prospect with feelings of acute economic inadequacy.

Others shared these feelings. The Royal Commission on Dominion-Provincial Relations, reporting to the government in early 1940, blandly reported that "Canada is one of the least self-sufficient countries in the world."[2] And, it could have added, in the words of the former Prime Minister R. B. Bennett, "one of the most difficult countries to govern."[3] It is not surprising that the King government reacted to this latest conundrum

⬦ N.F. Dreisziger, ed., *Mobilization for Total War* (Waterloo: Wilfrid Laurier University Press, 1981), 59–69.

by adopting a policy of rigourous restraint—much as it had done in time of peace. But after 1939, instead of resisting the claims of greedy provinces clamouring for welfare, it turned a deaf ear to cries for the development of a wartime economy.

King could hardly be blamed. For besides the economic strain, there was a political problem. In the late 1930s there were some belated attempts to establish some lines of war production in Canadian factories. Encouraged by the Minister and Deputy Minister of National Defence, a Toronto manufacturer had secured a British order for light Bren machine guns; a Canadian order followed. It was a sensible arrangement, and one that later contributed notably to Canada's war production capacity, but it was badly handled and turned into a messy political scandal. In response, war production contracts were divorced from the Department of National Defence and handed over to a civilian board—"an expert advisory group of competent business men."[4] Embodied as the Defence Purchasing Board, this turned out to be a part-time solution to a full-time problem. On the outbreak of war, the Defence Purchasing Board was replaced by a War Supply Board, organized along the same lines, to direct the government's war purchasing requirements. It only remained to identify those requirements.

Initially, the King government believed that its war supply organization would serve as the local purchasing agency for the British government, in the first instance, and for the Canadian military in the second. Basic items of military equipment would, of course, follow British designs; in order for Canadian production to be economically feasible, it would be necessary to secure orders from Britain as well as from Ottawa. When a British supply mission arrived in Ottawa early in September 1939, it was immediately besieged by anxious Canadian manufacturers looking for orders. In the view of the King government, that was as it should be.

Working with the best will in the world, however, Britain's supply representatives could not satisfy Canadian expectations. The expectations themselves were exaggerated, unfortunately: the British had a very short shopping list and a bank account to match. In any case, Canadian industry was notoriously undeveloped, and in the opinion of some of the British mission, it should continue to be. Britain, one of them explained, "could not maintain her munitions industry in full operation unless there was fighting on the western front. England was filled up with shells, bombs and so on."[5] Even if something was required, nobody in Ottawa seemed to know where to get it; as a result the visiting Englishmen launched themselves on tours of the countryside to compile a rudimentary industrial survey.[6]

o

Out of this confusion there nevertheless emerged, slowly, a war production policy. Though it was hesitant to use them, the Canadian government possessed almost absolute emergency powers under the War Measures Act. In September 1939 it had specifically applied these powers to the problem of war supply, passing through Parliament a Munitions and Supply Act, modelled on the British Supply Act. The Act remained dormant and unproclaimed through the winter of 1939–40. Thanks to the Bren gun scandal of 1938–39, control over munitions supply remained firmly in civilian hands. Equally important, the King government reinforced its political position by calling and winning a snap election at the end of March 1940. Its political rivals, the Conservatives, emerged from the election decimated and shaken. In appearance and spirit they were no longer an alternative government.

The several disasters to allied arms in Europe in the spring of 1940 further strengthened the political position of the Liberal cabinet. Public panic at the prospect of imminent defeat solidified opinion behind the government. In extreme circumstances nothing less than extreme measures would do. The fiscal restraints of the fall of 1939, and the cautious husbanding of what the government conceived to be very limited resources, were jettisoned virtually overnight. The stage was set for a sweeping expansion of administrative power.

The first step was the abolition of the government's board of business experts, the War Supply Board, in April 1940. The second was the recruitment of business advisers from across Canada to serve in the new Department of Munitions and Supply, at the top of which sat a serving politician, C.D. Howe. A fifty-five year old engineer, Howe had a reputation inside Ottawa for being "quick in making decisions," though it took some time to persuade Canadian business leaders of this.[7] Howe needed to be quick in making decisions, for in his new department were concentrated all the supply functions of the Canadian government. When the Department was established, these functions consisted largely of pious promises and large expectations, but within six weeks the spectre of defeat in Europe permitted firmer policies to take hold. As Howe told his officials in June 1940, "If we lose the war nothing will matter . . . if we win the war the cost will still have been of no consequence and will have been forgotten." Howe instructed his officers to "take such steps as we felt proper in the interest of getting on with the war effort. . . ."[8]

These were large steps, because the Munitions and Supply Act allowed the Minister "in his absolute discretion" to "mobilize, control, restrict or regulate" whatever he thought was necessary for war supply and production. Armed with Howe's orders, and equipped with their own extensive knowledge of the Canadian economy, the government's "dollar-a-year men" fanned out across the country to mobilize, organize, control, restrict, and regulate essential supplies, and to provide for their transformation into necessary war equipment.

Orders, with Howe's sanction, were no longer any problem. "In some cases," a contemporary observer wrote, "plants were actually constructed and production undertaken before the legal contracts had been signed."[9] The minister asked to be informed of the problems, not of the details. As a system, it was the only feasible as activity spread out in ever-widening circles. Howe's businessmen-executives followed their own instincts and consulted their own judgment. In effect, a highly centralized structure of authority permitted a highly decentralized system of management and direction. Howe's department was compared, accurately, to a thirty-ring circus. Less accurately, the Minister was described as the ringmaster in each ring.[10] Considering the high degree of autonomy each production act enjoyed, Howe's role was closer to that of an impresario—an impresario, moreover, of a circus spread out over three thousand miles.

The diversity of Howe's new enterprises contributed to a certain confusion when matters failed to move forward exactly on schedule. That was, to begin with, a public relations problem. Public opinion demanded quick results; inexperienced executives, with little idea of the complexity of their new tasks, promised them. The results could be embarrassing, and the process frantic. In the summer of 1940, as British purchasing exhausted the supply of airplane engines in the United States and British airplanes naturally failed to arrive on schedule for the air training programme, even Howe's legal adviser was conscripted into the hunt for an acceptable substitute.[11]

Howe's prominence mirrored the decline and then eclipse of the British purchasing mission. That unhappy group had become increasingly marginal in the eyes of its potential Canadian clients over the winter, as its impotence to produce substantial contracts became evident. Canadian manufacturers turned to their own government to influence what the local British representatives could not; the links thus established endured after June 1940 to make real production possible. Isolated from the course of events, uncomfortable in an alien political climate—as its reports home made plain—the British supply mission was terminated in the summer of 1940. Its responsibilities were taken over by the Department of Munitions and Supply, which thus became sole agent for British purchasing in Canada. An agreement between C.D. Howe and Henry Morgenthau, the American Secretary of the Treasury, in August 1940 completed the process. Henceforth foreign governments would procure access to Canada's industry and production only through one central bureau. The economic power Howe acquired enhanced his, and his department's, dominance over Canadian industry.

That dominance was increased further in the summer of 1940 by the imposition of allocation controls over strategic goods required for war production. Oil control, established 28 June 1940, was a case in point. The Oil Controller began his work by discovering that there were no adequate statistics that would allow him to appraise his new job as a whole. While a survey of Canada's oil and gas industry proceeded, the Controller established an advisory committee of industry representatives (presidents of the several refineries and a representative of the largest importer of gasoline) and toured most of the provinces to brief the provincial premiers and their civil servants on what would be required. That turned out to be the redirection of refineries to produce fuel oil rather than gasoline, and the consolidation of production was encouraged, but in its best year, 1942, the Canadian oil industry only managed to produce 18 per cent of the country's total requirements.[12]

Canadians grew used to the restrictions imposed by scarce petroleum— shortages became uncomfortably apparent in the summer of 1941 and rationing was imposed in April 1942. They also grew used to the limitations imposed by the necessity for electrical conservation. The Munitions and Supply Power Controller promptly banned electric space heaters, enforced daylight saving time the year round, and undertook to expand local capacity for hydroelectric production. New hydroelectric plants received the highest priority for materials from other Munitions and Supply controllers, whose jurisdiction, by the middle of the war, extended over aircraft, chemicals, coal, construction, machine tools, motor vehicles, ship repairs and salvage, timber, transit and transport.

o

Running the system were officers recruited from the private sector. To begin with, Howe turned to men experienced in the areas they were intended to regulate. H.R. MacMillan, a lumber baron and one-time Chief Forester of British Columbia, became Timber Controller, on the plausible ground that no one else knew as much about the lumber business. An executive of the Aluminum Company of Canada was recruited to dismantle his company's carefully constructed sales empire, and to redirect its product to war purposes. There was no evidence that these men, or others like them, exploited their position to favour their own companies. With war production running flat out, with profits carefully controlled, and with every square foot of usable space profitably occupied, it would have been strange if economic favouritism had prevailed. Yet inevitably there were rumours and uneasiness, and later in the war Howe seems to have made it a policy to appoint his controllers from outside the industry they were to supervise. It was understood that within their own bailiwicks they could recruit all the expert advice they might need.

The choice of the senior advisers to direct the Canadian war economy was largely Howe's. His selections were necessarily eclectic, especially on the production side, where expertise was notably lacking, and sometimes eccentric, as when he selected E.P. Taylor to be his trouble-shooter in April 1940. In one case, a Halifax fishpacker with a reputation as a private flyer was appointed to head Canada's aircraft production programme; to build Anson trainers for the Commonwealth Air Training Program, Howe selected a London printing and graphics magnate. Neither man knew much about producing aircraft, Howe reasoned, but then neither did anybody else. And as it proved, once technical problems were disposed of, aircraft production proved spectacularly successful. In the period from September 1939 to the end of 1940, 904 aircraft were produced; in 1941, 1699; in 1942, 3781. The expansion of Canada's aircraft industry was of such a magnitude as almost to represent a complete new industrial creation.[13]

The situation was different in steel. Canada had a steel industry in 1939, and one with considerable unused capacity. Moreover, about one-third of Canada's steel requirements was made up from imports from the United States and Great Britain. But the latter vanished altogether, and the former soon became a scarce and highly negotiable commodity. For a period in the summer of 1940, "steel was so scarce that . . . the whole Canadian war program was threatened."[14] To meet the crisis, Munitions and Supply decided that it had little alternative: wholly new plants would spread existing resources too thin. Instead, the department would reinforce such strength as existed, and construct additions to facilities already in being. At Algoma Steel in Sault Ste Marie, a blast furnace capable of producing 1000 tons of steel a day was installed, one of five new blast furnaces distributed around the industry. To handle the large labour force (employment in the steel industry almost doubled between 1939 and 1943), the Department of Munitions and Supply went into the housing business, constructing 200 new units to accommodate Algoma's expanding workforce.[15]

Across the country we may still see the evidence, in these wartime housing projects and in elderly factories, of wartime industry. Between 1939 and 1944 the government poured $490 000 000 into 98 new war production plants that it ran for itself. Besides that, some $166 million went into additions to private plants, the largest single amount being at Algoma Steel.[16]

o

The effect of war production on such a scale was clearly to reorient the allegiance of Canadian business away from regional or provincial preoccupations towards a single, national centre. Full order books and government marketing had much to do with this new attitude, of course. The central government was fulfilling the fondest dreams of the country's industrialists, and they responded in kind. The chairman of Algoma Steel, Sir James Dunn, explained in August 1944 that he was "in favour of continuing Steel Control when the war is over for as far into the future as I can see. I feel that Steel Control has done a splendid job for the country."[17]

From business's point of view the transition to national control was made easier by several prominent factors. First, of course, there was the military crisis that exercised so strong an influence in every aspect of Canadian life. Next, there remained a sense that controls were being administered by a relatively congenial and compatible group, the "dollar-a-year-men." Their existence guaranteed a sympathetic hearing, if not satisfaction. Moreover, their attitudes, as labour leaders complained, closely resembled those in Canadian business as a whole. They had the ear of the Minister, and the Minister, most of the time, thought as they did. Lastly, the Canadian war effort was almost ostentatiously free of partisan taints. A cursory survey of the professed political allegiances of Howe's controllers and directors-general shows a distinct postwar Conservative preponderance. Suspicious businessmen, hating Liberals and the King government as an article of faith, could now visit Ottawa and see for themselves. Leaving, they made an exception of Howe in their general denunciations and jeremiads. And after the war they would join in the chorus led by H.R. MacMillan, praising "the greatest organizer Canada has ever seen." Part of the failure of the Conservative party and the National Government movement during the

Second World War must surely be ascribed to the co-opting of their business support by Howe early on. Such criticism of Canada's industrial front as did reach Conservative leaders would as likely as not be muted if not suppressed altogether: there would be no threat to the government, or to Howe, from that quarter.

o

The Chief opposition to the Department of Munitions and Supply, therefore, came from within the government itself. There were political complaints from ministers or Members of Parliament jealous of the war contracts handed out to Tory firms. There was grumbling among the Members of Parliament and Liberal provincial governments about the functioning of the gas rationing programme. There were jurisdictional and personality conflicts between Munitions and Supply's controls programme and that of the Wartime Prices and Trade Board, which administered the civilian economy. Though occasionally difficult, these were hardly serious. Howe invariably backed up his executives and administrators: any necessary housecleaning was done behind closed doors, sparing, as far as possible, the feelings of the failed subordinate. One, for example, was given the opportunity to ply his talents as lieutenant-governor of British Columbia, while another went to an easy rest as president (but not managing director) of Polymer. In still another case, an incompetent controller found himself transformed into an "associate controller"; but no one insisted that he leave.

The only serious limitations on Munitions and Supply were financial. These problems dated from Howe's decision, supported by the rest of the cabinet, to take the dollar sign off war production in the summer of 1940. At that time there were serious deficiencies, not only in Canada's productive capacities, but in the way in which those capacities were reported and recorded. With the new production managers and controllers given a virtual *carte blanche*, it was inevitable that the system would get worse before it got better. "Nobody knew what anybody else was doing." A.F.W. Plumptre wrote soon after the event. "Everyone was overworked. The Department was largely staffed by persons unfamiliar with government regulations and government routine, most of them business men and engineers, so that there was a good deal of misdirected effort." Among other things, nobody had a clear idea of just how much government money had been committed (over $300 million for capital projects in 1940 alone, it turned out), nor exactly where.[18] But as Howe once told an anxious executive, "Hell, who's paying for anything these days."[19]

Much of Munitions and Supply's money was spent out of the country. To give one modest example, the president of Federal Aircraft, discovered that he needed machine tools and parts unavailable in Canada. Paperwork to get them would take months. Only direct action would do, he decided. And so, every few weeks, the president drove from Toronto to Cleveland for a "business" trip, bringing back trunks full of parts that were, in this case, more precious than gold.[20] Repeated on a vast scale, these purchases helped to produce an exchange crisis for the Canadian dollar in the fall of 1940. Searching for the cause of the drainage of American dollars, the Finance Department soon concentrated its attention on Munitions and Supply. In November 1940 a Wartime Requirements Board was constituted to look into the matter with Clifford Clark, the Deputy Minister of Finance, sitting prominently on the board.

The Wartime Requirements Board in the end became a disappointment to its creators, losing itself in an internal power struggle inside Munitions and Supply. But it did concentrate attention on Howe's statistics, and it resulted in a substantial improvement in the kind of information being fed to Ottawa, and in the rate of information flow.[21] If the first year of the war resembled a thirty-ring circus, the second and subsequent years took on the sober character of an excursion through acres of balance sheets, which were placed daily on Howe's desk. This information, coupled with a growing personal acquaintance with his industrial managers, contributed to Howe's personal ascendancy in the Munitions and Supply hierarchy.

The dollar war between Finance and Munitions and Supply was eventually resolved by the conclusion of the Hyde Park agreement between Mackenzie King and Franklin Roosevelt in April 1941. The controversy over scarce dollars was succeeded by a controversy of a different order: scarce manpower. Relations between the Department of Munitions and Supply and the other agencies responsible for shepherding Canada's manpower during the war were never entirely satisfactory. Howe looked out for his interests as far as he could, sponsoring the admission to the cabinet and election to Parliament of a Minister of Labour who was unlikely to contradict him at any point, but it was the Department of National Defence that was his major adversary. Through the summer of 1942 Howe's statisticians and those of National Defence laboured to discover a mutually acceptable means of computing how many eligible soldiers and war workers there might be in Canada. It was a controversy that was never satisfactorily resolved from the point of view of either department; its effect on Munitions and Supply may be discerned in the diminished figures of base metals and coal production after 1943.

o

Munitions and Supply's accomplishments during the Second World War were substantial, whether they were in the realm of primary production or secondary industry. They placed Canada fourth in importance as a supplier among the allied countries. They reinforced Canadians' sense of self-esteem and permanently altered their perceptions of what might be achieved by government. But one of the most remarkable achievements must be the establishment and functioning of what E.P. Taylor described as "the best of all" supply departments on the Allied side. The creation of a Department of

Munitions and Supply with unique powers over war production was less the product of foresight than of accident, but it had the beneficial by-product that most of Canada's struggles during the Second World War were outside the country, rather than inside the government.

Naturally, Munitions and Supply's accomplishments were stressed over its limitations. But these were substantial. There were, as we have seen difficulties with labour supply, a matter over which the department could exercise very limited control. There were problems of finance, though those were largely resolved through an agreement with the United States—supplemented, it must be pointed out, by a myriad of happy production relations across the forty-ninth parallel. Since Munitions and Supply was forced to cram its capital investment into a very narrow compass, it necessarily reinforced existing industrial structures. Geographically, this means that Canada's war production was concentrated overwhelmingly in central Canada, in Ontario and Quebec.[22] There was no time or opportunity to allow expensive and doubtful experiments with an abstract redistribution of capital, labour, supply or market.

The very success of Canada's industrial expansion helped to create another illusion, which since the war has been a very active delusion in certain circles. Between 1939 and 1945 Canada was a massive importer of technology of all kinds. As the British had feared in 1939, this left Canada less dependent at the end of the war than before on foreign suppliers of high technology. And to a degree it permitted the establishment of an autonomous Canadian research and development programme. In the process, however, the derivative origins of that programme were forgotten, as were the special circumstances that persuaded Canada's allies freely to give up their technological secrets. It encouraged, as well, the belief that Canada's economic development would after the war be export oriented. The first delusion led in a straight line to the Avro Arrow; the second was quietly discarded inside Howe's department even before the end of the war. Outside, however, it continued to exercise great influence.

Politically, the experience of the war bequeathed a legacy that lasted for a generation. Those who ran the war ran Canada for twenty years after. On the economic front, the federal government's refinement of the techniques of economic forecasting and fiscal inducement guaranteed it a place larger than before in the direction of the economy. The experience of war and administration encouraged many businessmen to accept this turn of events as perfectly natural. Some even went beyond the federal government's own wishes and asked for an indefinite continuation of the wartime millennium of guaranteed supply, demand, price, and profits. Briefly, the war gave the Liberal party an entrée into the Canadian business community, so that those businessmen who obstinately continued to vote Conservative nevertheless rejoiced that each successive election victory kept C.D. Howe in office—even if that meant a continuation of Liberal rule. It is only within the past five or ten years that the personnel of Canada's government was finally changed and that central power, such as it is, has passed to a newer, if not fresher, generation.

NOTES

1. O.D. Skelton memorandum, 10 September 1939. Skelton Collection, External Affairs Papers. These records, originally with the Department of External Affairs in Ottawa, have been transferred to the Public Archives of Canada.

2. Canada, Royal Commission on Dominion-Provincial Relations, *Report of the Royal Commission on Dominion-Provincial Relations*, Book 1, Canada: 1867–1939 (Ottawa, 1940), 179.

3. Paul Martin interview, 1974. The texts of this and other interviews cited in this paper have been deposited in the Bothwell Collection of the Archives of the University of Toronto.

4. Cited in C.P. Stacey, *Arms, Men and Governments* (Ottawa, 1970), 101.

5. Grant Dexter memorandum, 30 November 1939, Dafoe Papers, Public Archives of Canada (hereafter PAC).

6. James Crone, "Canadian Reminiscences," AVIA/22/3170/ERD/7525, Public Record Office.

7. "Backstage at Ottawa," *Maclean's*, 1 February 1940.

8. Henry Borden interview, 1976.

9. A.F.W. Plumptre, "Organizing the Canadian Economy for War," in *Canadian War Economics*, ed. J.F. Parkinson (Toronto, 1941), 9.

10. J.P. Moffat to the United States Secretary of State, 26 February 1941. State Department Papers, 842.002/107, United States National Archives. Cited in R. Bothwell and W. Kilbourn, *C.D. Howe: A Biography* (Toronto, 1979), 140.

11. Henry Borden interview, 1976.

12. J. de N. Kennedy, *History of the Department of Munitions and Supply*, vol. 2 (Ottawa, 1950), 157.

13. On this subject see Bothwell and Kilbourn, *C.D. Howe*, ch. 10, esp. 137–39, 155.

14. Ibid., 135f.

15. Department of Munitions and Supply, *The Industrial Front* (Ottawa, 1944), 290.

16. H. Carl Goldenberg, *Government-Financed Expansion of Industrial Capacity in Canada* (Ottawa, 1945).

17. Quoted in Bothwell and Kilbourn, *C.D. Howe*, 179.

18. Plumptre, "Organizing the Canadian Economy," 9.

19. Report by W.A. Harrison and E.A. Bromley, early 1940, C.D. Howe Papers, vol. 5, PAC. Also, Henry Borden interview, 1976. For a further discussion see Bothwell and Kilbourn, *C.D. Howe*, 132f.

20. Ray Lawson interview, 1977.

21. H. Carl Goldenberg interview, 1976. Goldenberg was the secretary of the Wartime Requirements Board.

22. John R. Leslie, "Summary Record and History of Allied War Supplies Corporation," 26 July, 1945, Allied War Supplies Corporation Records, PAC.

THE AMERICAN INFLUENCE ON THE CANADIAN MILITARY, 1939-1963 ✧

J.L. GRANATSTEIN

o

On Armistice Day in 1927, officials of the Canadian and United States governments dedicated a monument at Arlington Cemetery near Washington D.C., to commemorate the service of those Americans who had fought with Canadian forces before their country became a belligerent in the Great War. The occasion, stage-managed by Vincent Massey, Canada's first Minister to the United States, was a glittering ceremonial featuring permanent force infantry of the Royal Canadian Regiment and the Royal 22^e Regiment in their British-pattern scarlet tunics, as well as the pipes and drums of the 48th Highlanders, a well-known kilted Toronto militia regiment. Everyone was on their best behaviour, and the event was a great success, even the review of the infantry at the White House by the taciturn, if not comatose, President Calvin Coolidge.

So what? What made this visit by the RCR and the "Vandoos" to Washington so noteworthy was that it was completely unprecedented, the Canadians being the first units in red coats to appear in the American capital since Washington was burned by the British (as retaliation for the torching of York) in 1814.[1] The undefended border and the "century of peace" that followed the War of 1812 were already staples of after-dinner speeches in both countries, but the military relations between the two countries were almost non-existent. Even during the Great War there had been little direct co-operation, other than some limited naval and air efforts to counter German U-boats in North American waters and minor combined anti-Bolshevik operations during the Allied intervention in North Russia and Siberia. Worse still, there was some resentment in Canada at the long delay before the US entered the war and shrill American claims after the Armistice that the Americans had won the war virtually single-handed.

✧ *Canadian Military History*, 2 (1993).

Strikingly, the military staffs in both countries continued to plan for war against each other. In Ottawa, Colonel J. Sutherland Brown, the Director of Military Operations and Intelligence at National Defence head-quarters from 1920 to 1927, had the responsibility for producing war plans. His Defence Scheme No. 1, slated to become operational in the event of a breakdown in relations between the US and Britain, was based on the assumption that the major threat to Canada was an invasion by American forces. Such a threat, Brown maintained, could best be countered by launching Canadian "flying columns" at key points in the United States. Brown and other officers even undertook reconnaissance trips to spy out their objectives and the best routes toward them, and Brown himself avidly collected postcards showing strategic American locations.[2] In the US, similar studies took form at much the same time at the Army and Navy War Colleges and at the War Plans divisions of the two services, both as training exercises and as contingency plans. The American army's "Plan Red" for war with Great Britain foresaw the occupation of Canada by four armies with the intention to "hold in perpetuity all . . . territories gained. . . . The Dominion government will be abolished." "Buster" Brown's scheme was cancelled in 1931, war with the United States seeming utterly impossible to contemplate in Ottawa; the US plans remained on the books until 1937 and war college students did additional studies in 1938.[3] Until a few years before the outbreak of the Second World War, in other words, conflict between the North American nations was at least a possibility for the defence planners in Ottawa and Washington.

o

The prospect of another world war changed everything. The threat posed by aggressive Nazis, Fascists, and Japanese militarists was all too real, one that had to be faced. Hesitantly, Canada and the United States began to come together in their own interests. The process is well known: President Franklin Roosevelt's growing concern about the defencelessness of Canadian coasts and his increasing desire to secure naval bases in Canada and land access to Alaska; his regular conversations on defence with Prime Minister Mackenzie King after 1935; the first ever meetings of the Canadian and US chiefs of staffs in January and November 1938 to discuss Pacific Coast defences; and Roosevelt's pledge at Kingston, Ontario in August 1938 that the United States "will not stand idly by if domination of Canadian soil is threatened by any other empire," followed a few days later by a reciprocal promise from Mackenzie King.[4]

Canada went to war "at Britain's side" in September 1939 and the US remained neutral. There was no co-operation of a military nature between them until after the disasters of May and June 1940 drove France from the war and left Britain all but alone in the face of Hitler's Nazi legions. From being a minor partner in the war effort, Canada had suddenly become Britain's ranking ally.

But Britain was desperately weak, barely able to defend its home islands let alone to protect Canada. By itself Canada could not have defended its territory against either a German attack after Britain was occupied or a sudden Japanese raid on the Pacific coast. The logic of the situation was clear, and Prime Minister Mackenzie King was quick to seize the opportunity when President Roosevelt invited him to a meeting at Ogdensburg, New York in mid-August 1940. The resulting agreement, which included a Permanent Joint Board on Defence, guaranteed Canadian security during the war and allowed the fullest possible military support to Britain; at the same time, it marked the establishment of a permanent defence relationship between the two countries and the real beginnings of the close and complex military links that persist to the present. The United States had definitively replaced the United Kingdom as Canada's senior defence partner.[5]

Certainly, there was no doubt who was the junior in the relationship. With a population of just over ten million, with most of its effective military forces in the United Kingdom and with more on the way there, Canada was a virtual supplicant, seeking the help of the 125 million-strong United States to defend its homeland in a world suddenly made unaccustomedly perilous. There was every sign of this in the negotiations over a strategic defence plan for the continent. The Americans proved to be as difficult to deal with as the British had ever been. The US generals' demands for unity of command seemed to translate into complete administrative and operational control over all Canadian forces involved in the defence of North America, an aim that was only partly defeated and only with enormous difficulty.[6] Washington, for its part, resisted Canada's efforts to establish a military mission in the US capital,[7] and when it sent troops into Canada in 1942 to build a highway to Alaska, they sometimes acted more as an army of occupation than as allies.[8] There were also serious difficulties in working out command relationships with the American services operating in Newfoundland.[9]

Still, the wartime co-operation with the United States was generally good. Some 8864 Americans served in the Royal Canadian Air Force during the war, and more than five thousand continued their service after the United States entered the war in December 1941.[10] RCAF squadrons participated in the defence of Alaska and Canadian Army troops shared in the reconquest of the Aleutian Islands.[11] A combined Canadian–American unit, the First Special Service Force, participated in action in the Aleutians, Italy, and France,[12] and American army divisions at various points served under command of the First Canadian Army during the campaign in Northwest Europe in 1944–45. Canada had a division prepared for service in the invasion of Japan, one that was to be organized on American lines and equipped with US weapons. Why? As General A.G.L. NcNaughton, the Defence minister at the time and always a cautious man in dealing with the Yanks, said, "One of the primary reasons . . . was to obtain experience with the United States system of Army organization and US equipment in view of the obvious necessity for the future to co-ordinate the defence of North America."[13]

As that suggested, post-hostilities planners in Ottawa clearly foresaw the increasing tension between the US and the USSR, understanding that this obliged Canada to consider carefully its role in the defence of the continent.[14]

The events of the war had marked a historic change in Canada's place in the world. Curiously, however, there was as yet little sign of this in the attitudes, equipment, and training of the Canadian forces. The direct influence of the American military on their Canadian counterparts was still relatively limited. The methods and models for Canadian soldiers, sailors, and air force without question remained British in 1945.

o

Britain's military weakness in 1940 had pushed Canada into its first defence alliance with the United States. The mother country would never recover its military power in a postwar world that had room only for two superpowers with their globe-girdling reach and massive nuclear arsenals. Ideological similarities and geography both made certain that Canada's policy generally moved in tandem with that of its neighbour, and the military marched to a different beat in the postwar years. In North America, the Military Cooperation Committee, a joint planning body spun off from the Permanent Joint Board on Defence, began the process of integrating continental defences in the spring of 1946.[15] The Canadian military planners worried about their cousins' practices and occasionally alarmist assessments. General Guy Simonds, arguably the most successful Canadian wartime commander, complained in 1947 that the American "military authorities made plans based entirely on potential enemy capabilities, whereas it was the practice in Canada to take into consideration not only capabilities but probabilities."[16] Officers like Simonds tried to maintain the filial links with the British forces (Simonds even created a regiment of Canadian Guards when he was Chief of the General Staff a few years later), but it was increasingly a vain struggle, as the Cold War wore on and American power increased.[17]

Signs of this began to be evident in the joint Canadian–US training exercises that started in the late 1940s and even more so during the Korean War, the first major armed struggle of an ideologically driven era.[18] The 25th Canadian Infantry Brigade served as part of the Commonwealth Division in Korea for purely practical reasons. In 1950, even though the government had already made the decision to switch to American-pattern equipment, the Canadian equipment in use was still primarily of British pattern (steel helmets, battle dress, 25-pounder field guns and .303 Lee-Enfield rifles, for example). The use of US equipment, the Chief of the General Staff noted, would have required "major changes in . . . minor tactical doctrine."[19] Moreover, the inherited dogma was that American battlefield doctrine was, as befitted a country with a huge population, wasteful of lives and treasure; like Britain, Canada had few of its sons to spare. Brooke Claxton, the Minister of National Defence, shared this received wisdom. After a visit to

Korea, he returned unimpressed with the American commanders and appalled by the "lying" of staff officers who gave the briefings. He wrote a friend that "American expenditures of lives and ammunition are high according to our standards, higher than our people would be willing to accept."[20]

The Royal Canadian Navy was equally critical of its American cousins. In his memoirs, Admiral Jeffry Brock recalled that the fleet signal book employed by the RCN had one code for going into action: "Enemy In Sight. Am Engaging." The comparable USN code translated as "Request Permission to Open Fire on the Enemy," something that Brock was convinced was part and parcel "of the determined resistance of American officers to make any move at all without the written and signed authorization of someone senior."[21] Even so, there was a pressing need "to share the United States Navy's technical know-how, especially in anti-submarine warfare weaponry." As one commentator noted, "The coming change was first detected in the new terminology"—the British term "asdic" was superseded by the American word "sonar."[22] The establishment of NATO's Supreme Allied Command Atlantic (SACLANT), with headquarters at Norfolk, Virginia also meant that the RCN had its place with the USN and not the Royal Navy.[23]

Moreover, in Korea the Canadians had learned that everything about the US forces was not bad. Brigadier John Rockingham, the first commander of the 25th Canadian Infantry Brigade, called on General Horace Robertson, the Australian officer serving in Japan as commander-in-chief Commonwealth forces. Robertson wanted the Canadians to use Australian rations "as this would be to Australia's financial advantage." Rockingham flatly refused: "I explained that my cooks had been trained to cook American rations and my soldiers had become used to them and liked them very much."[24] The RCN, operating three destroyers in Korean waters, found much the same thing. "The Commonwealth base at Kure [Japan] . . . had the right kinds of ammunition and machinery spares for the Canadian ships." Commander Tony German wrote, "but . . . British provisions were terrible. Canadian ration scales were much better than RN now, but in Kure they mainly got tough mutton. . . . From the Americans in Sasebo there was first-rate beef . . . ice cream, milk, fresh fruit and vegetables and such magic as frozen French fries. . . ."[25] Armies march on their stomachs and, whenever they could be secured, the Canadians now simply refused to march without American rations.

American equipment too was increasingly coveted. Sometimes this was because US equipment was both more comfortable to wear, better designed for protection, and simply more effective than the Second World War-pattern British material used by the Canadian forces. For example, the steel helmet used in the Second World War and in Korea by British and Canadian forces offered no cover for the back of the neck, weighed a ton, and was so awkward that it was almost impossible to run while wearing it. "The less said about the present helmet the better," wrote one infantry batallion commander. The American helmet, by contrast, offered better protection and, because it had a liner that was removable, could even be used for cooking over an open fire in a pinch. No Canadian wept when the UK helmet was scrapped in the late 1950s. Even the Americans' mess tins, eating utensils, and cups were better designed than the comparable Canadian equipment issued to soldiers in the field.[26]

The desire for American equipment was especially evident at the expensive end—aircraft and missiles. The sky-high costs of developing technologically sophisticated weapons systems were driving smaller nations out of the market. Canada's experience with the Avro Arrow fighter aircraft is too well known to need comment,[27] but even larger powers like Britain were having their problems in paying the bills. For the UK that meant abandoning efforts to develop missile systems of its own. For Canada, the logic of the calculus of expenditures was equally clear. As General Charles Foulkes, the powerful Chairman of the Chiefs of Staff Committee, told the Minister of National Defence in 1958, "It appears apparent that we are reaching a stage in Canada where it will not be possible for us to develop and produce further complicated weapons purely for Canadian use." That meant, he said, that "the defence of North America will have to depend upon the joint development and production of these weapons and weapon systems for our joint use."[28]

In fact, even that was a hope, certainly not a certainty. The best Canada could expect was to purchase American air-to-air missiles, Bomarc SAMs and aircraft like the Voodoo, the CF-104, the CF-5, the Aurora, the Boeing 707, and others under schemes that often saw parts built in Canada or offsets for Canadian industry included in the deal. American equipment was not always the very best available, but it was invariably close to it. Moreover, in contrast to the increasingly impecunious Canadian and British armed services, the United States military had the goods of modern warfare in lavish profusion,[29] and the officers and members of the Canadian Forces inevitably and understandably wanted their small share of it. In effect, this equipment envy, this military *amour-propre*, often was a driving force for policy.

Nowhere was this more true than in the Royal Canadian Air Force. Even before the NORAD agreement formally integrated the air defence systems of the US and Canada in 1957, the RCAF was already tightly linked with the USAF on both a personal and a professional level.[30] "The two air forces," Joseph Jockel wrote of the early 1950s, "came to see the air defence of North America more and more as a problem to be tackled jointly with resources that were becoming more and more intertwined. Direct and permanent links were forged between the two air defence commands, by-passing the more formal channels of communication. . . ."[31] He added that the

> two air forces had every reason to co-operate. They were faced with a common military threat. As airmen, they shared an outlook which created a similar identity and even an emotional bond. They were interested in convincing civilians of the danger to the continent. Both . . . were locked in struggles with their sister services for defence funds. Finally, for the RCAF, the USAF was a source of

funding for radar stations and a source of pressure on Ottawa to recognize the importance of air defence.[32]

Leonard Johnson, who retired as a major-general, put it more succinctly: "The US Air Force was a big and powerful cousin with whom we identified."[33] Moreover, until the early 1970s in NATO, members of the Canadian air force served with the Americans, unlike the army brigade that operated in the British sector of West Germany.[34] And because the Canadian aviation industry (even after the cancellation of the Arrow) was relatively well developed, much more so than was necessary to support Canada's relatively small air force needs, export sales to the United States were essential. That gave the air force a community of interest with the industry and gave both an effective joint lobby with the government.[35] The result from the early 1950s onwards was that the RCAF received the lion's share of the defence budget and the greatest influence on defence policy making.[36]

This became especially apparent when the Conservative government of John Diefenbaker decided in 1959 to arm the Canadian Forces with nuclear weapons. The RCAF would have nuclear missiles for its aircraft flying in defence of Canada, a nuclear ground attack role for its CF-104s serving in NATO, and nuclear-armed Bomarc missiles at two bases in Ontario and Quebec. By 1960, however, the government began to waver in its commitment to nuclear weapons and, some might say, in its support for Canada's alliances. One obvious sign of this came during the Cuban missile crisis of 1962 when Diefenbaker resisted the entreaties of his Defence minister and delayed putting Canada's NORAD aircraft on alert. The Americans were justifiably furious, but in fact the aircraft had been prepared for war, the Defence minister informally authorizing the action and every fighter aircraft being armed and put on readiness.[37]

The RCAF was not alone in going to war readiness. The Royal Canadian Navy cancelled leaves and put every ship it had (in 1962, it had some!) to sea, relieving the United States Navy of responsibility for a critical sector of the western Atlantic. One authoritative account summarized the reasoning of Admiral Kenneth Dyer, in command on the east coast: "North America was under direct immediate threat. That included Canada. All the long-standing arrangements and government-to-government agreements said if one partner boosted its defence [readiness] condition, the other followed." When Ottawa did nothing, Dyer simply acted. His relationship with his US NATO commanders at SACLANT had been formed over the years in countless NATO exercises and was so close and so trusting, his assessment of the Soviet threat so fearful, that he felt obliged to put to sea to assist his ally. "That 'band of brothers,' Nelson's basic way of running things at sea, by mutual understanding and a firm grasp of the basic aim," Commander Tony German wrote, "was alive and well in North America in 1962. The navy . . . honoured Canada's duty to stand by [its] North American ally,"[38] even if the Prime Minister had wished otherwise.

This same attitude, this placing of the service's assessment of the threat and its particular needs ahead of government policy, was even more apparent

in the next few months. Badly battered by its ineptitude during the Cuban crisis, the government was now about to be brought down by its anti-
Americanism and its reluctance to arm the Bomarc missiles it had secured
with the nuclear warheads they needed to become effective. A key player in
the process was Wing Commander Bill Lee, the head of RCAF public relations in Ottawa. Lee had graduated at the top of his class from the USAF
Public Relations School ("renowned as one of the leading institutions in its
field," Paul Hellyer, Defence Minister after 1963, wrote).[39] Now Lee, other
RCAF officers, and USAF senior officers worked together to lobby the media
against the Diefenbaker policy and in favour of nuclear weapons. As Lee
admitted later,

> It was a flat-out campaign, because Diefenbaker was not living up
> to his commitments. Roy Slemon [Canadian Deputy Commander
> of NORAD] was going bananas down in Colorado Springs. We
> identified key journalists, business and labour people, and key
> Tory hitters in Toronto, and some Liberals, too, and flew them out
> to NORAD. It was very effective.

So it was, as the American view of the defence needs of the North American
continent was put forcefully and clearly to key Canadian opinion makers,
not least by Air Marshal Slemon. Significantly, Lee's campaign had the full
support of his superiors, he says, the Air Council having authorized it, and
the Chief of the Air Staff having secured the permission of the Chairman of
the Chiefs of Staff Committee, Frank Miller. "You go ahead," Lee quotes
Miller. "Just don't tell me the details."[40] The campaign worked, undoubtedly helped mightily by Diefenbaker's incompetence. In February 1963 the
government was defeated in Parliament over its nuclear policy.[41] Within a
few months, the incoming Liberal government had accepted the nuclear
weaponry.

The significance of this affair and the Cuban missile crisis ought to be
clear. The military in Canada in the past had occasionally disagreed with
government policy and made known its feelings to the media. That was certainly true during the conscription crisis of 1944, to cite only one example.
But these issues in 1962 and 1963 were different, directly involving as they
did the policies of a foreign government and its armed services. Whether the
actions of John Diefenbaker were malign, stupid, or simply wrong-headed,
his administration had been duly and properly elected and was entitled to
expect the armed forces to put the government's policy interests ahead of
those of the United States. The military's defence undoubtedly would be that
they were doing just that. But in fact, over the years, what had happened
was that the formal and informal links between the Canadian and American
military—the transgovernmental relations, or so political scientists call
them—had grown so close that the senior officers placed their service interests and their assessment of the situation ahead of their government's. As
Jocelyn Ghent put it, "the Canadian military perceived and acted on a threat
that was defined not by their government, but by the transgovernmental
group to which they felt much closer, the Canadian–American military."[42]

This situation was probably inevitable. No small country living cheek by jowl with a superpower could expect to retain full military independence and especially not if it were entwined in defence, economic, and political alliances to the extent Canada and the US were. Still, the impropriety of these actions is evident, and their effect on civil–military relations in Canada has never been confronted or even acknowledged by the political leadership. So far as is known, no attempt was ever made to impede or reverse the entrenchment of the transgovernmental linkages that pushed Canadian policy far ahead of the government's wishes in 1962 or that helped to bring down a government in 1963; no serious or sustained effort was made to reverse or even check the Americanization of the armed forces. Indeed, the process of unification, launched by Defence Minister Paul Hellyer in the mid-1960s with the intention, among other things, of giving Canada distinctive military forces, may have speeded the trend by dealing a killing blow to the Army's system of corps and its distinctive and much-loved uniforms, buttons, and badges. The army had been the least Americanized of the forces before the mid-1960s,[43] proud of its regiments and their traditions. The dark green uniform that came with unification homogenized the Canadian military and weakened the land forces' psychological defences against Americanization. The budget cuts of the Trudeau and Mulroney years and the sometimes slavish adherence of the latter government to Washington's policy have effectively completed the process. If Quebec becomes independent in the early 1990s and if the Canadian Forces in consequence abandon their bilingualism, there will be very little remaining other than Canada's self-professed expertise in United Nations peacekeeping to differentiate the military forces of the two countries.

NOTES

1. Vincent Massey, *What's Past is Prologue: The Memoirs of Vincent Massey* (Toronto, 1963), 141–44.

2. James Eayrs, *In Defence of Canada: From the Great War to the Great Depression* (Toronto, 1964), 70ff. I have one of Brown's postcards, showing Seneca Mountain in Allegheny State Park, NY.

3. R.A. Preston, *The Defence of the Undefended Border: Planning for War in North America, 1867–1939* (Montreal, 1977), 217ff.

4. There is a brief account of this process in P. Roy et al., *Mutual Hostages: Canadians and Japanese During the Second World War* (Toronto, 1990), 35ff.

5. See J.L. Granatstein, "Mackenzie King and Canada at Ogdensburg, August 1940," a paper presented at "The Road from Ogdensburg: Fifty Years of Canada–U.S. Defence Cooperation," St. Lawrence University, August 1990. One sign of the growth of defence relations with the US can be found by comparing the number of pages devoted to defence in the Department of External Affairs' *Documents on Canadian External Relations* series: the *1936–1939* volume does not even list defence as a topic under the rubric of Relations with the US; the *1939–1941* volume has 212 pages; the *1952* volume has 120 pages for its one year period.

6. See, for example, General M.A. Pope's memorandum after the PJBD meeting of 28–29 May 1941 in D.R. Murray, ed., *Documents on Canadian External Relations*, vol. 8, *1939–1941*, *part II* (Ottawa, 1976), 217ff.

7. C.P. Stacey, _Arms, Men and Govern- ments: The War Policies of Canada, 1939–1945_ (Ottawa, 1970), 354ff.

8. See, for example, essays by Curtis Nordman and R.J. Diubaldo in _The Alaska Highway: Papers of the 40th Anniversary Symposium_, ed. Kenneth Coates (Vancouver, 1985).

9. On the difficulties in Newfoundland, see the demi-official correspondence files in Royal Military College Archives, Kingston, Ontario, General W.H.P. Elkins Papers. Elkins was in overall command of Canadian Army forces on the Atlantic Coast during the critical period, and Newfound- land ground force commanders reported to him. On the RCN–USN relationship, see W. Lund, "The Royal Canadian Navy's Quest for Autonomy in the North West Atlantic: 1941–43," in _RCN in Retrospect, 1910–1968_, ed. J. Boutelier (Vancouver, 1982), 138ff. and, more generally, Marc Milner, _North Atlantic Run: The Royal Canadian Navy and the Battle for the Convoys_ (Toronto, 1985).

10. F.J. Hatch, "Recruiting Americans for the Royal Canadian Air Force 1939–1942," _Aerospace Historian_, 18 (March 1971), 17.

11. See S.W. Dziuban, _Military Relations Between the United States and Canada, 1939–1945_ (Washington, 1959), 242ff.

12. See Robert Burhans, _The First Special Service Force: A Canadian/American Wartime Alliance: The Devil's Brigade_ (Toronto, 1947, 1975).

13. John Swettenham, _McNaughton_, vol. 3, _1944–1966_ (Toronto, 1969), 171.

14. J.L. Granatstein, _Canada's War: The Politics of the Mackenzie King Government, 1939–1945_ (Toronto, 1975), 323; D. Munton and D. Page, "The Operations of the Post- Hostilities Planning Group in Canada, 1943–1945," CHA paper 1976.

15. For planning for Arctic defences, see Shelagh Grant, _Sovereignty or Security?: Government Policy in the Canadian North 1936–1950_ (Vancouver, 1988), chaps. 7, 9.

16. Quoted in D.J. Bercuson, "'A People So Ruthless as the Soviets': Canadian Images of the Cold War and the Soviet Union—1946–1950," a paper presented at the Elora Conference on Canada–USSR relations, 1989, p. 12.

17. Simonds' views on the US are detailed briefly in D.J. Bercuson, "The Return of the Canadians to Europe: Britannia Rules the Rhine," in _Canada and NATO: Uneasy Past, Uncertain Future_, ed. M. MacMillan and D. Sorenson (Waterloo, 1990), 20ff.

18. The impact of Communist deceit— and subsequent support for the US position—was strong on military officers who served in Indo-China with the International Control Commissions from 1954 to 1972. The Poles, one of the three countries on the ICC, were difficult to work with, and the North Vietnamese were gen- erally considered ruthless and duplicitous. Information provided to the author and, more generally, Douglas Ross, _In the Interests of Peace: Canada and Vietnam 1954–73_ (Toronto, 1984) and James Eayrs, _In Defence of Canada: Indochina: Roots of Complicity_ (Toronto, 1983), especially chaps. 7–8.

19. Quoted in Bercuson, "Rhine," 17. On defence procurement, see Danford Middlemiss, "Defence Co-opera- tion," in _Partners Nevertheless: Canadian–American Relations in the Twentieth Century_, ed. G.N. Hillmer (Toronto, 1989), and especially pp. 171–72 which discuss the PJBD deci- sion of October 1949 and the subse- quent agreement of May 1950 on the necessity of American defence pur- chases in Canada to provide the American dollars Canada needed for procurement of US-type equipment.

20. Claxton to G.V. Ferguson, 27 May 1953, Brooke Claxton Papers, vol. 31, National Archives of Canada (here- after NAC).

21. Jeffry Brock, _The Dark Broad Seas_, vol. 1, _With Many Voices_ (Toronto, 1981), 199, 220–1.

22. John Harbron, "The Royal Canadian Navy at Peace 1945–1955: The

Uncertain Heritage," *Queen's Quarterly* (Fall 1966), 317.

23. Ibid., 319.

24. John Melady, *Korea: Canada's Forgotten War* (Toronto, 1983), 82.

25. Tony German, *The Sea Is At Our Gates: The History of the Canadian Navy* (Toronto, 1990), 223.

26. Lt.-Col. K.L. Campbell, "Summary of Experiences: Korean Campaign, 25 Mar 53–25 Mar 54." This document was kindly provided by Professor David Bercuson.

27. See J.L. Granatstein, *Canada 1957–1967: The Years of Uncertainty and Innovation* (Toronto, 1986), 105ff.

28. "Aide-mémoire for Minister of National Defence . . . ," 10 July 1958, Dana Wilgress Papers, vol. 2, NAC, att. to Foulkes to Wilgress, 23 July 1958.

29. "By Canadian standards," wrote one officer of his experience at the US Armed Forces Staff College in 1965, "the scale of even modest American military operations is staggering." Leonard Johnson, *A General for Peace* (Toronto, 1987), 64.

30. James Eayrs, *In Defence of Canada: Peacemaking and Deterrence* (Toronto, 1972), 59–60 on Air Marshal Robert Leckie (who sometimes fought with the USAF but was basically friendly toward it) and Air Marshal W.A. Curtis. It is also worth recollecting that senior RCAF officers had been short-changed in their search for high command in World War II by the RAF; feelings of bitterness toward the British may well have persisted.

31. Joseph Jockel, "The Creation of NORAD (1954–1958)," unpublished paper, 4.

32. J.T. Jockel, *No Boundaries Upstairs: Canada, the United States and the Origins of North American Air Defence, 1945–1958* (Vancouver, 1987), 56.

33. Johnson, *A General for Peace*, 33.

34. Bercuson, "Rhine," 15ff. demonstrates that the CGS, General Foulkes, intended Canadians to serve in the US sector for practical reasons, but that it was General Eisenhower who determined that the brigade would operate in the British sector.

35. See on this J.V. Allard and Serge Bernier, *The Memoirs of General Jean V. Allard* (Vancouver, 1988), 225, 251–2.

36. See Joseph Jockel, "From Demobilization to the New Look: Canadian and American Rearmament, 1945–53," CHA paper, 1981, 26.

37. Information given the author in confidence. For an account of the Canadian response in the crisis, see Granatstein, *Canada 1957–1967*, 113ff.

38. German, *The Sea is at Our Gates*, 260ff. See also J. Sokolsky, "Canada and the Cold War at Sea, 1945–1968," in *RCN in Transition 1910–1985*, ed. W.A.B. Douglas (Vancouver, 1988), 218–19.

39. Paul Hellyer, *Damn the Torpedoes: My Fight to Unify Canada's Armed Forces* (Toronto, 1990), 28.

40. Knowlton Nash, *Kennedy and Diefenbaker: Fear and Loathing Across the Undefended Border* (Toronto, 1990), 145–46.

41. On the politico-military crisis of 1963, see Granatstein, *Canada 1957–1967*, 116ff.

42. Jocelyn Ghent, "Canada, the United States, and the Cuban Missile Crisis," *Pacific Historical Review*, 18 (May 1979), 172. There were likely earlier cases of transgovernmental linkages between Canadians and British officers—for example, the decision to send troops to Hong Kong in late 1941. The relations between the two governments in wartime, not to say between the British and Canadian armies who were serving together, were scarcely comparable to that prevailing between Canada and the US two decades later.

43. Writing in 1962, General M.A. Pope noted of the 1920s "how closely . . . our army was modelled on that of the United Kingdom; to a considerable extent it still is today." *Soldiers and Politicians: The Memoirs of Lt.-Gen. Maurice A. Pope* (Toronto, 1962), 53.

section

2

1945–PRESENT

---

anada's emergence from World War II as a "middle" power, anxious to avoid a catastrophic conflict between the new superpowers, involved it in international commitments that earlier generations could scarcely have imagined. The new strategic and technological realities of the atomic age, as well as self-interest as to how the major powers managed their relations, have seen Canada take leading roles in the creation and work of the United Nations and the North Atlantic Treaty Organisation, and active support of the concepts of war avoidance through deterrence, peacekeeping, and various aid efforts. Managing the dialectic of national and international needs became increasingly complex, especially in light of Britain's supercession by the United States as Canada's main partner in continental defence and military production sharing. Successive Defence White papers have outlined the substance of defence policy in terms of periodic re-orderings of international, alliance, and home defence priorities as determined by shifting circumstances or governments. The processes of policy development, through the traumas of unification, the "civilianization" and modernization of the defence bureaucracy, and persistent underfunding, have themselves been revolutionized in the name of economy and, perhaps, efficiency.

The first four papers in this section explore the substance of major policy developments since 1945. In "A Seat at the Table," Joel Sokolsky demonstrates how Canada's alliance commitments have formed the bedrock of defence policy throughout the Cold War period, contributing as well to other foreign policy goals and largely determining the shape and nature of the Canadian Forces. Joseph Jockel turns more specifically to the issue of joint Canadian–American continental defence. In describing the creation of NORAD, including the RCAF's role in securing a formal agreement, he highlights the importance of institutional and personal links as prime movers in policy making that have sometimes tested the boundaries of civilian control of the military.

Peacekeeping has become a major defence commitment of which Canadians are justifiably proud. Rod Byers's article, in tracing Canada's efforts in this area, also highlights the ambiguities, difficulties, and costs they have entailed, including their impact on the Forces' resources and professional attitudes. His questioning, before his own untimely death, of how seriously these operations stretch an already dangerous commitments–capability gap has been poignantly underscored by more recent events in south-east Europe and elsewhere around the globe where Canadian involvements are increasing.

With respect to sovereignty protection, particularly in the North, Harriet Critchley demonstrates how the interplay of changes in superpower strategies, international law, military technology, resource potentials, and domestic political concerns have, and have not, effected assessment by policy makers of the strategic importance of this difficult defence role.

The next two articles investigate the machinery of policy development. Douglas Bland critically assesses the roles of defence ministers and the bureaucratic system that has produced three White Papers since 1945.

Related to this, Harriet Critchley examines the myth that unification was seen as an instrument for the political management of military influence through increased numbers of civilian experts within the Defence Department.

Turning to the economics of defence, Dan Middlemiss surveys industrial preparedness policy since 1945. He explains how the military's needs for secure supplies have continued to conflict with wider national economic and political goals, with the chronic realities of comparative costs and economies of scale, and the extent to which further integration of Canadian and US defence industries has been justified. Rod Byers looks to defence spending levels and their impact on the forces' capabilities. His criticisms of all governments' lack of clear procurement policies and commitment to using defence buying as a tool of regional development have been overtaken in some specific respects by recent events, such as the 1987 White Paper. However, it seems unlikely that the uncertainties of a still unfolding post-Cold War world will in any way diminish the general importance of the questions raised.

A SEAT AT THE TABLE:
CANADA AND ITS ALLIANCES *

JOEL J. SOKOLSKY

o

Canada's alliance relationships, the North Atlantic Treaty Organization and the North American Aerospace Defense Command, constitute nearly the sum total of Canadian defence policy. From the weapons acquired and the forces deployed, to the very strategic and tactical assumptions under which the Canadian Armed Forces (CF) operate, the needs and perceptions of the Allies are dominant.[1] Since both NATO and NORAD are American-led pacts, Canadian defence policy complements and is closely coordinated with U.S. global strategic interests and postures.

This situation represents more than the yielding of a small power to the harsh realities of the nuclear age and the interests of its superpower neighbour. Apart from its own national interest in supporting collective defence, Canada has joined and actively participates in these alliances as part of a global foreign policy. In the economic sphere in which it is a "principal"[2] if not a major power, Canada claims an important seat in such global councils as the Organization for Economic Cooperation and Development, the World Bank, and the group of seven leading industrial democracies.

Alliance membership complements these other relationships but is harder to understand or even justify, given Canada's somewhat unique situation. Unlike the European countries, it is not directly threatened; unlike the United States, it cannot be decisive in the common defence and its defence contribution is marginal compared with that of the larger European Allies.[3] Nevertheless, Canada has put great emphasis upon having a seat at the table in the councils and organizations that deal with global strategy and security in the nuclear age.

* *Armed Forces and Society* 16, 1 (Fall 1989), 11–35.

AMERICA'S "CLOSEST ALLY"

In U.S. global strategy, Canada does not figure prominently. Few Americans, even those well-informed on foreign policy issues, devote much thought to Canada; even fewer are aware of the military dimensions of American–Canadian relations. To be sure, most Americans would be surprised if Canada were not an ally, but most would reject as fanciful the notion that Canada was America's closest ally.

In reality, while Canada may contribute very little materially to the common defence of the West, the nature of that contribution has made military ties with the United States almost intimate in comparison with other American security relations. Canada is the only U.S. ally who participates directly in the defence of the American homeland. It is the only U.S. ally with a formal defence production-sharing arrangement, which provides Canadian defence contractors privileged access to the U.S. defence market. Canada was a founding member of NATO and, like the United States, maintains land and air forces in Europe. Canada's maritime forces are almost entirely postured to work with the U.S. Navy in the protection of the approaches to North America and in securing the North Atlantic in conjunction with other NATO Allies.

The bureaucratic and personal ties between the Canadian Department of National Defence and the CF, and the U.S. Department of Defense and all three American services are exceptionally close and are reinforced through almost daily contact. While not without problems, especially in the economic sphere, the overall relationship is excellent and untroubled compared with America's other international relationships, especially in defence matters.

The evolution of this alliance relationship was by no means the result of American policy initiatives alone. Although from time to time Canada has felt uncomfortable with the high level of co-operation between the two countries and the extent to which that co-operation has determined its defence posture, membership in NATO and NORAD has been viewed as fundamentally consistent with Canadian self-interest. Geographic location, historical ties, common democratic values, cultural affinity, economic interests, and finally the exigencies of security in the nuclear age have all contributed to the broad complementarity of Canadian and American strategic outlooks. As a global nuclear superpower, America has security interests and commitments that go well beyond those of Canada. However, as far as Canadian security interests extend, they are not at variance with those of the United States.

Canada regards the Soviet Union as the only power that threatens Canadian physical security—albeit indirectly, as Canada would only become a target in a general nuclear war as a result of its geographic location. In this sense, anything either the United States or the USSR did anywhere in the world that might lead to an armed conflict between the two superpowers is of concern to Canada.

Yet Canada's strategic outlook is more than the fatalism of a small power caught between two giants. Since the cold war, successive Canadian governments have regarded some form of Soviet aggression or threat of aggression as the likely initial cause of the next global conflict. In this largely Eurocentric view, the most immediate threat to the West is in Europe. Thus, Canada draws its strategic perimeter along the iron curtain.

To dissuade the Soviets from aggression in Europe, Canada has subscribed to collective defence. The U.S. strategic nuclear arsenal is fundamental to the Western deterrent posture, and since 1949, Canada has held that the linking of this arsenal to the defence of Europe is crucial for the prevention of war. The importance of American nuclear retaliatory forces for Canada is reflected in its participation in the defence of North America.

In the most general sense, by supporting the fundamental strategic outlook of the United States, Canada seeks peace and stability. If that stability is based upon collective deterrence, then it is also in Canada's interest to preserve the cohesion of the Western Alliance. The Alliance not only enhances Canadian security but also engenders a transatlantic political environment in which Canada is spared from having to take sides in a major American–European rift.

The complementarity of Canadian and American strategic outlooks has shaped the CF but offers no precise guide to "how much is enough." No NATO commander would turn away the CF's roughly 6000 personnel, 77 tanks, and 44 fighter aircraft currently stationed in Germany because it was too small a force to make a difference. On the other hand, even if this force were doubled (bringing it to pre-1969 levels), it would hardly change the balance along the Central Front. Canada contributes personnel, forces, and facilities to the defence of North America, yet the level of this contribution is determined more by political and budgetary factors than by any military consideration. In theory, Canada could let the United States assume the entire North American air defence role if it were willing to have American troops stationed in Canada.

All of this has left Canada relatively free to determine the level of its contribution to collective defence. To be sure, Ottawa is not impervious to the criticism that its defence expenditure as a percentage of gross domestic product is the second lowest of actively contributing nations, and the government has heeded the Allies' calls for various annual increases. Yet without a direct threat to its security and with the understanding that it will not play a decisive role in the overall balance of power, Canada has tended to answer the question "how much is enough?" by spending *just* enough: just enough to keep the forces somewhat compatible with those of the Allies, just enough to reduce the need for an American presence in Canada, and just enough, as well, to secure the Canadian seat in Allied councils. . . .

Using its Alliance position has been a consistent theme in Canadian external policy since 1945. Indeed, contributions to collective defence have traditionally been justified on the grounds that they enable Canada to have its voice heard not only in Allied councils, but elsewhere. According to one analyst, "the main and overriding motive for the maintenance of a Canadian military establishment since the Second World War has had little to do with our national security as such . . . [and] everything to do with underpinning our diplomatic and negotiating position vis-à-vis international organizations and other countries."[4]

A particular aspect of Canadian diplomacy has been the various efforts to perform a peacekeeping, peacemaking, or moderating function. Thus Canada has been a regular contributor to all major international peacekeeping undertakings, from Suez in 1956 to Afghanistan and Iraq in 1988. As an ally of the United States, Canada also argues that it has the right to be consulted on arms control issues and to have its counsel taken into account. Canada has been an active participant in United Nations (UN)-sponsored disarmament efforts.

For the most part, Canadian global activities neither enhance nor harm American interests, and they leave the military relationship between the two countries untouched. It is difficult to know what influence Canada has had over U.S. strategy through NATO councils, but on the surface there appear to have been no fundamental disagreements. In fact, on occasion, Washington has benefited from Canada's efforts to smooth over Allied disputes, such as the one over the Siberian oil pipeline. However, in the event of a serious rift between the Americans and the Europeans, there is little expectation that Canada would be able to broker a settlement.

The desire to play a more independent global role and the belief that Alliance ties prevent such a role are ever present. Witness the wide popularity that the Trudeau "Peace Initiative" in late 1983 gained among Canadians. At a time when the United States wanted the Alliance to hold together in the face of a breakdown of Soviet–American arms talks and the impending intermediate-range nuclear forces (INF) deployment in Europe, Pierre Trudeau's call for a new approach caught the imagination of the country. Moreover, although public protest over the testing of unarmed U.S. air-launched cruise missiles in Canada had subsided, there was vigorous opposition to Canadian participation in the Strategic Defense Initiative (SDI).

At times it does seem that, as Henry Kissinger observed, Canada's "interest in favor of common defense" has been in conflict "with the temptation to stay above the battle as a kind of international arbiter."[5] Overwhelmingly, however, it has been the instinct for the common defence that has shaped Canadian security policy. Any misgivings or hesitations have certainly not been compelling enough to keep Canada above the battle. . . .

THE COLD WAR AND NATO

The results of [World War II] left Canada in a position to take an active role in the postwar diplomacy. The old powers of Europe—both victors and vanquished—had been left prostrate. Canada was physically untouched by the war and was in fact strengthened by the wartime boost to industrial strength and resource development. Most important, the direction of Canadian foreign policy devolved upon a new group that might be considered the Canadian counterpart to the fabled American cold war foreign policy establishment. Centred around soon-to-be secretary of state for external affairs (later prime minister) Lester B. Pearson, this group sought to use Canada's important wartime contribution and its new high standing in the world to secure an active role in the global councils that would determine the shape of peace and attempt to establish a stable and just world order. Canada was perceived both internally and externally as a distinct and major international actor; it used this perception to seek peace and security for itself and others but without joining any new alliances.

As Canada stepped out on the world stage in 1945, however, it did so in the shadow of the United States, no longer just a great and powerful neighbour but also an atomic superpower in an international system that soon became increasingly bipolar and dangerous. Looking back from the vantage point of 40 years, the era of the cold war and the Pax Americana were to be the halcyon days of Canadian foreign policy.

Canada's postwar foreign policy orientation as a Western ally was determined by a number of realities: the rigid bipolar structure of the international system, geographic fact, historical linkages, its political and economic structure, and public attitudes. Any other orientation, either of Soviet ally or even of neutral state, would have run counter to these realities. Canada was not only anxious to ally itself with the United States; it was also determined to play a leading role in the creation of a transatlantic alliance that would link North American and West European security and in so doing serve a multitude of Canadian strategic, political, and economic interests.[6]

In 1947, the then secretary of state for external affairs Louis St. Laurent became the first Western leader to publicly call for a transatlantic pact. Speaking before the United Nations, which had become mired in the growing bipolarity, St. Laurent served notice that Canada could no longer "accept an unaltered [Security] Council." Placing the blame on the Soviet Union, he announced that, if necessary, "Canada would seek greater safety in an association of democratic peace-loving states willing to accept more specific international obligations in return for a greater measure of national security."[7]

Canada did take an active role in the creation of NATO. Apart from its deterrence value, the Alliance would help to resolve a pressing Canadian foreign policy problem. With the decline of the Anglo-Canadian alliance, the rise of the United States to superpower status, and the strategic realities of the nuclear age, a multilateral security pact would allow Canada to avoid a bilateral security arrangement with the United States. For while Canada supported the continuance of the PJBD [Permanent Joint Board of Defence] into the postwar era and while it was largely supportive of the need for the United States to play an active role in the world, Ottawa was still concerned about its sovereignty and independence vis-à-vis the United States. As one senior Canadian diplomat cabled from London during the negotiations:

> A situation in which our special relationship with the United Kingdom can be identified with our special relationsip with the other European countries in western Europe, in which the United States will be providing a firm basis both economically and probably militarily for this link across the North Atlantic, seems to me such a providential solution for so many of our problems that I feel we should go to great length and even incur considerable risks in order to consolidate our good fortune and insure our proper place in this new alliance.[8]

Canada viewed NATO not only as a "proper" forum in which to influence transatlantic security issues, but also as a means to foster greater economic cooperation among the Allies. At Canadian urging, Article II of the North Atlantic Treaty was inserted, which pledged the Allies to "eliminate conflict in their international economic policies and . . . encourage economic collaboration." Subsequently, a number of other international organizations were established to deal with economic matters (in all of which bodies Canada assumed an important role). Thus, Canada's hopes for NATO in this area were not realized.

Canada participated actively in the Western rearmament, which took place after the outbreak of the Korean War. The argument was that Canada would have more influence if it contributed standing forces to NATO and deployed them in Europe. "We and our allies believe," the minister of national defence told the House of Commons in 1951, "that the fact of participation by the Canadian army will show more emphatically . . . that we stand together with our allies."[9]

Over the years, this commitment has, to a significant extent, shaped Canada's armed forces. A mechanized force of nearly 7000 men plus a tactical air division was deployed to Europe. In addition, almost all Canadian naval and naval air forces on the Atlantic coast were earmarked for NATO. Europe was Canada's front line, and its armed forces were to fight alongside the NATO Allies. While made in the interest of collective deterrence, this commitment, especially that of the standing forces in Europe, was basically a political one: to solidify Canada's standing within the Alliance, particularly as it related to ties with Europe and Germany. In time, as the relative strategic value of the Canadian contribution declined, its symbolic value increased. The forces in Germany came to symbolize Canada's commitment to NATO and the necessary minimum price for a seat at the table.

Active participation in NATO also provided the armed forces with justification for military expenditures. Canada's military leadership was able to ask the government for the funds needed to fulfill the new Alliance commitments. Without NATO, it would have been difficult to justify the maintenance of modern conventional armed forces, such as tanks, artillery, fighter aircraft, and a wide range of antisubmarine warfare (ASW) forces. The need for Canada to "live up to its obligations" became the rallying cry for proponents of increased military expenditures. When, from time to time, the government did initiate major military expenditures, they invariably were related to NATO.

NORTH AMERICAN DEFENCE

As Canada began to posture much of its armed forces to support the American-led NATO multilateral Alliance, its strictly bilateral security relations with the United States in North America became increasingly cooperative. This was despite a certain amount of political ambivalence about formalizing collective defence in North America through the creation of integrated military structures.

The PJBD was not disbanded, and American–Canadian military coop-eration continued into the immediate post-World War II period. The threat to North America was in fact greater than it had been during the two previ-ous periods of American–Canadian alliance. For the first time, the United States and Canada were vulnerable at home to more than maritime forces. Soviet long-range bombers could conceivably deliver atomic bombs on North American targets, especially the facilities associated with the then single U.S. retaliatory force, the bombers of the Strategic Air Command (SAC). More so than at the time of the joint U.S.–Canadian declarations in 1938, the United States had an interest in Canada not becoming a strategic liability. Yet it seemed clear that Canada on its own could not make itself reasonably immune from attack. The question then became how much and what kind of co-operation with the United States would be required.

Accordingly, a series of radar lines were built across Canada, partly funded and manned by the United States. These were the CADIN Pinetree Line in 1951, the Mid-Canada Line in 1954, and the Distant Early Warning (DEW) Line in 1955. The next step seemed to be a fully integrated opera-tional control and command organization that would combine U.S.- and Canadian-based radar networks and interceptors along with newer surface-to-air missiles. Working together, the U.S. Air Force and the Royal Canadian Air Force (RCAF) successfully lobbied for the creation of NORAD, which was established through a simple exchange of notes between Ottawa and Washington.[10]

Unlike NATO, NORAD is a purely military organization, headed by an American Air Force general, who simultaneously commands U.S. continental air defence forces. His deputy is always a Canadian officer. At the headquar-ters in Colorado Springs, American and Canadian personnel receive informa-tion supplied by the radar networks in Canada as well as U.S. surveillance systems in the continental United States, at sea and eventually abroad.

NORAD has four main missions: tactical warning and attack assess-ment of bomber or ballistic missile attack on North America; space surveil-lance, tracking, and cataloging of all human objects in space; peacetime surveillance, detection, and identification of aircraft; and operational con-trol of U.S. and Canadian air defence forces. NORAD is linked to the global U.S. surveillance system, including the Ballistic Missile Early Warning sta-tions in Alaska, Greenland, and Scotland, and has a responsibility to the Joint Chiefs of Staff to provide worldwide detection of missile launches and nuclear events. It provides SAC and the Pentagon with missile warning and attack assessments and is becoming increasingly involved with and depen-dent upon space-based systems.

With NORAD, Canada obtains not so much a seat at the table as a seat at the console. Canada gains surveillance over its own territory and the opportunity to participate in the initial deliberations at NORAD headquar-ters, in the event of a warning and/or attack. However, while Canadian offi-cials would be involved in the first attack assessments and overall threat assessments, missile attack conferences would involve only the U.S. govern-ment in Washington. Beyond the existing consultative mechanism, therefore,

NORAD does not offer the Canadian government any special influence over U.S. presidential decisions. NORAD does, however, reduce the need for American forces to be stationed in Canada, which has always been a concern of the Canadian government, and it does provide a credible role for the RCAF. Moreover, insofar as Canada has a stake in the credibility of the U.S. deterrent posture, contributions to North American defence, even if marginal, can be viewed as in keeping with Canadian security interests.

It would have been inconsistent at best for Canada to be an active and contributing member of NATO, yet reject a structured arrangement for North American defence. Canada could hardly have supported the need for European Allies to accept an extensive American presence on their soil and highly integrated and elaborate command arrangements while refusing a much lower American presence in Canada and a far less elaborate command arrangement. Ironically, had NORAD been fully integrated with NATO, the United States, with the support of other Allies, might have justifiably asked Canada to accept a greater American presence and a more highly integrated command arrangement. Such was the case with the Supreme Allied Commander, Europe and his subordinate regional commands.

NUCLEAR WEAPONS AND UNIFICATION

In 1957, the Liberal party, which had shaped Canadian postwar foreign and defence policy and brought Canada into its allied relationships with the United States, was voted out of office. The Progressive Conservatives, under John Diefenbaker, did not closely identify themselves with the internationalist outlook held by the Liberals and tended to be wary of American influence. The issue of nuclear weapons convinced Diefenbaker that Canada was not playing an influential role in NATO. Although it had collaborated in the development of the first atomic bomb, Canada had renounced the deployment of nuclear weapons in its own forces in the interest of nonproliferation and had become active in the UN Disarmament Agency. In the late 1950s, however, Canadian forces, in order to maintain the effectiveness of their Allied contributions, had acquired five weapons systems, all of which required nuclear warheads. The Diefenbaker government, anxious to expand Canada's role outside NATO, hesitated in accepting these warheads from the United States.[11] The United States began to pressure Canada; General Lauris Norstad, the American and NATO Supreme Commander in Europe, voiced his opposition to the Canadian government's position at an open meeting in Ottawa. The U.S. State Department went further and directly contradicted the prime minister, who had told the House of Commons that the United States and the other Allies also had doubts about a nuclear role for Canada.[12]

The nuclear fiasco resulted in the defeat of the Diefenbaker government, and the return of the Liberals under Lester Pearson brought about a quick acceptance of the nuclear role for Canada. The new government argued that Canada had to live up to its obligations as a member of NATO: "Having accepted responsibility for membership in a nuclear armed alliance, the question of nuclear weapons for Canadian Armed Forces is a subordinate issue."[13] Pearson was concerned both with the maintenance of Canadian influence in the Alliance and with Alliance unity, recently shaken by the withdrawal of France from NATO's integrated military structure.

Although the Liberals remained committed to Canada's alliances, they believed that changes had to be made in the structure of the CF if Canada was to continue to make an economically sustainable contribution to collective defence. The Liberals also wanted Canadian defence policy to serve the goal of national distinctiveness. In its 1964 *White Paper on Defence*, the government reaffirmed its commitment to collective security but sought ways for Canada to make a more identifiable and economical contribution. International peacekeeping was given a higher priority, with NATO second, NORAD third, and defence of Canadian sovereignty fourth. No committed forces were to be withdrawn from either NATO or NORAD, but the three services were to be abolished and a unified armed forces put in their place. Part of this force would be a 12 000-man mobile unit with the strategic mobility to move quickly to meet emergencies anywhere in the world.

It was this mobile force that Canada offered as its distinctive contribution to NATO defence. However, its usefulness to the Alliance stemmed not so much from its ability to be moved to Europe but rather from its ability to perform peacekeeping functions. It was argued that Canada's greatest contribution to collective security could be made through UN peacekeeping activities. Canada's first active peacekeeping role began with the United Nations Emergency Force. The initial attack by Great Britain and France on Egypt drove a wedge between the British and American NATO partners, and the Canadian peacekeeping effort was as much an attempt to heal that rift as it was an attempt to end the hostilities.

Once again, during the Cyprus crisis in 1964, Canada was given the chance to prove the importance of its peacekeeping forces to NATO. By sending troops to Cyprus, Canada hoped to prevent a confrontation between two NATO members and to avoid exposing the eastern flank of the alliance to Soviet advantage and, perhaps, the island itself to overt Soviet intervention."[14]

THE TRUDEAU ERA

In the 1960s, it became evident that not only was the Alliance not serving as a forum for economic co-operation between the Allies, but the rise of the European Economic Community was effectively freezing Canada out of transatlantic economic relations. Indeed, the rise of the EEC and the Eurogroup within NATO seemed to be bringing into reality the "two-pillar" approach to the Alliance opposed by Canadians in 1949. The Europeans, who themselves were working toward greater unity, were not sensitive to Canadian concerns about the drift toward continental integration in North America and Canada's desire to appear as a more or less independent actor within NATO. According to Jean Monnet, Canada had to "recognize facts."

Militarily it was part of North American defence plans (the Canada–U.S. planning group in NATO).[15]

When he assumed office at the head of a majority Liberal government in 1968, Pierre Trudeau was determined to recognize the facts about Canada's defence policy and its Alliance role. After a review, he cut the Canadian military presence in Europe in half. This decision was the most significant change in Canada's NATO policy since 1949. Trudeau defended his action on the grounds that the essential nuclear deterrent rested with the United States and that, on the conventional level, "the magnificent recovery" of Western Europe had given those countries the ability to provide for their own conventional defence. Although reducing its standing commitment in Europe, Canada would maintain forces in Canada to be sent to Europe in case of emergency. The prime minister believed that "NATO continues to contribute to peace by reducing the likelihood of a major conflict breaking out in Europe" and also rejected a complete withdrawal from the Alliance on the grounds that Canada could continue to play a "vital role . . . in the promotion of détente and arms reduction," only by remaining within the Allied councils. However, he rejected the view, held by previous Canadian leaders (with the exception of Diefenbaker) and voiced by the Europeans, that "our acts . . . will have profound international consequences." His decision to leave some troops in Europe was mainly for symbolic purposes— to give "visible evidence of Canada's continuing commitment to the alliance."[16] His decision to remain in NATO would allow Canada to exercise its traditional policy of seeking to moderate U.S.–Soviet relations, but Trudeau was not willing to maintain costly forces in Europe merely to have a seat at the table.

To a certain extent, the prime minister had judged correctly. Despite the reduction in the German forces, the 1970s were especially active years for Canada in NATO and in European diplomacy. Ottawa participated in the Conference on Security and Cooperation in Europe and the Mutual and Balanced Force Reduction talks, and continued a high level of involvement in intra-Alliance politics—for example, in the INF decision later in the decade. Trudeau was right: Canada could have its seat at the table and all the diplomatic activity it wished for at half the price.

The 1971 Defence White Paper did more than ratify the steps already taken. It stated that the first defence priority would henceforth be the protection of Canadian sovereignty; next would come NORAD, then NATO, and finally peacekeeping.[17] The White Paper ushered in several years of expenditure cutbacks, from which, as of 1989, the CF have yet to recover.

While the prime minister's view of the external environment did influence his decision, the main impetus was domestic. Foreign policy was to be the extension abroad of domestic priorities, and defence policy would be shaped accordingly. The domestic priorities were economic growth, sovereignty and independence, social justice, peace and security, and a harmonious natural environment.[18] In deciding to reduce the Canadian commitment to Europe and place more emphasis on the sovereignty protection role of the military, Trudeau was responding to the second priority. For example,

Canada had to have the capability to patrol its own coastal waters, including the Arctic regions, in order to protect its mineral and fishing resources.

In 1972, the Trudeau government turned its attention to ways to offset the growing continental integration. It developed a conceptual approach, called the "Third Option." Rejecting the first two options, status quo and increased integration (essentially straw men), the government stated that henceforth Canada would seek to strengthen its ties to other areas of the world beyond North America.[19] In concrete terms, this meant expanding trade and other economic links internationally. Aside from Japan, the only other region capable of serving this objective was Europe and the EEC. In other words, the government wanted the Europeans to serve as a counterweight to the United States. The original decision to join NATO had been made on the grounds that Canada could influence Allied decision making and expand its economic ties through the Alliance. Having only partially achieved the former and having failed to achieve the latter through the pact, Canada would not pursue its economic goals outside NATO. When it came time, however, to actually negotiate a contractual link with the EEC, Canada found the Europeans, and the Germans in particular, tying economic links with the upgrading of Canadian forces in Europe. German chancellor Helmut Schmidt reportedly told Trudeau "no tanks, no trade."[20]

In 1975, the United States joined the European Allies in urging Canada to upgrade its forces in Europe. Speaking in Ottawa, U.S. Defense Secretary James Schlesinger said, "The basic premise, I believe, is that unless we are prepared to defend parts of the world other than the North American continent, we will soon have nothing more than the North American continent to defend, and that would be a calamity from the standpoint of both our nations."[21] This was in line with the long-held U.S. view that Canadian contributions to North American defence did not offset Canada's obligations to Europe. Moreover, by the mid-1970s, as the rate of growth in U.S. defence expenditures declined in the post-Vietnam reaction, Washington was increasingly looking to the Allies to assume more of the burden.

Despite the reduction of Canadian forces in Europe and the declaratory emphasis upon sovereignty protection, the Trudeau government maintained all Canadian commitments to NATO, with the exception of a limited nuclear role. Thus, in the mid-1970s, when pressure from the Allies and the seriously deteriorating condition of the equipment compelled Canada to begin major capital replacements, Alliance requirements largely determined which weapons were bought. Budgetary constraints and the continuing Trudeau ambivalence about defence meant, however, that weapons were not replaced on a one-for-one basis. A total of 18 Aurora Long-Range Patrol/ASW aircraft replaced 36 Argus aircraft. Canada bought 128 German Leopard tanks for its forces in Europe. Further, the government decided to replace all fighter aircraft in Europe and North America, but the final choice of the F-18A was not made until 1981, with full delivery not completed until later in the decade. It was also decided that a new class of ASW frigate would be built. It took until June 1983 for the Trudeau government to decide to go ahead with only six made-in-Canada frigates. Four of the existing destroyers would be refitted, leaving Canada with only 10 surface ships, capable of fulfilling NATO roles by the mid-1990's.

As a further indication of Canada's continued commitment to collective security, Ottawa assumed a new NATO commitment to send land and air reinforcements to Norway in the event of war or crisis. This commitment was undertaken in part to fulfill earlier reinforcement pledges to the Central Front that could not be met. Yet in the absence of substantial improvements in Canadian air- and sealift capabilities, the Norwegian commitment was equally untenable.

The rearmament of the mid-1970s allowed Canada to claim that it was meeting the Allied objective of 3 percent annual real increases in overall defence expenditures. Furthermore, Canada became a contributor and participant in NATO's Airborne Early Warning and Control program and also allowed a number of Allied nations to use bases in Canada for training purposes. Nevertheless, by the early 1980s, Canada's defence expenditures as a percentage of gross national product hovered at about 2 percent, almost the lowest in the Alliance. Despite the handful of equipment projects, all of which experienced delays, a serious gap had emerged between capabilities and commitments.

THE 1987 WHITE PAPER

During the 1984 summer election campaign the Conservatives, let by Brian Mulroney, had pledged, among many other things, to rebuild the CF and to restore Allied confidence in Canada as a reliable partner in collective defence. Soon after taking office, the Conservatives initiated reviews of both defence and foreign policy and took steps to show that they were serious about these issues. One of the first decisions was to augment the ground forces in West Germany by 1200 men, bringing the total to 4200, which was close to war strength. The idea of having the battalion that was meeting Canada's commitment to NATO's Allied Mobile Forces-Land (AMF[L]) eventually join the Canadian Air-Sea Transportable brigade group in Norway to meet that commitment was abandoned, and a decision was made to conduct a full-scale exercise of the CAST commitment in 1986. The government proceeded with the acquisition of a low-level air defence system for the forces in Europe and reaffirmed the need for more new ships, ASW helicopters, and submarines. In March 1985, at a summit meeting between Prime Minister Mulroney and President Reagan, a bilateral agreement was signed for the modernization of NORAD's air defence facilities. A year later, at their second summit, the two leaders renewed the NORAD agreement for another five years. Finally, to help boost the morale of the armed forces, the government brought back distinctive uniforms for the three services, but it did not restore their old names.

As welcome as these steps were, the central problem of funding remained. Faced with a federal deficit proportionally higher than that of the United States and a continuing demand for spending on social programs and regional development, the Mulroney government was unable to increase defence spending significantly. Indeed, in its first two years, it fell short of the 3 percent real increases achieved by the Trudeau government.

Initially (and surprisingly, in view of the subsequent White Paper), the Mulroney government sought to close the commitment-capability gap by suggesting to NATO that Canadian air and ground forces be withdrawn from West Germany. In compensation, Ottawa would improve its reinforcement commitment to Norway.[22] Put forth in the fall of 1985, this proposal found no support among the Allies. West Germany was determined to keep Canadian forces on its territory, and Oslo did not want to be accused by Bonn of "stealing" the Canadians. Washington opposed the move on the grounds that it would set a bad precedent. More importantly, it appeared too much like the Trudeau moves of the early 1970s, when Canada cut its forces in West Germany in half, pledging to maintain adequate reinforcement capabilities but never doing so. This view was bolstered by the fact that by the autumn of 1985, it had become clear that Canada was not going to increase its defence spending dramatically. From the American perspective it was better for Canada to stay in West Germany, thus helping to sustain Allied unity, than to restructure in favor of Norway. In the summer of 1986, Canada conducted the first full-scale exercise of CAST, but the results did not impress the new minister of national defence, Perrin Beatty, who had initiated a new defence review.

As the government was examining options regarding Canada's contributions to Western collective defence, the issue of national sovereignty, especially in the Arctic, re-emerged. In the summer of 1985, a U.S. Coast Guard icebreaker, the *Polar Sea*, sailed through the Northwest passage, raising the issue of Canadian sovereignty—given the continued U.S. claim that the passage was an international strait and not, as Ottawa contended, internal Canadian waters. The Mulroney government, responding to public outcry, announced that it would be building a giant icebreaker and issued a declaration of straight baselines around the archipelago. The issue of Arctic sovereignty arose again in 1986, when it was learned that several U.S. Navy nuclear attack submarines (SSNs) had held a rendezvous at the North Pole.

Along with sovereignty, new questions were being asked about Canada's relationship to U.S. strategic nuclear plans. Was the United States shifting from a doctrine and posture based on stability through a secure second-strike capability (the mutual assured destruction approach) toward a nuclear warfighting doctrine that sought superiority over the Soviet Union? The Carter administration's "Presidential Directive 59" and the Reagan strategic modernization program seemed to suggest the latter. The launching of the SDI in March 1983 further encouraged belief in a shift in American nuclear strategy. SDI raised particular concerns for Canada because of the country's continued involvement in NORAD as well as the close links between the Canadian defence industry and the U.S. military. In the spring of 1985, the Reagan administration issued an invitation to Canada to become involved in SDI research. The Mulroney government decided not to engage in government-to-government participation in SDI although it did allow Canadian private firms and researchers to compete for contracts.

After receiving Cabinet approval, the White Paper entitled *Challenge and Commitment: A Defence Policy for Canada* was brought before the House of Commons on 5 June 1987.[23] Minister of National Defence Perrin Beatty declared that it represented a "made in Canada" defence policy. Reflecting the pessimistic international mood of the early 1980s and the government's desire to strengthen Canada's Allied ties, the White Paper stressed the continued Soviet threat to Canadian security and hence the continued importance of the NATO and NORAD commitments. Canada had no choice but to seek its security in a "larger family of like-minded nations." The White Paper rejected as "naive and self-serving the arguments of those who promote neutrality or unilateral disarmament." Given Canada's values, traditions, and political and economic interests, neutrality would by hypocrisy. More important, opting out of collective defence would mean relinquishing Canada's valued seat at the table.

The desire to maintain a seat at the table was most evident in the White Paper's plans for Canada's NATO ground and air commitments. Citing the continued conventional imbalance along the Central Front as one of four "geopolitical and geostrategic" factors that Canadian defence policy had failed to take into account, Beatty proposed a consolidation and improvement in Canada's commitments. The CAST commitment to Norway was dropped in favor of the Central Front role. The brigade group heretofore committed to CAST would now be given the task of reinforcement of the armored brigade already in West Germany. Together, they would constitute a new Canadian division dedicated to the Central Front. The two air squadrons that had been earmarked for Norway would also assume a reinforcement role as part of a five-squadron air division with three squadrons permanently based in West Germany. Canada would retain its commitment of a single battalion to the AMF(L) with existing pre-positioned stocks being left in Norway. The air division would use currently deployed CF-18s, but the ground forces would require new tanks and a revitalization of the reserves.

The decision to sustain and improve Canada's German commitment was consistent with Canada's approach to NATO since the early 1950s, which views forces dedicated to and stationed in Germany as the most tangible symbol of its support for the Alliance and the key to securing Canada's seat at the table. "The presence of Canadian armed forces in Western Europe," announced the White Paper, echoing a familiar theme, "ensures that we will have a say in how key security issues are decided."[24]

But while the White Paper reaffirmed the importance of NATO Europe, its main thrust appeared to be geared toward the improvement of Canada's North American and maritime capabilities, reflecting concern about both security and sovereignty. In addition to the situation along the Central Front, Beatty drew attention to three other changes in the international environment. The first of these was the growing importance of the Pacific Ocean in international political, economic, and strategic relations. Soviet naval power had also expanded in the Northern Pacific, and this coupled with Canada's growing trade and political ties in the region, had to be reflected in defence policy. Another change noted was the emergence of the Arctic as a region of strategic importance, particularly for the operation of "foreign submarines," which raised "security and sovereignty concerns for Canada." "Technology . . . is making the Arctic more accessible. Canadians cannot ignore that what was once a buffer (between the two superpowers) could become a battleground."[25] Finally, there was the development by the Soviets of air- and submarine-launched cruise missiles (ALCMs, SLCMs) and "Canada's vulnerability to attack by these weapon systems from our three ocean approaches."[26]

Of all changes mentioned, three related to the direct defence of Canada in its air and especially maritime dimensions. It is not surprising, therefore, that the centerpiece of the White Paper's acquisition proposals was the creation of a "three-ocean navy." Such a navy, the government argued, would be able to fulfill Canada's commitments to NATO while meeting the new sea-based threats to Canadian security and sovereignty.

To meet the bomber/ALCM threat and to support the maritime forces in countering the SLCM threat, Canada would continue with the modernization of NORAD's facilities. But the White Paper also stressed that Canada would be promoting research into new space-based and airborne air surveillance systems, as well as co-operating with the United States in American air defence research programs such as those being conducted under the Air Defence Initiative (ADI).

To a certain extent, the weapons acquisitions contained in the 1987 White Paper were actually consistent with the thrust of the 1971 White Paper in that they would greatly enhance Canada's ability to patrol its own waters and airspace and thus assert its sovereignty. Indeed, there was a new emphasis on territorial sovereignty, especially in the north. New units, as well as reserve units, were "to deal with threats in any part of the country." In fact, the government planned to construct a "northern training centre" at the tip of Baffin Island to train regular and reserve forces for territorial defence. Responding to public concern, the White Paper had an unmistakably nationalist character.

There was, however, a significant difference between the 1987 and 1971 approaches. In the early 1970s, protection of Canada's sovereignty was seen as protection from nonmilitary threats, such as fishing violations or pollution. There seemed to be a declining direct military threat to North America from Soviet air forces and general purpose maritime forces. Ballistic missiles constituted the only threat and one against which there was no defence.

In contrast, the 1987 White Paper assumed that North American defence would become more important in the larger strategic context as the Soviets develop new generations of cruise missiles, as the sophistication of their SSNs increases, and as the Arctic becomes more of a strategic arena. This will bring about a heightened American concern for continental defence (consistent with U.S. thinking on strategic defence).

The Mulroney government described a threat to sovereignty originating from a lack of Canadian capabilities, which would, in time, result in an overwhelming U.S. dominance of North American defence at sea and in the air. In explaining the government's decision to research space-based air defence systems and participate in ADI, the White Paper warns that "failure" to meet the challenge of new technologies "could mean forfeiting the responsibility for surveillance of Canadian airspace to the United States." As Beatty told the House of Commons, Canada would not contract out its defence. While the government was "prepared to discuss co-operation in all aspects of the defence of North America," it would not "allow Canada's sovereignty to be compromised. We will be a partner with our allies and not a dependent."[27]

Its differences notwithstanding, the 1987 White Paper shared the same fate as its predecessor; within two years it was all but repudiated by the very government that had introduced it with such promise and fanfare. Faced with a growing federal debt that consumed one out of every three tax dollars in interest payments, a recently reelected Mulroney government raised taxes and cut government spending in its April 1989 budget. Defence spending was hit particularly hard. With the exception of the six additional frigates and the minesweepers, for which funding decisions had already been made, every other major program contained in the White Paper was either canceled or postponed. This included the SSN program, which was dropped entirely. Plans to create a new armored division for Germany are in flux, and the projected tank purchase was cut in half and postponed. The recommitment of three air squadrons for Germany was retained, but the government said it would not be replacing any CF-18s lost through accidents. Also cancelled were six additional long-range patrol aircraft and equipment for the reserves. Canada would maintain its forces in Germany and honour its NORAD commitments, but it was clear that the gap between these commitments and its capabilities would continue to grow. The only area in which the government pledge to expand was kept was in peacekeeping, a role that had received much favorable support but that the 1987 White Paper had downplayed.

There was more to the about-face on the 1987 White Paper than money. Although in the early eighties there was a growing public consensus that the CF had to be improved, the arrival of a new era of détente and arms control by mid-decade had cast doubt upon the need to increase defence spending, especially if such expenditures were related to a renewed Soviet threat.[28] Canadians were worried about sovereignty, and the Mulroney government tried to play to this with such programs as the SSN. But there was never any strong support for meeting the sovereignty challenge with greater defence expenditures, especially for nuclear submarines. As for Canadian Alliance commitments and the global influence they bring, here too there was some questioning. The public and the government wanted to remain in NATO and NORAD, but there was an assumption that the seat at the table could be had at a lower cost. After all, for Canada the political aspects of its Alliance commitments had always mattered more than the military ones, and its influence in Allied councils had been less a function of military capabilities than of diplomatic skill. Because of this, Ottawa has always sat comfortably at the table.

CONCLUSION: SITTING UNCOMFORTABLY AT THE TABLE

In the absence of an improvement in Canada's military capabilities, it will take all the diplomatic skill Ottawa can muster to cope with the problems expected to arise in both NATO and NORAD in the coming years. In Europe, the economic integration of 1992, coupled with a plethora of transatlantic problems over burden sharing and nuclear weapons, is likely to encourage the emergence of a real European pillar within NATO. The United States will remain in the Alliance although the level of its involvement in terms of forces on the ground and management of Allied affairs will most likely decline. What Canada's position will be under these circumstances is unclear.

Following the logic of the 1987 White Paper, Ottawa could try to stay in NATO but concentrate military efforts in the North Atlantic and in the maritime and air defence of North America. Yet here, too, a lack of commitments may hamper Allied co-operation, principally with the United States. Unless more frigates and conventional submarines replace the now scuttled SSN program, Canada will have only 16 surface ships and an aging fleet of 18 patrol aircraft by the mid-1990s and no under-ice capabilities. Thus, stated intentions to support NATO at sea, to patrol the ocean approaches to North America for which Canada is responsible, and to assert Canadian sovereignty will be hard to maintain. Similarly, a declining interceptor capability and lack of funding for space-based air defence research may well leave Canada behind in future air defence efforts. Moreover, despite the current slowdown in SDI, strategic defence is not going to disappear from the American security agenda. This will, in turn, complicate the future of bilateral relations in NORAD.[29] A Canada becoming more of a dependent and less of a partner in North American defence, while at the same time drifting further away from European security affairs, is bound to find its seat at the table increasingly uncomfortable. The only thing more unsettling to Canadian foreign and defence policy would be not to have any seat at all.

NOTES

1. John English and Norman Hillmer, "Canada's Alliances," *Revue internationale d'histoire militaire*, Edition Canadienne 51 (1981), 31.

2. On this point, see David B. Dewitt and John J. Kirton, *Canada as a Principal Power* (Toronto: John Wiley and Sons, 1983).

3. Henry Kissinger, *White House Years* (Boston: Little Brown, 1979), 383.

4. James Eayrs, "Military Policy and Middle Power: The Canadian Experi-

ence," in *Canada's Role as a Middle Power*, ed. J. King Gordon (Toronto: Canadian Institute of International Affairs, 1966), 70.

5. Kissinger, *White House Years*, 383.

6. John W. Holmes, *The Shaping of Peace: Canada and the Search for World Order, 1943–1957*, vol. 2 (Toronto: University of Toronto Press, 1982), 29.

7. Robert Mackay, ed., *Canadian Foreign Policy, 1945–1954: Selected Speeches*

and Documents (Toronto: McClelland and Stewart, 1971), 97.

8. As quoted in Escott Reid, *Time of Fear and Hope: The Making of the North Atlantic Treaty, 1947–1949* (Toronto: McClelland and Stewart, 1977), 132.

9. John Gellner, *Canada in NATO: A Documentary History* (Toronto: Ryerson Press, 1970), 26.

10. For an account of the origins of NORAD based upon declassified documents, see Joseph T. Jockel, "The Military Establishments and the Creation of NORAD," *The American Review of Canadian Studies* 12, 3 (Fall 1982), 1–16; and Joseph T. Jockel, *No Boundaries Upstairs: Canada, the United States and the Origins of North American Air Defence* (Vancouver: University of British Columbia Press, 1987).

11. Jon B. McLin, *Canada's Changing Defense Policy, 1957–1963: The Problems of a Middle Power in Alliance* (Baltimore, MD: Johns Hopkins University Press, 1967), 136.

12. Peter C. Newman, *Renegade in Power: The Diefenbaker Years* (Toronto: McClelland and Stewart, 1963), 366.

13. Canada, Department of National Defence (hereafter DND), *White Paper on Defence* (Ottawa: Queen's Printer for Canada, 1964), 13.

14. Arthur Blanchette, ed., *Canadian Foreign Policy 1955–1965: Selected Speeches and Documents* (Toronto: McClelland and Stewart, 1977), 42.

15. Elliot R. Goodman, *The Fate of the Atlantic Community* (New York: Praeger, 1975), 145.

16. On the NATO decision, see Bruce Thordarson, *Trudeau and Foreign Policy: A Study in Decision Making* (Toronto: Oxford University Press, 1972), chap. 5; and Larry R. Steward, ed., *Canadian Defence Policy: Selected Documents, 1964–1981*, National Security Series, no. 1/82 (Kingston, ON: Centre for International Relations, Queen's University, 1982), 10–21.

17. Canada, DND, *Defence in the 70s* (Ottawa: Information Canada, 1971), 16.

18. Canada, Department of External Affairs, *Foreign Policy for Canadians* (Ottawa: Information Canada, 1970), 14.

19. Mitchel Sharp (then minister of state for external affairs), "Canada–U.S. Relations: Options for the Future," *International Perspectives* (Autumn 1972), 1–24.

20. Gerald Porter, *In Retreat: The Canadian Forces in the Trudeau Years* (Montreal: Deneau and Greenberg, 1978), 164.

21. As quoted in D.W. Middlemiss and J.J. Sokolsky, *Canadian Defence Decisions and Determinants* (Toronto: Harcourt Brace Jovanovich, Canada, 1989), 38–39.

22. On the Canadian approach to NATO regarding Norway, see Joseph T. Jockel, *Canada and NATO's Northern Flank* (Toronto: York University Centre for International and Strategic Studies, 1986), 26–28.

23. Canada, DND, *Challenge and Commitment: A Defence Policy for Canada* (Ottawa: Minister of Supply and Services, 1987).

24. Ibid., 6.

25. Ibid.

26. Canada, DND, "Tabling of the Defence White Paper in the House of Commons," 5 June 1987, 4.

27. Ibid., 10.

28. On changes in Canadian opinion toward defence, see Don Munton, "Canadians and Their Defence," *Peace and Security*, 3 (Winter 1988/89), 2–5.

29. On current issues in Canada–US defence relations, see Joel J. Sokolsky, *Defending Canada: U.S.–Canadian Defense Policies* (New York: Priority Press, 1989); and David G. Hadlund and Joel J. Sokolsky, eds., *The U.S.–Canada Security Relationship: The Politics, Strategy and Technology of Defense* (Boulder, CO: Westview, 1989).

THE MILITARY ESTABLISHMENTS
AND THE CREATION OF NORAD◇

JOSEPH T. JOCKEL

o

Few aspects of Canadian defence policy, or of Canadian–American relations have been as debated in Parliament, examined by scholars, or attacked by critics as the 1957 creation by the United States and Canada of the North American Air Defense Command (NORAD). It might thus be expected at first glance that there is not much more to be said about it. In fact, however, one major aspect of the events leading to NORAD's establishment has never been dealt with satisfactorily, largely due to shrouds of military secrecy. That aspect is the role in the Canada–U.S. negotiations of the two national military establishments.

Melvin Conant once summarized the suspicions surrounding NORAD: "Rightly or wrongly, fairly or unfairly the impression exists in Canada that the real decision to establish NORAD was made by Canadian military officers in discussion with U.S. Air Force officers and that this took place without the prior, full and continuing comprehension of either the Liberal Cabinet, under which the discussions began, or the newly installed Conservative government, which took responsibility for implementing the military agreement."[1] This suspicion was voiced most dramatically perhaps, by James Minifie, a CBC Washington correspondent, in his widely-read *Peacemaker or Powdermonkey: Canada's Role in a Revolutionary World*, published in 1960. Minifie claimed that there had been collusion between the United States Air Force and the Royal Canadian Air Force. A joint command, in his view, had been the USAF's pet project, but the U.S. Defense Department and the Joint Chiefs of Staff did not share their Air Force colleagues' enthusiasm. Having been blocked at the highest military levels in the United States, the proposal was "forwarded surreptitiously by the USAF and brought to fruition in Canada by the Royal Canadian Air Force."[2]

◇ *American Review of Canadian Studies* 12, 3 (Fall 1982), 1–16.

Certainly there was a good deal of circumstantial evidence to support Minifie's charge. After the U.S. Secretary of Defense and the Chairman of the Joint Chiefs of Staff publicly poured cold water on the idea in 1955 it popped up again in a speech in Montreal by the Chief of Staff of the RCAF. The Chairman of the Canadian Chiefs of Staff committee would later confess to pressuring the government. No doubt Minifie and others had access as well to the usual off-the-record journalistic sources. Yet he was forced to conclude that the origins of NORAD "cannot yet be traced with any sureness."[3] Conant concurred, writing in 1962 that "no verdict can be given now."[4]

The picture has now changed. The relevant American documents, including the records of the Joint Chiefs of Staff are almost all open, to a large extent as a result of the Freedom of Information Act. It is now possible to trace in substantial detail the military-to-military contact which led to the establishment of the joint command. Similarly, a greater grasp can be had on the manner in which the Canadian military presented the proposal to the Canadian political authorities.

The idea of a post-war Canada–United States air defence command was proposed for the first time in 1946, and arose sporadically thereafter. The military, though, had little interest, due to several obstacles. First, a necessary prerequisite was the establishment of an *American* unified continental air defence command. The U.S. Army, Navy, and Air Force all had responsibility for different aspects of continental air defence. In particular, the Air Force established an Air Defense Command (ADC) responsible for most interceptor aircraft while the Army maintained its Anti-Aircraft Command (ARAD-COM) with responsibility for anti-aircraft artillery and surface to air missiles. Not until 1954 was the Continental Air Defense Command (CONAD) brought into existence and given operational control over all U.S. continental air defence forces. Operational control was defined as the "authority to direct the tactical air battle, control fighters, specify conditions of alert, station early warning elements, and deploy the command combat units."[5] The situation in Canada was far simpler, for continental air defence was almost exclusively in the hands of the RCAF's Air Defence Command (ADC).

Of equal importance in inhibiting military interest in a joint Canada–United States air defence command was the simple lack of necessity. Both countries had been very slow to erect continental air defences. The Soviet Union seemed, in the early post-war years, a long way from developing a concrete air atomic threat to North America. This perception changed, of course, as the Russians revealed that they had acquired not only atomic weaponry, including hydrogen bombs, but the capability as well to deliver these weapons to targets in the United States and Canada. By the mid-1950s, the RCAF was deploying nine squadrons of advanced CF-100 Canuck interceptors for the defence of Canada. The United States programmed eighty-six squadrons of interceptors for active service and sixty-one battalions of Nike surface-to-air missiles. Extensive radar networks had been developed in the continental United States (the Permanent system) and in southern Canada (the Pinetree radars). Decisions had been taken, as well, in 1953 and 1954 to deploy two great early warning systems. The Mid-Canada Line, stretching along the fifty-fifth parallel, would be built by Canada. The Distant Early Warning (DEW) Line would be built by the United States across northern Canada and Alaska, with seaward extensions.

These two early warning lines would, it was realized in the mid-1950s, place a premium on tactical co-operation between the two growing national air defence systems. First, given the extra warning time which the lines would provide, the battle areas would be pushed northward. This would entail more American interceptors entering Canadian airspace. Secondly, the extra early warning would allow for a more rationalized air defence strategy across the entire continent. Successive waves of enemy bombers could be detected. This would encourage the judicious use of "forward" and "reserve" interceptors. Given that the "forward" squadrons could be Canadian and the "reserve" American, tactical co-operation between the Royal Canadian Air Force and the U.S. Air Force would become still more important.

In fact, co-operation between the two air forces grew steadily in the early 1950s as they developed more interceptors and built more radar stations. Formal methods of co-operation were arranged under the auspices of the Canada–United States Permanent Joint Board on Defense (PJBD). In 1951 the U.S. and Canada established procedures whereby fighter aircraft could cross the international boundary to carry out interceptor operations. PJBD Recommendation 51/4, approved by both governments, provided that either country could send interceptors into the airspace of the other to intercept a hostile aircraft if that aircraft were headed towards the boundary and if the air defence forces of the country in which it was located were obviously unable to intercept it before it crossed the boundary. It was clearly stipulated, though, that while in the airspace of the second country, the interceptors would act under the operational control of that country, exercised through its air defence command and ground control intercept stations. This entailed that the interceptors would be guided on target by the "host" country's ground stations and could destroy the hostile aircraft only with the permission of the host country.

PJBD Recommendation 51/4 was supplemented by Recommendation 51/6, approved in early 1952. It filled an important gap. As Secretary of State Acheson wrote President Truman, "there is no provision at present for the deployment of United States Air Force air defence forces in Canada and Royal Canadian Air Force air defence forces in the United States except for training exercises. . . . In the outbreak of war against a common enemy, circumstances might occur which would require the rapid reinforcement of forces of one country by another."[6] Under the provisions of Recommendation 51/4, American interceptors could be deployed to Canadian air bases, and Canadian interceptors to American bases. Once again, though, all interceptors would be under the operational control of the country in which they were located.

The stipulation in Recommendation 51/4 concerning the interception of only those aircraft approaching the boundary entailed a severe limitation on the ability of the two countries to co-ordinate their air defence efforts. In the wake of growing Soviet offensive capabilities and in the wake of the growth in Canadian and American defensive capabilities, this limitation was removed in 1953 with the approval of PJBD Recommendation 53/1. This provided that interception of aircraft by fighters of one country could be undertaken in the airspace of the other when the second country's air defence command would be unable to execute the interception with its own forces. Once again, though, solely national operational control was to be retained. All interceptors in Canada would fall under Canadian operational control.[7]

Operational control over continental air defences was a sensitive matter in Canada, for it touched on two concerns. The first was command and control of Canadian armed forces. Canada pursued, in its military relations with Britain during both world wars, what C.P. Stacey has described as "a sturdy policy of autonomy."[8] Canadian resistance was even stiffer when it came to command and control of Canadian forces in Canada during the Second World War. The problem here, of course, was with the United States. Canada had been prepared to accept U.S. direction of continental defences in early plans which assumed a defeated Britain and a retreat to Fortress North America. But acrimony between the U.S. and Canadian governments was the result of Canadian refusal in 1941 to accept a U.S. proposal for American control of all forces in the Maritimes and British Columbia. Thus many Canadians assumed in the post-war years that Canadian control of Canadian forces in Canada represented "a basic and unalterable tenent of policy."[9]

The second concern resulted from the wartime presence of U.S. forces in Canada. There had been some unpleasant experiences. As Vincent Massey commented, "The Americans . . . have apparently walked in and taken over power as if Canada were unclaimed territory inhabited by a docile race of aborigines."[10] It is thus not surprising that the Advisory Committee on Post-Hostilities Problems recommended that in the future, "Canada should accept full responsibility for all such defence measures within Canadian territory as the moderate risk to which we are exposed may indicate to be necessary."[11] This proved, however, in the early post-war years to not always be possible. Canada could not bear the expense of the entire Pinetree radar systems, or of the DEW Line. Thus the latter was built and manned by the Americans, as were some Pinetree stations. Strict conditions were set, though, by Canada on the activities of these ground-based Americans. Furthermore, as had been seen, Canada retained national operational control of not only its own interceptors in Canada, but as well of the American interceptors which entered Canadian airspace.

It is clear, and perhaps somewhat surprising that by the mid-1950s the highest levels of the American military were aware of Canadian sensitivities. In the spring of 1954 Representative Sterling Cole of New York became a public advocate of the establishment of a joint Canada–United States command for continental defence. Sterling also privately wrote Secretary of Defense Charles Wilson, putting forth the idea. Wilson turned to the Joint Chiefs of Staff for guidance. The Chiefs advised the Secretary in June that "The desireability of considering the North American continent as a strategic entity for defence is recognized and organizational arrangements which would further this concept are under continuous study within the Department of Defense."

Nonetheless, a negative recommendation was in order: "Existing arrangements provide for the necessary close collaboration and afford greater simplicity and freedom of action than would be probable under an organization of the type proposed." The Chiefs reviewed for the Secretary the channels of Canada–U.S. defence co-operation, including the PJBD, the Military Co-operation Committee and direct command-to-command contact. Two problems would arise if an attempt were made to replace or complement these channels with a joint command. First, Canadian political authorities could be expected to impose restrictions:

> Our experience in military planning with the Canadians has been that Canadian military planners are unable to arrive at negotiated positions without agreement on a governmental level. A combined U.S.–Canadian command would in all probability be equally restricted, on the Canadian side. Thus a combined U.S.–Canadian command would not seem sufficiently effective to warrant the expense in money and personnel involved.

Secondly, the Chiefs warned the Secretary, the Canadians might want to link a North American command to NATO. Such linkage would force the disclosure of certain American plans to the Europeans and "might impose upon continental U.S. defence forces restrictions which would be militarily unacceptable."[12]

The Secretary took the Chief's warning to heart. Wilson poured cold water on the idea of a joint command when Representative Cole repeated his suggestion at a January 1955 session of the House Armed Services Committee. As Wilson put it, "Our Canadian friends have a parliament and they are a little fussy about their southern neighbor—whom they are fond of—but they don't want us to think that we're running their country."[13]

To the Royal Canadian Air Force and the United States Air Force, however, Canadian operational control was not a paramount concern. The two air defence establishments grew more intimate in the wake of the growing Soviet threat, the increased size of the continental forces deployed to meet that threat and the expanded authority military commanders were granted to cooperate. They came, thus, to see the air defence of the continent more and more as a joint effort, and less and less of one in which sovereignty should play a role. In the spring of 1954 the Commander of the RCAF Air Defence Command and the Commander of the USAF Air Defense Command set up a joint planning group, which established itself, with permanent staff at USAF/ADC headquarters. That fall the two commanders directed the joint group to draw up one plan for the air defence of the continent. The group found that the greatest deficiency was "that forces deployed to defend against attack from one direction (for instance from the North) are not now under one commander, which imposes severe practical limitations in day-to-day training and in our capability to conduct a properly co-ordinated air battle in case of actual attack."[14]

The conclusion was significant, for it precipitated action by the RCAF and USAF to secure the designation of such a commander, despite Canadian sensitivities and despite the wariness of the U.S. Joint Chiefs of

Staff and Secretary of Defense. In February 1955 RCAF and USAF personnel briefed the Canada–United States Military Study Group (MSG), a body charged with considering various aspects of continental air defence, on the conclusions reached by the joint planning group. MSG agreed to bring the matter of a single air defence commander to the attention of higher military authorities in both countries. In Canada, the Chiefs of Staff Committee was convinced by the report, and no doubt by the strong arguments made in its favor by the RCAF. The Canadian Chiefs approved the joint air defence command in principle, but did not submit the matter to the Cabinet for approval.[15]

The Chief of the Air Staff did float a trial balloon. On 2 June 1955 Air Marshal Roy Slemon told the Aviation Writers Association in Montreal that a joint Canada–United States air defence command was "inevitable."[16] The balloon was promptly shot down the next day when the Minister of National Defence, Ralph Campney was asked in the House of Commons about Slemon's remarks. Campney told the House that Slemon's statement was "not a declaration of government policy" and that "he was simply stating a trend in military thinking." Campney added, "I may say in relation to this matter that there has been no pressure by our neighbors [for the establishment of a joint command] and that our relations continue as they were when Secretary of Defence Wilson made a statement, I think it was in January, that we were working closely and harmoniously together and there was no need for any change."[17] Secretary Wilson himself chimed in a few days later, pointing out that Slemon was "on his own."[18]

In fact, though, Slemon was not entirely on his own, for the United States Air Force, in turn, pressed from its end. On 9 September 1955 the Director of Plans in USAF headquarters, acting on behalf of the Chief of Staff, sent General Earl Partridge, who was both Commander of USAF-ADC and Commander-in-Chief of CONAD a draft proposal for a joint command. "This action," wrote the Director, who was no doubt aware of the RCAF's enthusiasm for the project, but who was obviously not up to date as to the political climate in Ottawa, "anticipates the eventual receipt of a Canadian proposal for the integration of the air defense systems under a combined command."[19] General Partridge replied on 21 October that

> It is my firm conviction that the integration of the Canada–United States air defense system and the establishment of a combined command would permit deployments of Canadian and American forces which would provide much better air defenses for both Canada and the United States than would otherwise be possible. Therefore, I feel there is an urgent necessity for an early decision on this matter.

Partridge warned, though, that Canada could not be expected to take the first step, as USAF headquarters seemed to believe. Partridge revealed:

> Informal information that we have just received from Canada indicates that the Canadian Chiefs of Staff have no firm plans for presenting a combined air defense proposal to the Canadian Cabinet.

In fact, our informal information indicates that for various reasons this is somewhat of a dead issue in Canada at this time and that impetus for action on a combined command might have to come from the United States rather than from Canada. Canada does not look with disfavor on such an arrangement but is not prepared at this time to take the initiative in making the proposal.[20]

The implications of Partridge's memorandum are clear. The Canadian Chiefs of Staff Committee was in favor of a joint command, but due to Canadian sensitivities over operational control they were unable to present the matter to the Cabinet. However, if a joint command were presented to the Canadian government as an *American* proposal, the Chiefs could provide the necessary backing. "Informal information" consisted, no doubt, of material passed on by the RCAF.

The Chief of Staff of the U.S. Air Force, General Nathan F. Twining turned to the task of convincing his colleagues on the Joint Chiefs of Staff that such an American proposal was now in order. On 5 December 1955 he brought the JCS up to date on recent events, including the approval in principle by the Canadian Chiefs adding: "It was expected that, as a result of this action, the Canadian Chiefs would formally indorse the proposal and submit it to the Canadian Cabinet for consideration. However, to date this latter action has not been taken." Twining also shared with his colleagues the complete text of Partridge's "informal information" memorandum. He concluded that it was "opportune" for the Joint Chiefs of Staff to

a. Approve in principle the desireability for (sic) the establishment of a combined Canada–United States Air Defense Command.

b. Inform the Canadian Chiefs of Staff Committee, through its Chairman, of recognition by the Joint Chiefs of Staff of the necessity and urgency for combining the air defense systems of Canada and the United States into a single integrated combined command.[21]

Twining's fellow Chiefs were largely, though not entirely, convinced by the weight of the Air Force's arguments. They agreed on 16 December 1955 that operational integration of the two air defence forces was desirable, and that they were prepared to make a proposal to Canada. However, they remained convinced that "A Combined Canada–U.S. command is probably not acceptable to the Canadians at this time and should not be proposed."[22] There would have to be operational integration, but no command. In short, what the U.S. Chiefs were proposing was a commander, who would exercise operational control over continental air defences, but who would not have a formal command, in the sense that CONAD or Allied Command Europe existed as commands.

The distinction was not entirely tenable, and, as shall be seen, eventually faded. A commander would have a headquarters staff which would in practice, if not in legal form, resemble a command. Moreover there were commands in existence which only exercised operational control. Such was the relationship between, for example, CONAD and the U.S. Army's air defence forces. By avoiding the use of the word, though, the Chiefs could accomplish two goals. First, the proposal might be more palatable to Canadian politicians and public. It would underline the fact that only operational control would be integrated, not other military functions such as initial deployment of forces, training, and discipline, all of which would remain in national hands. Secondly, Canada could not suggest making the non-command into a NATO command.

On 18 January 1956 the U.S. Chiefs informed Secretary Wilson that they had concluded that "centralization of authority during peacetime for operational control" of U.S. and Canadian air defence forces was necessary, and requested permission to formally approach the Canadians.[23] Wilson agreed,[24] whereupon on 14 February 1956 the JCS caused a letter to be sent to General Charles Foulkes, Chairman of the Canadian Chiefs of Staff Committee. In it, the U.S. Chiefs outlined their conviction that integrated operational control, including "authoritative direction" was necessary, but carefully excluded integration of other military functions and avoided entirely the use of the word "command":

1. In their continuing study of military measures essential to the strengthening of the air defenses of the North American continent, the United States Joint Chiefs of Staff have considered the desireability of peacetime integration of operational control of the continental elements of the air defense systems of Canada and the United States, including the continental portions of the warning systems. As a result of their examination of this subject, the U.S. Joint Chiefs Staff have approved in principle the need for such integration of operational control.

2. The U.S. Joint Chiefs of Staff consider it would be desireable to reach agreement in such areas as the composition of subordinate forces, the assignment of tasks, the designation of objectives, and the authoritative direction necessary to accomplish the mission of providing air defense for Canada and the continental United States. Operational control in this respect would not include such matters as administration, discipline, international organization, unit training, and logistics.

3. In order that the necessary details of such an arrangement may be developed at an early date, the U.S. Joint Chiefs of Staff have directed the undertaking of immediate studies in this field. In taking this action the United States Joint Chiefs of Staff are fully aware that any studies made unilaterally on the subject in the United States will be without substance if not responsive to Canadian views.

4. With this in mind the U.S. Joint Chiefs of Staff request the views of the Canadian Chiefs of Staff committee on the desireability of integrating in peacetime the operational control of the continental elements of our two air defense systems. Suggestions on the procedure by which combined study and planning in that direction might be undertaken also are solicited.[25]

The Canadian Chiefs of Staff Committee caused a reply to be sent on 27 February:

> The Canadian Chiefs of Staff agree that it would be desireable to study methods of integrating in peacetime operational control of the continental elements of the air defence of North America. They suggest that this study be undertaken by an ad hoc group of U.S. and Canadian Air Force officers reporting separately. The Canadian Chiefs of Staff further suggest that to avoid raising delicate political problems the ad hoc group should limit their discussions and recommendation to the problems of operational control. They point out that this subject is very sensitive politically in Canada and that it is important there should be no leakage of informations regarding the proposed group or the subject of its discussions to the press.[26]

The American Chiefs, in turn, assured their Canadian counterparts on 27 March 1956 that "The political sensitivity of this matter is recognized by the U.S. Joint Chiefs of Staff and every precaution is being taken to preclude any leakage of information."[27] There was further discussion between the two sets of Chiefs about the composition and nature of the study group, but it was finally decided in June that the Military Study Group would organize an ad hoc study group.

Of course the overall result of any such study was a foregone conclusion. After all, the Chief of Staff of the USAF and the Chief of Staff of the RCAF (the latter publicly) were strong supporters of the designation of a single commander. The two air defence commands were in favor. The matter had already been studied by a joint RCAF/USAF planning group. Approval in principal had been granted by the U.S. Joint Chiefs of Staff and the Canadian Chiefs of Staff Committee. An arrangement would have to be worked out, though, which avoided the creation of a formal command. Thus to no-one's surprise the ad hoc study group and the MSG recommended that the two national air defence efforts be operationally integrated and authority over them be delegated to a "Commander-in-Chief, Air Defense, Canada–United States" (CINCADCANUS), who would report to both the U.S. and Canadian Chiefs.[28]

The U.S. JCS endorsed the report on 6 February, and final approval on behalf of the United States was granted by Secretary Wilson the next month.[29] In a private conversation on 4 March Foulkes told Twining that, as the latter later paraphrased it, "no difficulty was expected" in securing the necessary political approval in Canada, on the assumption that Deputy CINCADCANUS would be a Canadian. Twining assured Foulkes that the JCS certainly had that in mind, for the ad hoc study group had recommended such an arrangement.[30] The Minister of National Defence also informed the Americans that he expected the Cabinet Defence Committee to grant its approval at a meeting to be held on 15 March 1957.[31] Despite these military and ministerial assurances, difficulties soon arose. Jon B. McLin has outlined what occurred:

The item was placed on the agenda of (the Cabinet Defence Committee) . . . for its consideration on March 15 (but was) removed from the agenda and was not considered by the committee. To explain this to the Americans a mission was sent to Washington on March 24, to convey to the Joint Chiefs of Staff this account: while there was no disagreement about the substance of the agreement, the question at the moment was potentially a political one, and formal approval would be withheld until such time as it ceased to be so. Finally, on April 26 Campney notified the U.S. Joint Chiefs of Staff that the Canadian government's decision—which was not likely to be negative—would not be taken before June 15.[32]

The obstacle was the Canadian general election, scheduled for 10 June. "There is . . . every reason," writes McLin "to suppose that the Liberal government chose to postpone briefly the NORAD issue so as to deal with it in the less politically charged atmosphere after the election."[33]

As it turned out, of course, the Liberals were relieved of the task of approving the agreement, for they lost the election of 1957. Foulkes pressed the new Progressive Conservative Prime Minister, John Diefenbaker, and the Minister of National Defence, George Pearkes, hard to grant quick approval to operational integration. As Foulkes would later put it, "Unfortunately—I am afraid—we stampeded the incoming government with the NORAD agreement."[34] Foulkes put forth two arguments to Diefenbaker and Pearkes. First, the agreement had already been approved by the United States. It would be embarassing, and detrimental to the overall state of Canadian–American relations if Canada now demurred. Secondly, Foulkes claimed that the agreement was in fact fairly minor, being simply the next logical step in the evolution of Canada–U.S. defence co-operation. Foulkes would later publicly explain this point of view:

> There were no boundaries upstairs; and the most direct air routes to the United States major targets were through Canada. Therefore, air defence was to be a joint effort from the start. It is important to keep this point in mind: that the decision for a joint air defence was taken in 1946, not 1958 as some of the critics claim when discussing NORAD.[35]

Thus, as McLin puts it, "the advice upon which Diefenbaker was acting held that it was a relatively unimportant agreement."[36] Consequently Diefenbaker acted fairly swiftly in endorsing the agreement. In so doing he ignored the formal Canadian constitutional practice in these matters (in particular, approval by the Cabinet Defence Committee and by Cabinet) which the Liberals had been following, and simply authorized the Minister of National Defence to grant approval.

On 1 August 1957 the Minister and the U.S. Secretary of Defense announced the agreement. The statement, as might be expected, emphasized operational integration and that command would remain in national hands. In a bow to reality, or perhaps, just a slip, there was a reference to "an integrated command," though "integrated headquarters" was also used. The term CINCADCANUS was omitted:

> The Secretary of Defense of the United States, the Honorable Charles E. Wilson and the Minister of National Defence of Canada, the Honorable George R. Pearkes, announced that a further step has been taken in the integration of the air defense forces of Canada and the United States. (The two governments have agreed to the setting up of a system of integrated operational control of the air defense forces in the Continental United States, Alaska and Canada under an integrated command responsible to the Chiefs of Staff of both countries.) An integrated headquarters will be set up in Colorado Springs and joint plans and procedures will be worked out in peacetime, ready for immediate use in case of emergency. Other aspects of command and administration will remain the national responsibility. This system of integrated operational control and the setting up of a joint headquarters will be effective at an early date. This bilateral arrangement extends the mutual security objectives of the North Atlantic Treaty Organization to the air defense of the Canada–U.S. region.[37]

Within days, though, the two countries bowed to the inevitable and further blurred the distinction between an integrated headquarters and command. On 12 August 1957 General Partridge, who would become the new CINCADCANUS (Air Marshal Slemon was designated Deputy CINCADCANUS) complained about the awkwardness of that title and pointed out that there would be in fact a joint command located at Colorado Springs, given the presence of a U.S. Commander-in-Chief, a Canadian Deputy, and an integrated staff. His suggestion that his new operation be entitled the North American Air Defence Command was thus quietly approved by both governments.[38] The May 1958 exchange of notes between the U.S. and Canada, which formalized the agreement, squared the circle by referring to "the integrated headquarters known as the North American Air Defence Command."[39]

CONCLUSION

NORAD has always had the faint air of illegitimacy about it. That reputation resulted not only from the clumsy manner by which the Diefenbaker government granted Canadian approval and from the equally clumsy manner with which it defended its action during the NORAD debate in the House of Commons. It arises as well from first, the suspicion voiced by Minifie and others, that the military establishments of the two countries had entered into much too close an embrace, and secondly from the perception that the Canadian military provided the incoming Diefenbaker government with misleading advice as to the nature of the draft agreement.

At the time it was made, Minifie's assertion that NORAD originated in the U.S. Air Force, that it was rejected by the Joint Chiefs of Staff and by the Department of Defense, after which it was forwarded surreptitiously by the USAF and brought to fruition in Canada by the RCAF may have seemed a bit shrill. In retrospect, though, it can be said that Minifie's claim, while not entirely accurate, contained some substantial elements of truth.

Certainly there was a close relationship between the two air forces. At the heart of that relationship was the jointly perceived air defence mission and the ensuing intimate contact. The creation of a joint command was a pet project of both air forces, not primarily of the USAF, as Minifie implied. The RCAF/USAF planning group located at Colorado Springs endorsed the idea the Canadian Chiefs approved in principle, whereupon A/M Slemon attempted to prepare the public and the politicians. That attempt failed, whereupon the USAF Chief of Staff, General Twining, took the matter up in the Joint Chiefs of Staff. Twining was armed with "informal information" while the RCAF had passed on to Colorado Springs, indicating that the Chiefs of Staff were prepared to endorse operational integration, but only if the Americans acted first.

Minifie was certainly correct in asserting that the U.S. Joint Chiefs of Staff were opposed to the creation of a joint command. That opposition, however, was not based on hostility to the concept itself, but rather on scepticism as to whether it could be sold, in proper form, to the Canadians. The Chiefs, and the Secretary of Defense were aware of Canadian sensitivities concerning command and control, and were worried that a proposal for a joint command would lead to the type of arrangement which would be unacceptable to the United States: a command hemmed in by limitations, and linked to NATO. Thus in 1954 the Chiefs flatly rejected Representative Cole's proposal. Once the USAF convinced the JCS in 1956 that the Canadian government just might go along, opposition by the U.S. Chiefs lessened, to the point where they were prepared to propose to Canada a commander without a command. The USAF did not, as Minifie implied, step around the JCS and Secretary of Defense to bring NORAD into existence. Rather the Chiefs and Wilson were signed on, once military logic and political assessment led in the same direction.

That close interaction between agencies, other than foreign ministries, of two different states has, of course long ceased to be even mildly surprising. Extensive studies of transgovernmental behavior, especially in the case of the Canada–United States relationship, have been undertaken. "Transgovernmental" as defined by Robert O. Keohane and Joseph S. Nye, Jr. "refers to direct interaction between agencies (governmental subunits) of different governments where those agencies act relatively autonomously from central governmental control."[40] That the NORAD concept was developed and championed by the RCAF and USAF, acting transgovernmentally, should not in itself be alarming. To expect Canada–U.S. lines of communication to pass solely through diplomatic channels is, at best, less than realistic.

However, the question remains whether in the manner of developing and supporting NORAD the two military establishments deprived Canadian political authorities of the opportunity to make a proper decision based, in particular, on access to all the pertinent facts. The Liberal government was misled by the military on one aspect: that the proposal for operational integration was American in its origins. Clearly both the RCAF and USAF wanted integration. The Canadian Chiefs of Staff Committee had approved it in principle. Indeed, USAF headquarters in 1956 was expecting to be called upon to react to a Canadian proposal until Partridge passed on his "informal information" from Ottawa. Here, too, the RCAF committed a technical sin: it transmitted information concerning the disposition of its military superiors and political masters.

There were thus two transgovernmental transgressions, the passing on of information and the orchestration of an "American" proposal to the Canadians. It still cannot be said, however, that the Liberals lacked the necessary information on which to make a judgement. They may not have known how the proposal came to be made by the U.S. JCS. They were not, though, led to believe that the proposal was American *in nature*, as opposed to one desired by both the Canadian and American military. The Canadian military did not argue that Canada had to accept integration in order to satisfy the Americans. They argued, rather, that there were strong strategic reasons for creating an integrated headquarters: First, an active air defence of Canada was necessary, which Liberal governments had agreed to since 1948. Secondly, co-operation with the Americans was a necessity in order to best utilize air defence resources. Having accepted these premises, and being aware of the growth in Soviet capabilities, the St. Laurent government could logically accept operational integration. The government was fully aware, though, of the significance of the proposal insofar as it touched on the sensitive issues of national operational control, for the Liberals were unwilling to deal with it until they were safely returned to power.

The same awareness had not dawned on John Diefenbaker. It is perhaps surprising that Diefenbaker did not grasp the delicacy of the issue, for he had, if anything, well-developed political instincts. However he was new to power and called upon to make a judgement in an area, national defence, which was to him, given his background, fairly arcane. Thus he relied upon the advice tended to him by the military. As James Eayrs has put it, "The decision to create a North American-Air Defence Command owed much to the Chairman of the Chiefs of Staff Committee, General Charles Foulkes, turning his powers of persuasion upon inexperienced ministers immediately after the change of government."[41] Conant was not especially charitable, though nonetheless accurate when he pointed out that "in retrospect, it seems clear that neither he (Diefenbaker) nor his principal cabinet members understood the full political import of placing integrated Canadian and U.S. forces under a U.S. commander-in-chief."[42]

Foulkes was certainly faultless in asserting that from a military point of view, operational integration of the two air defence systems under a single commander-in-chief was the next logical move in a series stretching back to 1946 when joint post-war defence planning began. By 1957 co-operation between the United States and Canada had progressed to the point where interceptors, in an emergency, could enter the airspace of the other partner in pursuit of hostile aircraft, and could be temporarily deployed to the other partner's bases. There was a full exchange of radar data. There was contiguous radar coverage in many areas along the border. There was constant communication between CONAD and RCAF ADC, including a joint planning group at CONAD headquarters in Colorado Springs. Militarily the next step would be to designate one commander who could use the authority to move interceptors across the border as part of a continent-wide strategy. Foulkes could also point out that Canada would only be placing its forces under the control of a commander with the integrated headquarters, not under an integrated command.

It was, still, a step across an important political threshold. There can be no doubt that Foulkes knew this when he advised Diefenbaker that operational integration was a fairly minor affair. Foulkes was certainly aware of the history of Canadian political sensitivity. After all, the Canadian Chiefs had cautioned the U.S. Chiefs of this very point, writing them on 27 February 1956 that the subject of operational control "is very sensitive politically in Canada." Foulkes knew as well that the more experienced Liberals had approached the issue with a good deal of caution. Had he been more restrained in his advice, in all probability the Conservative government would have taken the traditional constitutional steps which the Liberals were preparing when they were defeated, and most of the furor which erupted over the governments' rapid granting of Canadian approval would not have occurred.

On balance, thus, the transgovernmental transgressions committed by the two military establishments were fairly minor compared to those committed by the Canadian military alone. NORAD deserves its air of illegitimacy, for aspects of its conception were sinful and it was rushed into birth. However, it deserves to be viewed as other illegitimate offspring properly are. The sins of the parents can be condemned, and steps can be taken to prevent their re-occurring. The child must be, in fairness, assessed on its own merits. Thus in judging NORAD one must determine whether an integrated headquarters or joint command with operational control over the air defences of both Canada and the United States was necessary in 1957 and whether it remains necessary today.

NOTES

1. Melvin Conant, *The Long Polar Watch: Canada and the Defense of North America* (New York: Harper and Brothers, 1962), 82.

2. James M. Minifie, *Peacemaker or Powdermonkey: Canada's Role in a Revolutionary World* (McClelland and Stewart, 1960), 98–99.

3. Ibid., 93

4. Conant, *Canada and the Defense of North America*, 83.

5. *Nineteen Years of Air Defense* (NORAD Historical Reference Paper No. 11) (Colorado Springs: Directorate of History, Headquarters NORAD, 1965), 34.

6. Acheson to Truman, 5 March 1952, Files of the American section, PJBD, Arlington.

7. The text of Recommendation 53/1 has not yet been declassified. The above account is based on secondary sources, including CONAD

operational plans, and interviews with Air Force officials.

8. C.P. Stacey, *Arms, Men and Governments* (Ottawa: Information Canada, 1970), 203.

9. Conant, *Canada and the Defense of North America*, 80.

10. Quoted in James Eayrs, *In Defense of Canada: Peacemaking and Deterrence* (Toronto: University of Toronto Press, 1972), 350.

11. Advisory Committee on Post-Hostilities Problems, "Post-War Canadian Defence Relationship with the United States: General Considerations," 23 January 1945. Files of the Department of External Affairs, Ottawa.

12. Memorandum, JCS to Secretary of Defence, 11 June 1954; approved in JCS 1541/94, 11 June 1954, "Proposed North American Continental Defense Organization," National Archives, Modern Military Branch, Record Group 218, Records of the Joint Chiefs of Staff (hereafter cited as NA.218.JCS).

13. US Congress, House, Committee on Armed Services, *Hearings on Sundry Legislation Affecting the Naval and Military Establishments* 84 Congress, 1st Session, 1955, 1:223–24.

14. Quoted in *Nineteen Years of Air Defense*, 48.

15. *Chronology of JCS Involvement in North American Air Defence* (Mimeo, Historical Division, Joint Secretariat, Joint Chiefs of Staff, 30 March 1976), 59.

16. Quoted in Jon B. McLin, *Canada's Changing Defense Policy 1957–1963* (Baltimore, MD: Johns Hopkins University Press, 1967), 39.

17. Canada, House of Commons, *Debates* (1957–1958) 2:1060–1061.

18. Quoted in McLin, *Canada's Changing Defense Policy*, 329.

19. Major General Richard C. Lindsay, Director of Plans, DCS/O, Hq USAF to General Earl Partridge, Commander-in-Chief, CONAD, 9

September; reprinted in Enclosure C to JCS 1541/102, 5 December 1955, "Report by the Chief of Staff, US Air Force to the Joint Chiefs of Staff on a Combined Canada–United States Air Defense Command," NA.218.JCS.

20. Partridge to Chief of Staff, USAF, 21 October 1955; reprinted in Enclosure C to JCS 1541/102.

21. Ibid.

22. "Note to Holders of JCS 1541/102," 16 December 1955, JCS files.

23. Memo, JCS to Secretary of Defense, 18 January 1956. Files of the office of the Secretary of Defense, The Pentagon.

24. Memo, Deputy Secretary of Defense to JCS, 10 February 1956. Files of the office of the Secretary of Defense, The Pentagon.

25. Secretary, JCS to Canadian Chiefs of Staff Committee (SM-126-56), 15 February 1956, NA.218.JCS.

26. Chairman, Canadian Joint Staff (Washington) to Secretary, JCS, 27 February 1956, reprinted in JCS 1541/104, 27 February 1956, "Note by the Secretaries to the JCS on Integration of Operational Control of the Continental Air Defenses of Canada and the United States During Peacetime," NA.218.JCS.

27. Secretary JCS, to Chairman, Canadian Chiefs of Staff Committee, 27 March 1956, NA.218.JCS.

28. The US Defense Department has refused to declassify the ad hoc group's report. It is partially summarized in the declassified "Eighth Report of the Canada–U.S. Military Study Group, 19 December 1956; reprinted as Enclosure C to JCS 1541/112, 25 February 1957, "A Memorandum by the Chief of Staff, U.S. Air Force, on Integration of Operational Control of the Continental Air Defenses of Canada and the United States in Peacetime," NA.218.JCS.

29. JCS decision on JCS 1541/112, 6 February 1957. NA.218.JCS, Memo, Secretary of Defense to JCS, 16

March 1957. Files of the office of the Secretary of Defense, The Pentagon.

30. JCS 1541/114, 8 March 1957, "Memorandum by the Chief of Staff, U.S. Air Force, for the Joint Chiefs of Staff on Integration of Operational Control of the Continental Air Defenses of Canada and the United States in Peacetime," NA.218.JCS.

31. McLin, *Canada's Changing Defense Policy*, 40.

32. Ibid.

33. Ibid.

34. Canada, House of Commons, Special Committee on Defence, *Minutes of Proceedings and* Evidence (1963), 510.

35. Charles Foulkes, "Canadian Defence Policy in a Nuclear Age," *Behind the Headlines* (May 1961), 5.

36. McLin, *Canada's Changing Defense Policy*, 45.

37. *U.S. Department of State Bulletin* (19 August 1957), 306.

38. "Chronicle of JCS Involvement in North American Air Defense," 76; *Nineteen Years of Air Defense*, 49–50.

39. *U.S. Department of State Bulletin* (9 June 1958), 979.

40. Robert O. Keohane and Joseph S. Nye, Jr., "The Complex Politics of Canadian–American Interdependence" in *Canada and the United States: Transnational and Transgovernmental Relations*, ed. Annette Baker Fox, et al. (New York: Columbia University Press, 1976), 4.

41. James Eayrs, *The Art of the Possible: Government and Foreign Policy in Canada* (Toronto: University of Toronto Press, 1961), 97.

42. Melvin Conant, "Canada's Role in Western Defence," *Foreign Affairs* (April 1962), 436.

PEACEKEEPING AND CANADIAN DEFENCE POLICY: AMBIVALENCE AND UNCERTAINTY*

ROD B. BYERS

○

Since the 1950s, peacekeeping has been acknowledged as a cornerstone of Canadian foreign and defence policies. The role has been officially designated as a priority for the Canadian Armed Forces (CAF); and, over the years, politicians, especially the late Lester Pearson, have vigorously advocated strengthening the international peacekeeping aspects of the United Nations. Canada has participated in all UN peacekeeping operations and has been one of a small minority of countries which has officially designed troops for such purposes. On the military side, the CAF have incorporated peacekeeping into their ethos, planning, training, equipment, and resource allocations. All these factors suggest a longstanding and serious commitment on the part of successive Canadian governments. While the reasons underlying this commitment are complex and numerous, it has often been argued that peacekeeping is a role that a middle power such as Canada can undertake in order to make a meaningful contribution to international peace and security.

The foregoing summary, while oversimplified and abbreviated, typifies the stereotype of the Canadian peacekeeping record.[1] On the surface, the overall record appears exemplary and, in many respects, the Canadian case can serve as a model for other states in the international system who wish to undertake the peacekeeping role. Yet this conclusion should only be drawn if it is understood that the Canadian experience has been fraught with uncertainty, ambivalence, and difficulty.

... In terms of officially declared policy, the fortunes of peacekeeping have hit both the top and the bottom of defence priorities. While this may

---

* H. Wiseman, ed., *Peacekeeping: Appraisals and Proposals* (Toronto: Pergamon Press, 1983), 130–56.

reflect the reality of international politics and the international environ-ment, it could also reflect uncertainty and ambivalence in the formulation of the country's foreign and defence policy objectives.

Needless to say, a relationship exists between declaratory policy and the implementation of such policy by the CAF and the Department of External Affairs. In the case of peacekeeping–training, the shifts in policy emphasis have complicated planning, training, equipment purchases, resource allocation, and personnel requirements. From the perspective of the CAF, it should be remembered that the composition of the Canadian contribution has varied with each operation and that the government's offer to supply integrated military units has not been taken up by the UN. One result has been the cannibalization of military units in order to supply the required number of communication and logistics specialists. A by-product of this type of arrangement has been a partial loss of expertise and opera-tional capability within the CAF land forces. Furthermore, it is not clear that senior military personnel attach the same degree of importance to peace-keeping as do their political masters. . . .

THE POLICY DIMENSION

. . . [Prior] to the mid-1950s, the Canadian government had not clearly artic-ulated which foreign policy objectives and interests were being served by peacekeeping; and no official government statement explicitly linked peace-keeping to other foreign and defence policy instruments. Yet, in fairness to the Canadian government, it should be remembered that, in the immediate postwar era, peacekeeping was not a major concern as attention was first focused on collective security and subsequently on collective self-defence via NATO, once it became clear that the UN was unable to fulfill its collec-tive security mandate. The contemporary regime of peacekeeping began to emerge only in early 1949 with the formation of the UN military observer group in India–Pakistan (UNMOGIP). Canada acceded to the UN request in the case of UNMOGIP but did so with minimum publicity and without declaring how participation related to Canadian foreign policy objectives. In fact, from 1949 through 1955, peacekeeping rated no mention in the annual reports of the defence department; and government policy remained unclear. During this period, the thrust of Canadian military activity focused primarily on NATO and participation in the Korean war. To some extent, the genesis of Canada's willingness to earmark and allocate military person-nel for UN duty on a continuing basis can be traced to the Korean war, but this was within the context of the 1950 "Uniting for Peace" resolution of the UN General Assembly which was deemed to fall within the framework of collective security. Initially, therefore, peacekeeping was a limited and rela-tively unimportant aspect of postwar Canadian foreign policy, and it is probably correct to argue that "the policy of the Canadian government toward the peacekeeping initiatives of the United Nations can only be described as uninspired in the early 1950s."[2]

Despite the undifferentiated approach to early peacekeeping activities, successive Liberal governments, over time, attached increasing importance to the role as an instrument of Canadian foreign policy. Here, the preferences and impact of Lester Pearson, as Secretary of State for External Affairs, influenced the actions of the Liberal government; and, as early as 1954, a pattern of continued participation emerged as the norm. For example, despite reservations over possible participation in the International Commission for Supervision and Control (ICSC) in Indochina, it has been argued that "there never had been a moment's doubt that this was an obligation that Canada had to accept."[3] During this early period, however, it should be remembered that participation required few personnel and only limited resources. In other words, a capability factor to perform the role was never a serious consideration. Nevertheless, with active Canadian involvement in early UN operations and the ICSC, the Liberal government of the day should have formulated a policy perspective which related peacekeeping to other defence roles and priorities. The lack of explicit objectives meant that Canadian participation in UNEF I in 1956 was undertaken without the benefit of a clear policy framework and with inadequate understanding of how large-scale participation would affect the future requirements and needs of the CAF. Unfortunately, UNEF I tended to become the "model" for Canadian peacekeeping activities and, for some time, constituted the norm for future operations. Yet, in actuality, Suez turned out to be the exception rather than the rule.[4]

After the 1957 general election, the Conservative government of John Diefenbaker made no immediate attempt to clarify the policy dimension. Observers generally argued that the Conservatives would be less receptive to peacekeeping than the Liberals, even though in principle an all-party agreement had emerged with respect to Canadian participation in UNEF I. Yet, the Diefenbaker government responded positively when asked to participate in the 1958–59 Observation Force in Lebanon (UNOGIL); and it was during this period that Canada officially earmarked a military unit as a standby force for possible UN operations. Such actions were entirely consistent with past Liberal practice.

In 1959, the Conservatives made a limited attempt to specifically link military activities to broader foreign and defence policy objectives. *Defence, 1959* stated that Canada would provide military forces for the defence of North America, for NATO in Europe and the North Atlantic, and for "the United Nations to assist that organization in attaining its peaceful aims." The annual report went on to outline the existing level of Canadian participation, but made no reference to future operations. Neither did the report explicitly address itself to the relationship between peacekeeping and the other major defence roles. Nevertheless, the Conservative government implied that NATO constituted the preeminent defence activity. This view remained consistent throughout the Diefenbaker era, despite a sizeable commitment to the Congo operation (ONUC) in 1960. On balance, the Conservatives agreed to participate in peacekeeping operations when the opportunity arose, but did not place particular emphasis on the role. It should be remembered, however, that the issue of nuclear weapons split the government and tended to push other foreign and defence issues into the background during the early 1960s.[5]

Prior to 1964, therefore, the lack of a clearly defined policy dimension for peacekeeping allowed Canadian participation to be interpreted from two rather different perspectives. One view held that peacekeeping could be perceived as a separate role in its own right to be pursued when the opportunity arose. Within this context, peacekeeping was accorded an unduly elevated status, almost a foreign policy objective, which enhanced Canada's stature within the international system. Alternatively, peacekeeping could be considered as one of a range of foreign policy instruments to be exercised as appropriate. In this case, it was considered less significant than the main defence commitments undertaken within the NATO and NORAD context. As a generalization, it could be argued that federal politicians tended to be motivated by the peacekeeping as "objective" perspective, while senior members of the armed forces were more concerned with matters of function and priority. Under these circumstances, it was not surprising that uncertainty from the policy perspective seemed the order of the day.

The Liberals formed the government after the 1963 general election, and it was widely assumed that Canada's foreign and defence houses would be put in order. On the defence side, the 1964 *Defence White Paper*[6] constituted an ambitious, and to that date most complete, attempt to articulate Canadian defence policy within the domestic and international environments, and then relate military roles to broader policy objectives. The white paper properly started with the premise that defence policy could not be divorced from foreign policy and listed three major objectives: "To preserve the peace by supporting collective defence measures to deter military aggression; to support Canadian foreign policy including that arising out of our participation in international organizations; and to provide for the protection and surveillance of our territory, our airspace and our coastal waters." In order to fulfill these objectives, the government agreed on the following set of priorities:

1. Forces for the direct protection of Canada which can be deployed as required.

2. Forces in-being as part of the deterrent in the European theatre.

3. Maritime forces in-being as a contribution to the deterrent.

4. Forces in-being for UN peacekeeping operations which would also be included in 1 above.

5. Reserve forces and mobilization potential.

Here, UN peacekeeping was ranked fourth, but then multi-tasked with forces allocated for the defence of Canada. The overall effect and impression was to upgrade peacekeeping in comparison to the priority allocated the role by the Conservatives.

The upgrading of peacekeeping was based, in good part, on the assumption that Third World "instability will probably continue in the decade ahead and called for containment measures which do not lend themselves to Great Power or Alliance action. The peacekeeping responsibilities devolving upon the United Nations can be expected to grow correspondingly."[7] In general, the Pearson government adopted the view that instability existed in the international system, that deterrence should be maintained across the entire range of the force spectrum, and that Canada could make a meaningful contribution at the lower end of the spectrum. Thus, the white paper claimed "it is essential that a nation's diplomacy be backed up by adequate and flexible military forces to permit participation in collective security and peacekeeping, and to be ready for crises should they arise." Based on this set of factors, the white paper stated that "It is the policy of the government, in determining Canada's force structure for the balance of the decade, to build in maximum flexibility. This will permit the disposition of the majority of our forces in Canada where they will be available for deployment in a variety of peacekeeping activities." At the same time, however, the white paper clearly indicated that Canada's main defence contribution would continue to be collective self-defence via NATO.

On balance, the Pearson government adopted the view that the peacekeeping role complemented other military activities, including NATO. Thus, the stated objective of reequipping the CAF in order to increase mobility and flexibility was considered compatible across the entire range of Canadian defence activities. Major reequipment expenditures for the army were to produce a mobile force with an air-sea lift capability for immediate deployment outside Canada. This objective dovetailed with the increased emphasis placed on peacekeeping and assumed a high degree of compatibility between existing NATO commitments and future peacekeeping operations. Both government and senior military personnel seemed to share the view that compatibility between the two types of commitments could and should be maintained.[8] This, of course, was based on the premise that the CAF had sufficient manpower to fulfill NATO commitments and concurrently participate in peacekeeping operations.

For peacekeeping, the 1964 *Defence White Paper* represented the high water mark in terms of declared government policy. According to one observer: "The white paper is a milestone in the development of Canada's peacekeeping policies, not because it marks the beginning of a radically new policy, but because it indicates the importance that the government attaches to peacekeeping responsibilities and to the necessity to integrate defence and foreign policies."[9] In more practical terms, the government set out a framework within which the defence department could undertake planning, training, and, to a lesser extent, equipment acquisition as well as budgetary allocations. This clearly indicated that, in policy terms, peacekeeping was deemed as one instrument, albeit one of the more important instruments, to support the country's foreign and defence policy objectives. Needless to say, in principle, the articulation of the relationship between objectives and goals constituted a major step in the right direction.

If 1964 represented a peak, then 1971 represented official disillusionment over the utility and future prospects for peacekeeping. By this time, Mr. Trudeau had been prime minister for three years and, to some extent, his personal stamp had been placed on the country's foreign and defence policies. At the same time, *Defence in the 70s*[10] could, with justification, give examples which pointed to the demise of peacekeeping. The rather ignominious withdrawal of UNEF I from the Middle East in 1967, the inability of the UN to reach an agreement on a general peacekeeping formula as well as the problems of financing peacekeeping operations, the ineffectiveness of the ICSC in Indochina, and the presumed reduced future demand for peacekeeping represented factors which caused the government to reevaluate the country's commitment to the role. In addition, the inability to move from peacekeeping to peacemaking contributed to a sense of frustration in government circles. The 1971 white paper argued that "the scope for useful and effective peacekeeping activities now appears more modest than it did earlier, despite the persistence of widespread violence in many parts of the world." Based on this set of premises, peacekeeping was deemed the least important defence priority as the Trudeau administration allocated greater priority to the protection of Canadian sovereignty, to the defence of North America, and to NATO in that order. In effect, this set of declared priorities reallocated the peacekeeping role to its pre-1964 position vis-à-vis other defence activities. Yet, the underlying basis of mobility and flexibility which had been outlined in the 1964 white paper was retained as the government stated its intention "to maintain within feasible limits a general purpose combat capability of high professional standard within the Armed Forces, and to keep available the widest possible choice of options for responding to unforeseen international developments." More specifically, the commitment for a standby peacekeeping battalion was retained and personnel continued to receive training for such operations. In addition, the necessary linkages between peacekeeping and other defence roles remained a key element in the 1971 white paper. That is, the practice of relating defence activities to broader foreign and defence policy objectives remained central.

Events since the publication of *Defence in the 70s* have tended to contradict the premises of the white paper as the scope for peacekeeping operations expanded during the mid-1970s. Canadian military personnel became involved with the abortive and short-lived ICSC in Vietnam, with UNEF II, UN Disengagement Observer Force (UNDOF) and UN Interim Force in Lebanon (UNIFIL) on the Syrian–Israeli border. This rather abrupt turn of events had to be taken into account when the defence department undertook its force structure review in late 1974. Upon completion of the first phase of the review in November 1975, the defence minister reaffirmed in the House of Commons the priorities of the 1971 White Paper and stated that "the structure of the Canadian Armed Forces will provide up to 2000 personnel to be available for United Nations peacekeeping purposes at any one time."[11]

In effect, the demand for peacekeeping during the mid-1970s meant that the Trudeau administration remained as, or even more, committed than had the Pearson government during the 1960s. From the policy perspective, the major difference was one of emphasis as the Trudeau government consistently maintained a more skeptical attitude toward peacekeeping. This clearly affected perceptions of the utility of peacekeeping. Here it could be argued that the policy was clear, the practice was clear, but the outcome was one of official ambivalence. Increasingly, therefore, during the late 1970s, members of the Trudeau administration adopted the position of reluctant participation—particularly in the case of Cyprus. Statements by spokesmen for the government of Joe Clark indicated that the Conservatives would retain the general policy posture of the Trudeau government on the question of peacekeeping even though no definitive position had been adopted by the end of 1979. To some extent, the current peacekeeping policy of the Canadian government seems to have come full circle to that of the early 1950s. Consequently, it became incumbent upon the newly elected Conservative government to clearly formulate its policy posture regarding the peacekeeping role.

But before that could happen, the Clark government was defeated and Trudeau returned to power. It was this government, which, somewhat reluctantly, acceded to the secretary-general's "insistence" that Canada contribute to UNIFIL—despite misgivings about the viability of the operation and the strain on logistical manpower in the CAF.

On an informal basis in the summer of 1981, the United States "sounded out" Canadian attitude toward participation in the non-UN multilateral peacekeeping force in the Sinai following the final and complete Israeli withdrawal from Egyptian territory according to the Camp David Accords. Middle East foreign policy concerns and the non-UN character of the operation strongly inclined Canada not to participate; however, because of the "informality" of the request, no formal refusal was made.

THE UTILITY OF PEACEKEEPING

The degree of uncertainty and ambivalence toward peacekeeping from the policy perspective can, in part, be explained by the difficulties which face any nation-state in terms of the interface between the formulation of declared policy and its actual implementation. Obviously, the policy process and conflicting priorities—both domestic and external—have had an impact on the ability of successive governments to implement declared policy. In the sphere of international relations, however, the major variables which affect policy implementation are often outside the control of individual governmental decision makers. Peacekeeping is a good example of this situation. Obviously, Canada and other peacekeepers have a considerable degree of latitude concerning the first order decision—that is, whether or not to agree in principle with the peacekeeping role as a foreign policy instrument. In terms of first order decisions, the policy discussion can be explicitly (and possibly rationally) articulated vis-à-vis other foreign and defence activities. Both the Pearson and Trudeau administrations attempted this process. Yet, the second order decisions, including the terms of commitment to a specific operation, tend to be outside the control of the peacekeeper. Not surprisingly, therefore, the policy dimension becomes difficult, at times impossible, to implement as originally anticipated. Under these circumstances, skepticism toward the utility of peacekeeping increases. Within this context, Canadian views could express common themes and concerns perceived by other states that undertake a peacekeeping role.

One set of factors which have adversely affected perceptions of utility stems from the inability of members of the UN to reach agreement on the establishment, administration, command and control, and financing of peacekeeping operations. The ad hoc approach forced upon the secretary-general and the ambiguous nature of his powers in this area along with disagreement among members of the Security Council have complicated most peacekeeping operations. The inability of the Special Committee on Peacekeeping Operations (the Committee of 33), set up in 1965, to establish principles and guidelines for peacekeeping reflects the inherent difficulties which exist within the UN. These factors have been discussed at length from a Canadian perspective by academics, by the House of Commons Standing Committee on External Affairs and Defence (SCEAND), and in Canadian submissions to the Committee of 33, which all indicate, in varying degrees, dissatisfaction and disillusionment with the UN situation.[12] However, it should be remembered that UN difficulties have been a continuing feature of the organization when security issues have arisen. This has been the case for collective security in Korea, peace observation missions, or larger peacekeeping operations in the Congo and Middle East. Obviously, the utility of peacekeeping would be enhanced if the UN resolved some of the problems which continually arise. Yet, peacekeeping operations have been mounted despite opposition from certain members of the UN. Thus, while each operation has been undertaken within a context of uncertainty at the UN level, this, by itself, has not been crucial in terms of Canadian support or disillusionment.

States that have expressed a willingness to undertake the peacekeeping role have come to realize that they do so with very little control over when, where, or how they will be called upon.[13] The fulfillment of the role is dictated by events and actors which leave the peacekeeper in a reactive rather than an initiatory position. The terms and conditions of involvement depend in part upon the superpowers and in part upon the participants to the conflict. This is to point to the obvious. Yet, the implications for Canada of Nasser's rejection of the Queen's Own Rifles for UNEF I were not sufficiently appreciated by either the Diefenbaker or the Pearson governments. It was only after the sudden rejection of the UN force in 1967 that the Pearson government realistically appreciated the inability of the peacekeeper to influence the outcome of such operations. In many respects, this constituted a rude and sudden shock to Canadian officials and members of the attentive public.

In addition, as a result of superpower politics, other operational conditions circumscribe the role of the peacekeeper. In 1954, Canada joined the three-member Indochina commission of the ICSC as a surrogate of the West, and while government spokesmen played down this criterion, it remained a central consideration. This emerged even more clearly with the formation of the short-lived ICCS in Vietnam. With the formation of UNEF II in 1973, the concept of "balance" between NATO and Warsaw Pact members established a new precedent within the UN context. Here, Canada clearly represented NATO while Poland represented the Warsaw Pact. In this particular instance, co-operation between the Canadian and Polish contingents [was] relatively effective. Yet, the precedent adds a further complicating factor for future UN operations. These examples indicate the extent to which the superpowers have affected the utility of peacekeeping irrespective of the views and wishes of those states that explicitly relate their foreign and defence policies to the peacekeeping role.

These factors accounted, at least in part, for the increased circumspection adopted by the Trudeau government on the peacekeeping front. *Defence in the 70s* referred to operations which have "been severely hampered by inadequate terms of reference and by a lack of co-operation on the part of those involved. . . .

Based on these considerations, the reader could be forgiven if the conclusion were drawn that Canadian decision makers were ready to forego the peacekeeping role. Yet this is not the case. Data collected in 1975–76 via the Canadian International Image Study (CIIS) indicated that the Canadian Foreign Policy Elite (CFPE), composed of some 251 federal politicians, officials, and academics, almost unanimously (97.5 percent) supported continued participation in peacekeeping operations.[14] In fact, the degree of support for peacekeeping was more extensive than support for NATO (92.5 percent) or for NORAD (85.9 percent). Furthermore, when asked to rank order ten possible defence roles for Canada, the CFPE allocated the top position to peacekeeping ahead of sovereignty, surveillance, NATO, and NORAD to name the most obvious. In other words, on the surface, the commitment appeared deep-rooted. This did not mean, however, that Canada should become involved in all peacekeeping operations as only 13.2 percent of the CFPE supported the proposition that Canada should *automatically* volunteer troops whenever the UN establishes a peacekeeping operation. Thus, while support in principle exists, this latter finding indicated that a certain sense of caution pervaded the Canadian foreign policy community.

Nevertheless, there would appear to be a real dichotomy or almost schizophrenia in Canada concerning the utility of peacekeeping. On the one hand, the Trudeau government officially downgraded the role as an instrument of Canadian foreign and defence policies, and repeatedly questioned its utility. The Conservative government of Joe Clark adopted a similar posture. On the other hand, members of the foreign policy community almost unanimously supported continuation of the role. Clearly, then, it becomes essential to try to explain this seemingly anomalous situation.

Perceptions of utility have, in part, been a function of the extent to which operations contributed to system and subsystem stability, to conflict avoidance between the major world powers (especially the superpowers) and, to conflict management at the regional subsystem level. The linkages among these factors are obvious in the sense that the second and third clearly enhance stability at the system and subsystem levels. Not surprisingly, therefore, the CFPE, when asked to explain their support for Canadian participation, attached greater importance to the stability rationale (nearly one-third of the respondents) than to any other single factor.

Going beyond this general response category, however, only a small number (2.6 percent) specifically referred to conflict avoidance, and conflict management did not emerge as a distinct category for the decision makers. This finding was rather surprising as the conventional wisdom contained in the literature cites numerous references to the linkage between peacekeeping and conflict avoidance.[15] Furthermore, Canadian politicians have referred to this rationale on any number of occasions. . . .

Since peacekeeping forces are normally deployed after the outbreak of hostilities, the contribution to conflict avoidance at the subsystem level is severely limited; but the conflict management aspect has been a continuing justification for the peacekeeping role. For Canada, the 1964 *Defence White Paper* referred to problems in the Third World and argued that "instability will probably continue in the decade ahead and call for containment measures which do not lend themselves to Great Power Alliance. The peacekeeping responsibilities devolving upon the United Nations can be expected to grow correspondingly.". . . While the Trudeau government tended to be less enthusiastic, the 1971 White Paper employed much the same rationale as it argued that peacekeeping "helps to prevent the outbreak or spread of hostilities in other areas of tension, so that underlying political problems can be settled through negotiation or a process of accommodation, and so that the probability of Great Power conflict is minimized."

The arguments supporting the conflict avoidance function are fairly persuasive, but critics have rightfully pointed out that the conflict management function, while introducing an element of greater stability into a crisis situation, all too often reinforces the status quo—political and/or territorial—and thus draws out the peacekeeping role: "All too often the negotiations never took place or got nowhere. Peacemaking almost never followed peacekeeping. The result was a decay of world morale, a lessening of faith in the UN. The disillusionment was felt in Canada, too. Suez 1956 had stirred hope, but Suez 1967 destroyed it. Suddenly Canadians began to question why peace had not followed peacekeeping. . . . A sense of futility was very sharp."[16] The mid-1974 outbreak of hostilities in Cyprus contributed to a hardening Canadian position on the question of utility. In his September 1974 address to the General Assembly,[17] the Minister for External Affairs claimed that "fighting has taken place on an unprecedented scale because the long-smouldering political problem remained unresolved. Moreover, it has been demonstrated once again in Cyprus that without the agreement and co-operation of the disputants, the constructive role of a peacekeeping force is severely circumscribed." On the more general issue of participation in future operations the Minister pointed out that "Canadians are today becoming less inclined to accept in an unquestioning way the burden of participation. Their concern springs mainly from the fact that peacekeeping endeavours often seem to do no more than perpetuate an uneasy status quo. . . . It must be accompanied by a parallel effort on the political level, especially by the parties most directly concerned, to convert the temporary peace . . . into something more durable. If this is not done . . . governments will be less willing to respond to future requests for troops." This view has been increasingly shared within Ottawa circles and only a small minority (3.9 percent) of the

CFPE mentioned the peacemaking function. While in opposition, spokesman for the Conservatives urged the Liberals to withdraw from Cyprus unless there was some progress towards a peaceful resolution of the conflict. . . .

Yet unilateral withdrawal from an operation such as Cyprus would raise the question of how the action would be perceived by other actors in the international system. After all, if Canada acquired an "opting-out" reputation there would be greater hesitancy to make an initial request in the case of future operations. But, within this context, withdrawal from the ICCS may have raised some doubts in the international community. Withdrawal from Cyprus would certainly add to this particular problem. . . . Yet, there could be repercussions in terms of Canadian reliability and credibility in the eyes of some UN members; and, if withdrawal were later linked to further serious problems on the southern flank of NATO, Canada's position within the Western alliance could be undermined.

The reliability factor should not be dismissed out of hand, as nearly one-quarter of the CFPE supported participation on the grounds that Canada is perceived by other actors as possessing the necessary foreign policy attributes to perform this particular role. Thus, Canadian reliability, credibility, detachment, neutrality, and pragmatism emerged as attributes which enhanced the country's peacekeeping ability. Here, the argument becomes somewhat circular as these terms are perceived important because Canada has actively participated and done so in a reliable manner. Should the opting out and/or nonparticipatory routes be adopted, then this set of factors would clearly be less significant in the eyes of other states in the international community and thus no longer constitute cause for Canada to maintain the peacekeeping role. It might be pointed out, however, that in international politics linkages generally exist between various components of a state's foreign policy posture and positive aspects of that posture should not be jeopardized without due cause.[18]

Having made this observation, however, it should be pointed out that Canadian decision makers do not attach a great deal of importance to national self-interest factors when expressing support for peacekeeping. Only a small minority referred to peacekeeping as augmenting Canada's prestige (7.8 percent), international role (7.8 percent), and benefits to the CAF (5.2 percent). Indirectly, of course, positive foreign policy attributes are beneficial to Canada, but self-perceptions of attributes supposedly held by others hardly constitute national self-interest criteria. Thus, from the perspective of most Canadian decision makers, participation in peacekeeping operations is not particularly relevant in terms of direct self-interest. That is, the tangible benefits to Canada and to Canada's role and position in the international system are not explicitly linked to continued support for peacekeeping. This suggests that the role, despite its first-order ranking in terms of preference for defence commitments, could be downgraded without adversely affecting Canadian self-interest.

In part, the utility of peacekeeping can be linked to the "voluntarist" component of Canadian foreign policy.[19] That is, Canada participates in peacekeeping out of a sense of responsibility to the international community, and actions in this policy sphere constitute an "ought" for the policymaker.

Thus, over one in ten (13.0 percent) members of the CFPE argued that there is a "need" for peacekeeping in the international system, while an almost equal number (12.5 percent) stated that Canada has an obligation to participate. When coupled with those respondents who indicated that it is important to maintain the UN as an organization and support its objectives (7.8 percent) and that the moral factor should be taken into account (5.2 percent), the voluntarist rationale for peacekeeping emerged as quite significant. Unfortunately, the altruistic component of a state's foreign policy can often be circumscribed by events beyond its control. . . .

On balance, therefore, Canadian rationales regarding the utility of peacekeeping indicate two rather different trends. On the one hand, declared government policy and statements have, particularly during the 1970s, called into question the role of peacekeeping as a central foreign policy instrument in spite of continued participation. On the other hand, support for peacekeeping would appear to be solidly entrenched among members of the foreign policy community. Yet, reasons for support tend to be in the sphere of intangibles (voluntarism, perceived attributes) rather than more tangible factors (self-interest). There are two exceptions. The first is that of system and subsystem stability. As long as peacekeeping is perceived as making a contribution to international stability, its utility will be enhanced— provided there is also progress in terms of peacemaking. . . .

THE MILITARY DIMENSION

The military dimension is directly and indirectly affected by both the policy dimension and perceptions of the utility of peacekeeping. Policy objectives, statements, and priorities require translation into specific commitments, force levels, equipment, and training. Indirectly, the policy dimension affects the Canadian armed forces in terms of military professionalism, the relationship between defence roles, and multi-tasking. If the policy dimension is unclear or a low priority is attached to peacekeeping, coupled with a sense that the utility is questionable, then the emphasis placed on the military dimension by the armed forces is bound to be circumscribed. If the opposite conditions exist, then the degree of emphasis from the military perspective is bound to be considerably higher. The latter circumstances existed during the middle and late 1960s when peacekeeping was given a high profile and considerable attention. During this period, a number of military uncertainties remained beyond Canada's control, but these did not affect Canadian support for the role. However, during the 1970s and into the 1980s, a combination of the change in defence priorities and increased doubts regarding the utility of peacekeeping led to greater uncertainty within the CAF.

One of the more important military concerns is the question of professionalism[20]—that is, the extent to which peacekeeping is considered to be, or can be equated with, the raison d'etre of the military establishment. As in most Western industrialized countries, a study by the CAF on professionalism[21] argued that "the raison d'etre of a professional military force is to apply, or to threaten to apply force on behalf of the state and at the lawful direction of the duly constituted government of the nation." The report went on to state that this function "quite rightly dominates the thinking and outlook of the military profession" and "is reflected in its development processes . . . and leadership style." *Defence in the 70s* had adopted the same philosophy with its objective to retain a high general purpose combat capability. . . .

Obviously, attitudes toward peacekeeping are affected by the value system of the military. Nevertheless, it was surprising that DND respondents in the CIIS of 1975–1976 ranked peacekeeping as the fourth least important defence commitment, i.e., seventh out of ten. . . . Peacekeeping normally does not require the application of force. The operations in Cyprus, the Middle East, the Congo, and Vietnam could all be classified as quasimilitary, while the peace observation missions could probably be more correctly labeled as nonmilitary. The fact that peacekeeping units make no concerted attempt to match the armament and force levels of the disputants in an interposition situation only further emphasizes the quasimilitary nature of the role. While various Canadian peacekeeping units have had to resort to the use of force on occasion, the primary emphasis has been, and is, on interposition between opposing sides in a conflict.

It could be argued that the Congo operation deviated from the norm in that UN forces became involved during various parts of the operation in war-fighting situations. This could have accounted for the brief employment of the term "peace-restoring" during the late 1960s. At that time, the Chief of Defence Staff explained to SCEAND that the philosophy behind Mobile Command was to provide a force which could operate independently within a peace-restoring context.[22] If implemented, this would have fundamentally altered the philosophy behind peacekeeping as the application of force would have become more central to any operation. In reality, of course, this concept was never feasible and, for the most part, constituted wishful thinking on the part of some senior defence planners. Yet, this example was important as the desire to move in the direction of peace restoration indicated ambivalence within certain sectors of the CAF.

There must be a wider range of factors, however, which can explain the low rank allocated to the peacekeeping role by some members of the CAF. A related issue is the extent to which peacekeeping can be considered as complementary to other defence roles. In the main, senior military personnel have adopted the view that peacekeeping can be successfully undertaken by units and personnel expected to fulfill more traditional military roles. For example, a positive spin-off is deemed to result from the training and experience gained within the NATO context which then allows Canadian personnel to be peacekeepers par excellence. The SCEAND report on peacekeeping accepted this view and noted that "the normal professional training and discipline of Canada's regular forces accounts for the effectiveness they have shown in past operations."[23] To some extent, this claim has been substantiated by the Canadian experience. Yet constraints on the use of force, the type of equipment and arms, the operational environment of the small squad, the need to exercise personal diplomacy, and the need for noncommissioned personnel to make rapid, on-the-spot decisions tends to be at variance with some aspects of "normal" professional training for military type roles. The operational environment of NATO's central front is rather different than that of the Green Line in Cyprus—with the exception of the Canadian defence of the airport at Nicosia during the attack by Turkey in 1974.

In other words, the peacekeeper has to be aware of the nature of the operational environment and realize that the skills and training can differ from the traditional norm of military professionalism. In some instances, military units and personnel who are tasked with an internal security role by their government may, in the first instance, more effectively perform the peacekeeping role. This being the case, the Canadian example may not necessarily be applicable to other military establishments.

Other aspects of the complementary issue involve the organizational and structural implications of the relationship between peacekeeping and other defence roles. The prevailing view within the CAF has generally supported the argument that armed forces should be structured to perform military tasks and then adapted as necessary to meet the contingencies of peacekeeping. This position was most clearly articulated in 1967 by the then retired Chief of Defence Staff: "I do not believe we should ever design forces for peacekeeping only. . . . We should design forces for war as we know it and then adapt them to the peacekeeping role or any other role that happens to fall to them in line with Canada's National Policy."[24] During the debate over unification of the CAF in the late 1960s, some senior retired military personnel expressed the fear that the Liberal government's emphasis on peacekeeping would adversely affect the CAF's capability to adequately fulfill the NATO and NORAD commitments.[25] While this view was rejected by government spokesmen, the discussion served to emphasize that peacekeeping can be perceived as a competitor to traditional military roles and that any serious attempt to upgrade peacekeeping at the expense of NATO and NORAD would be opposed by segments within the CAF. . . .

As a generalization, the larger the size of a country's armed forces the more likely it is that peacekeeping can be undertaken without adversely affecting the capability to carry out other commitments and, at the same time, retain a reserve capability. In the Canadian case, force levels have steadily declined from 126 000 in 1962 to 80 000 in 1981. Over this period, the decline of some 37 percent has been quite drastic and, while all elements of the CAF have been affected, the reduction for the land environment (army) has been of a greater magnitude than on the air and sea side. For example, in 1962, with a force level of 52 000, the army constituted 41 percent of the total military manpower. By 1978, with 29 300 personnel, the land environment had dropped to 36.6 percent of the CAF. In 1977, the Liberal government announced that the CAF would be increased by some 4700 personnel to bring them to the 1974 level of 83 000. Yet, most of the increase was to be allocated to the support arms, and the infantry component would be limited to an increase of some 600 personnel. . . .

In order to cope with the variety of requirements and commitments, multitasking has become the norm, and nearly all segments of the CAF are now multitasked.[26] In principle, the concept constitutes a rational approach to defence planning and resource utilization. Furthermore, most military establishments rely upon multitasking in order to meet the range of commitments which support a country's domestic and foreign policies. However, should multitasking become excessive or be based on insufficient force levels, then the capability to carry out primary tasks will be adversely affected.

The CAF is currently at a force level and posture which cannot realistically meet the extent of multitasking deemed necessary by government requirements should a crisis occur. This issue was extensively reviewed in the mid-1970s within the context of the force structure review which utilized the roles of the 1971 defence white paper. Some 55 tasks were identified, and apparently all of these were retained for planning purposes. A sense of the extent of multitasking was ascertained from the annual reports of the defence department. For example, the 1974 annual report outlined approximately 24 separate tasks which were assigned to the six commands in Canada with a high of eight being assigned to maritime command. Even though some tasks are more important than others, conflicting demands on manpower and equipment can hardly be avoided. In the case of mobile command, in 1976, some 5000 of 17 600 personnel were attached to secondary units, including headquarters and training centres, while the three major combat groups accounted for only 12 600 personnel. Even though mobile command was restructured during 1977, the situation, as of 1982, has not been appreciably altered.

Most of the multitasking for the land element is the responsibility of the two combat groups and the Special Service Force. At present, the land element is multitasked in five areas: NATO, Canada/U.S. Defense (CANUS), peacekeeping, internal security, and national development. Within these five areas, multitasking can be either explicitly specific commitments, as in the case of NATO and peacekeeping, or unspecified commitments, as in the other three areas. In the former category, approximately 4500 personnel are currently serving overseas in Europe or on peacekeeping assignments, that is, nearly one-sixth of the total land force. Furthermore, the government has explicitly committed a Canadian air-sea transportable combat group to the NATO flanks. The remaining explicit commitment is to provide up to 2000 personnel for peacekeeping. Taken together, these commitments represent a further 12 percent of the total land force; and, if drawn primarily from mobile command, the manpower problem would become even more acute. In other words, multitasking has serious implications for the land environment. If the overseas commitments are called upon at the same time as a major unspecified commitment turns into a specific commitment, only two-thirds of the total land element remains available as a manpower pool.[27] Since the major combat arms constitute only one-third of the land environment, the difficulties of coping with multitasking become obvious. While peacekeeping is not a direct cause of this problem, it is part of the overall context, and the uncertainty of the demand for peacekeeping forces contributes to ambivalence regarding its role.

The ad hoc approach to the formation, composition, financing, and operationalization of peacekeeping operations has complicated defence planning. . . . However, the Canadian case deserves greater elaboration as a serious gap has existed between the types of prior commitment recommended by Canadian governments and the final composition of Canadian peacekeeping contributions. Dating back to UNEF I, the Pearson government had offered a major operational combat unit to the UN as the preferred contribution. In every case except Cyprus, such offers have been rejected with the result that the Canadian land forces for UN operations have been drawn from administrative and support personnel. The 1964 defence white paper noted this situation as "requests from the Secretary-General for assistance have been for specialists of various kinds, mainly from the Canadian army and the RCAF." However, the white paper later stated that training for peacekeeping would still be undertaken at the unit level even though the possibility of this type of deployment remained remote.

The 1971 defence white paper adopted much the same position as the government continued to commit a battalion group for UN standby even though it seemed unlikely to be deployed. This was the situation in October 1973 with the offer to commit the Canadian airborne regiment, in the Middle East. This contribution was not deemed essential by the secretary-general who responded with the request for a logistics component—air support, transport, communications, and other administrative and support personnel. In order to comply with this request, the CAF had to draw personnel from a variety of units and commands. This resulted in a diminution of support elements within the combat groups, as well as a reduction of administrative personnel elsewhere in the system. This type of situation served as a double source of frustration for the CAF. First, most personnel realized that peacekeeping training at the unit level would not result in unit deployment even though it could serve as a useful function for contingency purposes. Second, the major support units tend to be cannibalized which, in turn, would adversely affect the overall combat effectiveness of the major units should they be involved in other multitasking situations. Even without active deployment, the normal training and operational effectiveness of combat groups tends to be impaired as key support and administrative personnel are serving elsewhere. On balance, therefore, there are solid grounds for being concerned with the military dimension.

THE CANADIAN EXPERIENCE: SOME CONCLUDING OBSERVATIONS

... The Canadian case can be considered illustrative of the types of problems faced by most nation-states that have undertaken the role of peacekeeper. Policymakers in other states may consider it of some value to assess their situation and aspirations in light of the Canadian experience.

In terms of the policy dimension, at least four observations come to mind: (1) The articulation of policy and policy objectives lagged behind the practice of peacekeeping, i.e., while Canada participated in its first peacekeeping operation in 1949, the policy implications were not clearly and officially spelled out until the 1964 defence white paper. (2) The priority and relationship between peacekeeping as an instrument of foreign policy and other instrumentalities has often been difficult to ascertain. Despite serious attempts to explicitly develop a coherent framework covering the range of Canadian roles and instruments, it has been difficult to implement policy as anticipated. (3) Despite the long-standing commitment on the part of successive Canadian governments to the principle of peacekeeping, the degree of support has varied considerably. (4) It could be pointed out that, upon the outset of the 1980s, the Canadian position on peacekeeping exhibited some of the characteristics of the mid-1950s. That is, the role and position of peacekeeping vis-à-vis other defence activities and foreign policy instruments required clarification. . . .

In part, problems in the policy dimension are a function of the perceived utility of peacekeeping. Not surprisingly, perceptions of utility have varied over time, and declared government policy has reflected these changing perceptions. According to policymakers and observers, three types of factors have shaped perceptions: (1) those that are rooted in the international system, such as the impact of superpower politics, as well as the preferences of disputants to a conflict on the selection and implementation of peacekeeping operations; (2) the internal difficulties within the UN system on the financing, establishment, support, etc. of peacekeeping; (3) the impact of the domestic environment including the commitments and capabilities of the Canadian armed forces. Thus, on the surface, the Canadian foreign policy community strongly advocated support for peacekeeping as an instrument of foreign policy but stressed the less tangible benefits of peacekeeping. Even though considerable emphasis was placed on the need for international stability, Canadian support for peacekeeping could dissipate rather quickly.

Linkages between the policy dimension and perceptions of utility must be taken into account when the military dimension is assessed. Obviously, the applicability of the Canadian experience will vary from country to country depending upon the structure, capability, and professionalism of individual military establishments. However, from the perspective of the CAF, peacekeeping has been undertaken with a certain sense of ambivalence. First, the professional ethic of the CAF is not entirely compatible with the peacekeeping environment. Furthermore, in terms of capabilities, the CAF is at a force level which calls into question its ability to successfully participate in major peacekeeping operations and also fulfill commitments in other areas. Third, there have been problems in relating the military dimension to the policy priorities and activities laid down by the government. This is particularly true when priorities appeared to deviate from declared policy.

In conclusion, it might be noted that the difficulties and problems which have arisen for Canada should be seen as illustrative of a nation-state that has made peacekeeping one of its major instruments of foreign policy and, in the process, has attempted to link policy and practice in as rational a manner as possible. This being the case, imagine the difficulties and problems faced by states in the international system that have not undertaken the peacekeeping role with the same sense of dedication and commitment.

NOTES

1. Alastair Taylor, David Cox, and J.L. Granatstein, *Peacekeeping: International Challenge and Canadian Response* (Toronto: Canadian Institute of International Affairs [hereafter CIIA], 1968).

2. Granatstein in Taylor et al., *Peacekeeping*, 107.

3. Ibid., 109.

4. James Eayrs, *Canada and World Affairs, October 1955 to June 1957*, vol. 9 (Toronto: CIIA, 1959).

5. P.V. Lyon, *Canada and World Affairs, 1961–63*, vol. 12. (Toronto: Oxford, 1968). For an American interpretation see Jon McLin, *Canada's Changing Defence Policy, 1957–1963* (Baltimore, MD: John Hopkins University Press, 1967).

6. For references to the 1964 white paper see Canada, *White Paper on Defence* (Ottawa: Queen's Printer, 1964).

7. Ibid., 11, 12, and 21 for quotes in this paragraph.

8. See David Cox, "Canadian Defence Policy: The Dilemmas of a Middle Power," *Behind the Headlines*, vol. 27, nos. 5–6 (Toronto: CIIA, 1968).

9. See Cox in Taylor et al., *Peacekeeping*, 49.

10. For references to the 1971 white paper see Canada, *Defence in the 70s* (Ottawa: Information Canada, 1971).

11. Canada, House of Commons, *Debates* (1975), 9503.

12. See Henry Wiseman, "Peacekeeping: Debut or Denouement?" *Behind the Headlines* (1972), vol. 31, nos. 1–2; Canada, House of Commons, Standing Committee on External Affairs and National Defence (hereafter SCEAND), Subcommittee on Peacekeeping, 21 May 1970, no. 31.

13. One of the best studies on peacekeeping remains Alan James, *The Politics of Peace-Keeping* (New York: Praeger, 1969).

14. For the background and explanation of the project see R.B. Byers, David Leyton-Brown, and Peyton V. Lyon, "The Canadian International Image Study," *International Journal*, 32, 3 (Summer 1977).

15. For example see C.S. Gray, *Canadian Defence Priorities: The Question of Relevance* (Toronto: Clarke Irwin, 1972), 109; and Alastair Buchan, "Concepts of Peacekeeping" in *Freedom and Change*, ed. Michael Fry (Toronto: McClelland and Stewart, 1975), 17.

16. Jack Granatstein in *Canadian Forum* (August 1974), 18.

17. Canada, Department of External Affairs, *Statements and Speeches*, 25 September 1974.

18. The case can be made that the Trudeau administration made this mistake with NATO in 1969. For example see R.B. Byers "Defence and Foreign Policy in the 70s: The Demise of the Trudeau Doctrine," *International Journal*, 33, 2 (Spring 1978), 312–38.

19. See Thomas A. Hockin in Lewis Hertzman, John Warnock and Thomas Hockin in *Alliances and Illusions* (Edmonton: Hurtig, 1969), 95–136.

20. For an overview of this issue see R.B. Byers and Colin S. Gray, *Canadian Military Professionalism: A Search for Identity*, Wellesley Paper 2 (Toronto: CIIA, 1973); and Adrian Preston, "The Profession of Arms in Post-War Canada, 1945–1970," *World Politics*, 13 (1971), 189–214.

21. Canada, Department of National Defence, *Report of Study on Professionalism in the Canadian Forces*, 1972.

22. House of Commons, Standing Committee on National Defence (1967), 275. Also see Canada, House of Commons, *Debates* (1966–67), 14559–60.

23. SCEAND, *Peacekeeping Report*, 28.

24. Standing Committee on National Defence (1967), 2327, 2315.

25. Ibid., 275. For example, Lieutenant-General F. Fleury argued that "If you are going to have a Canadian defence force . . . for peacekeeping duties, and peacekeeping duties only, then I think perhaps you ought to have another favourable look at unification, because this might be the one circumstance that I can see where unification might be a real boon." (p. 1305).

26. It should be pointed out that there had been reductions in other commitment areas. In the mid-1960s Canada's NATO land commitment consisted of a 6500 man mechanized brigade group allocated to the European central front with a standby commitment of two brigade groups. This was renegotiated by the Trudeau government and now the European land commitment consists of a 2800 man mechanized battle group to be deployed on the flanks of NATO with a standby commitment of one combat group stationed in Canada for flank deployment. Unfortunately the Canadian land forces have not been equipped with sufficiently modern equipment to effectively offset deployments within the Warsaw Pact armies. The acquisition of 128 Leopard I tanks constitutes a partial attempt to rectify this situation, but the effectiveness of the Leopard I will soon decline with the new generation of main battle tanks being deployed by the USSR. Furthermore, the CAF must acquire more rapidly the next generation of precision guided munitions or the decline in combat capability will continue.

27. An example of the problems posed by multi-tasking occurred during the FLQ crisis in 1970 when twelve of the fourteen major combat units stationed in Canada had to be deployed for internal security purposes. Only two units stationed with 1 Combat Group in Western Canada remained uncommitted. If the crisis had coincided either with the request for a peacekeeping contribution of the size deployed in Vietnam or later with UNEF II the Canadian government would have been unable to comply. This scenario indicates the difficulties which could arise from unrealistic multi-tasking.

THE ARCTIC◇

W. HARRIET CRITCHLEY

○

The Arctic first became a region of strategic interest for Canada during the
Second World War. In 1946, Lester B. Pearson succinctly described this
change: "Not long ago this vast Canadian Arctic territory was considered to
be little more than a frozen northern desert, without any great economic
value or any political or strategic importance. . . . We know better now. . . .
The reason is obvious. The war and the aeroplane have driven home to
Canadians the importance of their Northland, in strategy, in resources and
in communications."[1] Over the succeeding forty-odd years, the level of
Canadian interest, as expressed in foreign and defence policy and related
government activity, has waxed and waned—largely as a result of events
and factors exogenous to Canada.

Some of these changes were chronicled in an article by R.J. Sutherland,
written twenty years after Pearson's.[2] That article has long been regarded as
a classic statement of the strategic significance of the Canadian Arctic. In it,
Sutherland described the 1947 agreement with the United States to establish
five joint weather stations in the Arctic islands; the construction of the
Distant Early Warning (DEW) Line and the Mid-Canada Line in 1955–57 as
components of a continental air defence system to provide better protection
(for the NATO defense alliance) for the United States strategic retaliatory
force from attack by Soviet long-range bombers; the *relative* decline in the
strategic importance of the two lines, and consequently the Canadian
Arctic, with the advent of operational Soviet intercontinental ballistic missiles (ICBMs) in 1957–60. In each instance, the impetus for change—whether
an increase or a decline—in the strategic significance of the Arctic came
from technological improvements that were incorporated into United States
and Soviet strategic weapons systems, not from any indigenous developments in Canada.

◇ *International Journal* 42 (Autumn 1987), 769–88.

While technology was the impetus for change, cost was the principal factor that dictated the choice of the actual response from a range of feasible options and Sutherland predicted—quite accurately, it turns out—that "the future strategic significance of Canada's northern regions [will depend] in the first instance upon the general state of international tensions and the total size of defence budgets." He also concluded that with a downward trend in defence budgets and success in reaching some limited arms control agreements, "the primary determinant of the strategic importance of the area *will tend* to become the importance of [the natural resources of the area] to the nations which depend upon them."[3] Although he pointed to the benefits to Canadian sovereignty of the DEW Line agreements, sovereignty per se was not a major factor in his analysis.

Three years after Sutherland's analysis was published, the voyages of the United States oil tanker, *Manhattan*, clarified the link between the development of Arctic resources and sovereignty in a way that took Canadian public opinion by storm and caused the government of the day to engage in a flurry of legislative activity.[4] Shortly thereafter, a new defence white paper, *Defence in the 70s*, was tabled. The white paper seemed to extend the earlier legislative activity to defence policy when it gave pride of place to "the protection of our sovereignty" by listing it first among the four commitments of the Canadian Forces: that policy document clearly identified Canada's sovereignty in the Arctic as the sovereignty in need of protection. Within a few years, however, the actual tasking of the Forces and the equipment procured for them to carry out those tasks made it evident that sovereignty protection had a lesser priority than Canada's commitments to the North American Aerospace Defence (NORAD) Command and the North Atlantic Treaty Organization (NATO).

This situation was mirrored in the capabilities of the various branches of the Canadian Forces. Where naval forces were concerned, the tasking and deployment of Maritime Command vessels revealed that the Atlantic Ocean had first priority, while the Pacific Ocean was a distant second and the Arctic an even more distant third. After the icebreaker, *Labrador*, was transferred from Maritime Command to the Coast Guard, the navy had the capability to sail only in the southeastern fringe of Canada's Arctic waters and then only for a few months of the year. Mobile Command could deploy only small units north of 60 degrees for survival training until the late 1970s when units up to battalion size were exercising at selected points in the Arctic islands. Even these latter deployments required detailed advance planning and elaborate logistical support. Air Command's unarmed Aurora long-range patrol aircraft have flown some 16 missions annually over the high Arctic in recent years, but these missions are limited to *visual* observation of the land, sea, and ice surface and are therefore hampered by cloud and fog cover, not to mention the months of prolonged darkness during the Arctic winter. Unarmed Hercules aircraft of Air Command's Air Transport Group provide logistical support and resupply to CFS Alert and to Mobile Command units exercising in the high Arctic. They are also available for search and rescue duties in that region. Until recently, fighter aircraft could not take off or land from any airstrip north of CFB Cold Lake (Alberta). In exceptional circumstances, fighter aircraft could (and have) flown in the high Arctic, but only after detailed preparations have been made for their in-flight refuelling. In summary, by the early 1980s, with the 1971 defence white paper still in effect, Canada's defence capabilities in the Arctic were exceedingly modest: most of the country's armed forces could not operate in over one-third of its sovereign territory.

Some planned future improvement in this rather embarrassing situation was signalled in 1985 with the announcement of an agreement with the United States to modernize continental air defence. The DEW Line will be replaced by the North Warning System (NWS)—a series of thirteen long-range and thirty-nine short-range radars—with construction set to begin in 1989. This radar warning system will be supplemented with an improvement of some five airstrips in the high Arctic which will allow CF-18 fighter aircraft to land and take off on a contingency basis and two airstrips will be improved to allow for take off and landing of AWACS (airborne warning and control system) aircraft attached to NORAD.[5] The impetus for this change in capability was, as before, improvements in military technology. In this case the improvement was the development by the Soviet Union of long-range air-launched cruise missiles and the deployment of those missiles on its long-range bombers. The development and deployment of this new weapons system reversed the relative decline in preceding decades of the threat to North America of attack by hostile bomber aircraft. The option chosen to defend against this new development was, as before, dictated by cost: only the long-range radars will be manned, the cost of construction of the whole system will be shared by the United States and Canada, and the fighter aircraft will not be based in the north. Canadian sovereignty concerns, although apparently not the principal determinant in the choice of defence options, did figure in the decisions to man the long-range radar stations with Canadian Forces personnel, to assign the overall management of the Canadian portion of the system to Canada, and to delay—for at least fifteen years—deployment of a space-based surveillance system for continental air defence.

The impetus of technological advance and the constraint of cost were the major factors, once again, in the reassessment of the strategic significance of the Arctic and the decision to construct the NWS. Within a few months of the announcement of the NWS agreement, however, sovereignty concerns were paramount: the voyage of the United States Coast Guard icebreaker, *Polar Sea*, through parts of the Northwest Passage in August 1985 provoked a reaction among the Canadian public not unlike the response to the voyages of the *Manhattan* some eighteen years earlier. Once again, the government responded with a flurry of legislative activity, accompanied this time by a firm assertion of Canadian sovereignty over the waters in and around the Arctic islands. In a speech to the House of Commons announcing the measures that the government would undertake to assert Canada's sovereignty in the Arctic, the secretary of state for external affairs set the tone for the government's policy with the following strong statement:

Canada is an Arctic nation . . . the Arctic is not only a part of Canada, it is part of Canadian greatness. The policy of the Government is to preserve that Canadian greatness undiminished. Canada's sovereignty in the Arctic is indivisible. It embraces land, sea and ice. It extends without interruption to the seaward-facing coasts of the Arctic islands. These islands are joined, and not divided, by the waters between them. They are bridged for most of the year by ice. From time immemorial Canada's Inuit people have used and occupied the ice as they have used and occupied the land. The policy of the Government is to maintain the natural unity of the Canadian Arctic archipelago and to preserve Canada's sovereignty over land, sea and ice undiminished and undivided.

A list of the actions the government proposed to take followed: an order in council establishing straight baselines around the outer perimeter of the archipelago; introduction of legislation to extend the application of Canadian civil and criminal law to all offshore zones (including the Arctic); withdrawal of Canada's 1970 reservation to acceptance of the compulsory jurisdiction of the World Court; construction of a Polar Class 8 icebreaker; an increase in the number of Aurora northern surveillance flights and in the level of naval activity in eastern Arctic waters; and an offer to engage in bilateral discussions with the United States on "all means of co-operation that might promote the respective interests of both countries as Arctic friends, neighbours and allies in the Arctic waters of Canada and Alaska."[6]

While Mr Clark's statement and the list of proposed actions constitutes the strongest ever assertion of sovereignty in the Arctic by a Canadian government, the original cause for such action was official embarrassment over Canada's lack of capability to enforce its sovereignty in the face of even an oblique challenge by a close friend and ally: "During [the voyage of the Polar Sea], Canada's legal position was fully protected, but when we looked for ways to exercise our sovereignty we found that the Canadian cupboard was nearly bare."[7] The Polar Class icebreaker will certainly improve Canada's capability—once it is at sea some five to seven years from now. In the interim, however, marginal increases in surveillance flights by unarmed patrol aircraft and naval sorties in the eastern Arctic (when ice conditions permit) are merely symbolic gestures rather than any real increase in capability.[8]

Eighteen months after the government statement on Arctic sovereignty, the 1987 white paper on defence, *Challenge and Commitment*, was tabled in the House of Commons. As in 1971, the white paper initially gave the appearance of extending the government statement and associated legislative activity to defence policy. The speculation in the media leading up to the publication of the white paper reinforced the impression that the Arctic was to become a major priority of Canadian defence policy. For example, much has been made of the plan to acquire up to 12 nuclear-powered attack submarines for use on Arctic patrol. Yet, when the policy document is carefully examined, the priority attached to the Arctic and to enforcing Canada's sovereignty there is not at all clear. In terms of the major outlines of the new policy, maintaining "effective and stable" strategic deterrence and contributing to the conventional defence of Canada and the Atlantic alliance have precedence over "sovereignty."[9] In fact, the paper contains a statement to this effect: "*After the defence* of the country itself, there is no issue more important to any nation that the protection of its sovereignty."[10]

The Arctic, as a region of defence interest to Canada, is nonetheless incorporated in each of the first three components of defence policy. With respect to the maintenance of strategic deterrence, it is one of the regions of Canada where "we enhance deterrence to the extent that we are able to deny any potential aggressor the use of Canadian airspace, territory or territorial waters for an attack on NATO's strategic nuclear forces." With respect to conventional defence, the Arctic is one of the regions of Canada where "the Soviet conventional threat to Canada and Canadian interests is . . . not entirely absent wherever Canadian interests and Soviet capabilities overlap." Finally, with respect to sovereignty, the Arctic is one of the regions where "the Canadian forces have a particularly important, though not exclusive, role to play . . . [and Canada's] determination to participate fully in all collective security arrangements affecting our territory or the air or sea approaches to our country and to contribute significantly to those arrangements is an important affirmation of Canadian sovereignty."[11]

The planned acquisition of various weapons and surveillance systems reflects these priorities. With the exception of the plan to deploy fixed sonar systems in the Canadian Arctic and the intention to increase the size and improve the equipment, training, and support for the Canadian Rangers— both relatively minor items—all other acquisition plans serve the much broader goals of closing the commitment-capability gap in strategic deterrence and conventional defence. However, given the lengthy period of implementation of up to fifteen to twenty years for the acquisitions mentioned in the white paper, there is a distinct prospect of change and/or cutback even in the plans for improvement in broad capabilities, not to mention specific Arctic uses of those capabilities.[12] Over the next fifteen years, pressures to reduce the federal deficit and to expand social welfare and education programmes are likely to continue, while unforeseen requirements for defence expenditures on personnel, operations, or maintenance will almost certainly occur. The former pressures can have a significant impact on the defence budget as a whole; the latter requirements can, as in the past, seriously affect capital acquisitions. In this respect, it should be noted that the planned level, or base rate, of real increases in defence spending of two percent per annum is not only lower than the six percent or the four percent promised at various times by the current government, but *it is also lower than the three percent real growth in defence spending that Mr Trudeau's government put into effect for most years in the late 1970s and early 1980s!*

In short, the defence policy of the 1987 white paper proposes marginal adjustments to various aspects of previous policy rather than a radical departure that would attach a clear priority to the Arctic over other commitments.[13] In this sense, the new defence policy is not to be a straightforward extension of the Arctic sovereignty policy that the government announced in 1985. Moreover, the white paper addresses the commitments-capabilities gap by adjusting or consolidating some of the commitments and announcing long-range plans to improve some capabilities, but at a spending level that continues to be very low in comparison to the spending levels of most other NATO governments, and with planned real increases in the spending level that are distinctly "conservative" in comparison to those of many previous Canadian governments.

Once again, as Sutherland predicted, advances in military technology stimulated some of the marginal changes contained in the white paper—such as the acquisition of nuclear-powered attack submarines, increased efforts in space-based surveillance, construction of the North Warning System—and cost was the major constraining factor in choosing among the various options to close the commitment-capability gap. Each of these changes has some application for the Arctic region. In addition, the Arctic does figure—though not exclusively—in the sovereignty section of the white paper. What, then, are we to conclude about the government's view of the strategic significance of the Arctic? Has the Arctic's strategic significance increased or declined?

Any evaluation of the changing strategic significance of the Arctic involves an assessment of at least five major factors and their interaction: changes in the strategic doctrine of either superpower, changes in international law, changes in military technology, changes in the economic value of the renewable and non-renewable resources of the region, and political developments in the region. Let us first look at how each of these factors impinges on the Arctic.

Over the past thirty years, the Soviet Union's navy has grown from a local coastal patrol force to a large, general-purpose navy which ranges throughout all the world's oceans and rivals the United States Navy in strength and capability. As this transformation occurred, most of the attention of naval defence planners in a number of NATO countries focussed on the Soviet Northern Fleet with its home ports on the Kola Peninsula. A major cause for NATO concern was the concentration in the Northern Fleet of over half of all Soviet SSBNS (nuclear-powered submarines armed with long-range nuclear missiles) and of all other types of Soviet submarines—including attack and cruise missile varieties. Because of limitations in missile range and accuracy, the SSBNS from the Northern Fleet had to take up patrol stations close to the Atlantic shores of Western Europe and North America.[14] In response, NATO forces created anti-submarine detection grids stretching from Greenland to Iceland and the United Kingdom, along the northeastern coast of North America, and out from Norway's northern coast. However, as the range and accuracy of the Soviet submarine-launched ballistic missiles increased and as the capabilities of NATO's detection grids improved, the Soviet Union changed its strategy to avoid North Atlantic waters and concentrate its SSBN deployment in the Barents Sea and adjacent ice-covered and ice-infested waters. The adoption of this "bastion" strategy for the Northern Fleet has caused a new response from the United States and other NATO navies: defence "forward operations" by

NATO's Striking Fleet Atlantic in the Greenland and Norwegian seas and possibly even further north.[15] The upshot of these changes in naval strategic doctrine by the two superpowers is a significant increase in the use of the Arctic Ocean and its adjacent seas by nuclear-powered submarines of the Soviet Union, the United States, and, perhaps, the United Kingdom.[16] Thus, these changes in strategic doctrine have increased the strategic significance of the Arctic Ocean for both superpowers and several NATO allies—including Canada.

Over the past two decades, there have been major changes in the international law of the sea. The vehicles for these changes have been the negotiations at the Third United Nations Conference on the Law of the Sea (UNCLOS III), the document which emerged from those negotiations—the 1982 Law of the Sea Convention—and the subsequent practice of, and related domestic legislation enacted by, various states which participated in UNCLOS III. For the present analysis, the most important changes to the law of the sea were those measures which enhanced the jurisdiction of coastal states over their adjacent waters: the creation of 200-mile exclusive economic zones, the extension of the territorial sea to twelve miles, and the new, special responsibilities of coastal states where adjacent waters are ice covered or ice infested.[17]

The purpose of the entire UNCLOS III exercise was to modify the law of the sea to reflect the modern era. In particular, some of the measures, including those mentioned above, were an attempt to regulate the exploitation of increasingly valuable offshore renewable and non-renewable resources. Although some major states, such as the United States and the United Kingdom, have not signed the 1982 convention, a large number of others *have* and are putting various measures into effect for their offshore waters. Canada is one of this group. Because of the real or potential value of offshore resources, a number of bilateral and multilateral disputes have arisen over the boundaries of 200-mile zones. In the Arctic region, for example, there are disputes between Canada and the United States concerning the boundary in the Beaufort Sea, between Denmark (on behalf of Greenland), Norway, and Iceland concerning boundaries in the Greenland and Norwegian seas, between Norway and the USSR over the boundary in the Barents Sea, and between the United States and the Soviet Union over boundaries in the Bering and Chukchi seas. While the resources in question vary from different species of fish to oil and natural gas deposits, the effect of the changes in the international law of the sea and of the importance attached to the boundary disputes by the states concerned has been to cause each of the disputing states to attach a higher strategic value to its own coastal waters and to pay much more attention to other states' activities in the waters of the same general oceanic region. For all states in the circumpolar region, the strategic value of the adjacent seas of the Arctic Ocean unquestionably has increased as a result of recent changes in international law.

Over the past twenty years, there have also been major improvements in technology associated with nuclear-powered submarines and their missiles. As noted, these particular changes have had an effect on strategic doctrine and on the strategic significance of the Arctic. Highly accurate, long-range cruise missiles are the product of yet another advance in military technology.[18] These small pilotless aircraft fly at subsonic speeds and can be launched from other larger aircraft (ALCMS), from the surface of the land (GLCMS), and from the surface or subsurface of the sea (SLCMS). Regardless of the method of launch, the current generation of cruise missiles flies at very low altitudes, is highly accurate, and can be armed with either conventional or nuclear warheads. Moreover, if the launching vehicle is a long-range bomber or a nuclear-powered submarine, a large number of cruise missiles can be launched by a single vehicle, and the vehicle itself may launch its cruise missiles some 1500 miles or more from the intended targets.

Although defence against a cruise missile attack is possible, the missile's small size and low-flying characteristics make a successful defence quite difficult—unless the launch vehicle is the defence target. Even then, the long distances from launch location to target and the wide variety of locations available for launch against the same target make defence only somewhat less difficult. Canada and the United States agreed to modernize the DEW Line by constructing the North Warning System mainly because of the new threat posed by the possibility of a long-range cruise missile attack by Soviet bomber aircraft across the northern approaches to the North American continent. Indeed, this development has reversed the relative decline in the strategic importance of the high Arctic for continental air defence which had occurred in the 1960s and 1970s. Hence the construction of the NWS. As well, Canadian fighter pilots are gaining some experience in defending against a Soviet cruise missile attack by flying CF-18s in practice pursuits of United States ALCMS during tests of the latter in the Canadian north.

One result of this general development in military technology—that is, long-range cruise missiles—cannot yet be assessed with reasonable certainty. That is the degree to which sea-launched cruise missiles, particularly those launched by nuclear-powered attack submarines, may become a major defence problem for Canada. Both superpowers are developing this weapon and deployment is near. For Canada, the uncertainty has two aspects. First of all, it is not possible at this time to predict how many Soviet cruise-missile-carrying submarines will be deployed or what the pattern of deployment will be. Answers to these two questions would indicate the scale of threat such submarines will present to North America. Secondly, only the passage of time can show whether such submarines may be deployed in Arctic waters or, as is more likely, whether Soviet submarines will use Arctic waters as transit routes between home ports in the Kola Peninsula and either the Atlantic coasts of Europe and North America or the North Atlantic sea lanes of communication between North America and Europe.[19] The extent to which the detection capabilities of NATO's anti-submarine grids in the North Atlantic continue to improve may be, in large measure, the incentive for Soviet submarines to avoid those waters as a transit route. If such deployment patterns do develop in the near future, this application of advances in military technology will further increase the strategic significance of the Arctic—for both of the superpowers and for Canada.

Over the past two decades, exploration for petroleum hydrocarbons in Canada's high Arctic, that is, the Beaufort Sea, the Mackenzie Delta, and the

Sverdrup Basin, has led to the discovery of several very large deposits of natural gas and some large deposits of oil. Further exploration will likely yield even more such discoveries. However, the level of exploration activity and the eventual exploitation of petroleum hydrocarbons in the high Arctic is entirely dependent on world supply and demand for this resource which, in turn, dictates the price for natural gas and oil. The costs of exploration, production, and transportation of the resources to their likely markets are sufficiently high that no economic profit can be made unless the world price is at least US$25 per barrel of oil or its equivalent for natural gas. When oil prices increased in the 1970s, the level of exploration activity in the high Arctic rose to almost boom proportions: hundreds of millions of dollars were being spent annually. With the precipitous decline in prices in the early 1980s and their apparent stabilization at the US$15–US$18 level, exploration in the Canadian high Arctic has virtually ceased.

Petroleum hydrocarbons are non-renewable resources. In the long term the world supply is finite, and yet the long-term demand for these resources is relatively stable or inelastic. In consequence many observers predict that oil prices will eventually increase at least to levels comparable to those of the late 1970s. If these predictions are correct, the economic value of the petroleum hydrocarbons in Canada's high Arctic will increase dramatically. For the present, however, the immediate economic value of these resources has declined in recent years and remains comparatively low today.

The discovery of large deposits of these resources and the high potential for further discoveries in the Arctic certainly adds another facet to the strategic significance of the region. But the degree to which Canada may become dependent in the future on supplies of oil and natural gas from reserves in the high Arctic will be a much more direct stimulus to increased strategic significance from a national, domestic point of view. While it is impossible to predict with any certainty when—or if—such a dependence will develop, let alone the degree of dependence, it should be noted that Canada is not the only Arctic littoral state which may come to depend on its Arctic region (onshore and/or offshore) for fossil fuel supplies. All of the Arctic littoral states—Denmark (for Greenland), Norway, the Soviet Union, and the United States—are in a similar situation. There are variations in the size of discovered and potential oil and gas reserves as well as variations in each state's need for or interest in developing its Arctic reserves. Thus, there are also variations in each state's estimate of the strategic significance of the reserves and of the region in which the reserves are located. Nevertheless, the fact that all Arctic littoral states possess at least potential large hydrocarbon reserves has led to a general *international* perception of an increase in the strategic significance of the Arctic.

The political awareness and activity of the northern populations in several of the Arctic littoral states has grown in recent years. This has been evident at the local, national, and international levels of policies. . . .

In Canada, the federal government is gradually devolving more powers and administrative responsibilities to the territorial governments of the Yukon and Northwest Territories. Political awareness and participation in the political process at the local and national levels have increased markedly over the past two decades—particularly in the Northwest Territories. Native groups in both territories have formed organizations primarily for the purpose of negotiating land claims settlements with the federal government, but these organizations have also galvanized local and national attention on the issues of aboriginal rights, the possible division of the Northwest Territories into two separate units, eventual provincial status for the territories, other constitutional issues, regional economic developments, and a host of environmental issues. . . .

While the extent of this increase in political awareness and participation varies for each of these countries, the effects have been similar. Much more national attention is focussed on "northern" issues or issues that affect the state's northern region: the central government and the public at large are more aware of "their" north. Moreover, the northern populations in each state are less isolated from the national mainstream. Whereas in the past these populations may have paid little attention to, or had little involvement in, debates over national policies and programmes in energy, economic development, or defence, now these issues are covered more extensively in the northern regions and individual northerners as well as organizations which seek to represent them make direct contributions to the national debate. In short, to the extent that the strategic significance of a country's Arctic region is changing—whether increasing or declining—political awareness and participation by northerners have made that change more apparent at both the local and national levels. And, to the extent that northerners' political participation influences the national debate, their participation in and of itself is a factor in the changing strategic significance of the region.

This brief review of the five factors which are involved in any assessment of the changing strategic significance of the Arctic—changes in strategic doctrine, international law, military technology, economic value, and political development—shows that, for Canada, at least four of the five factors register changes that *increase* the strategic significance of Canada's Arctic. Only one, the economic value of the region's petroleum hydrocarbon resources, registers a decline for the past few years, but this may well be only a short-term phenomenon. If it is a short downward "blip" in a longer term trend of increasing economic value, then all five factors register changes that contribute to an across-the-board increase in strategic significance. This conclusion should come as no surprise to those readers who are familiar with the Canadian Arctic as it was forty years ago, when Lester Pearson wrote his article, and can compare it to the Canadian Arctic today. However, very few Canadians have that appreciation for the extent and pace of change in the Arctic. Too many of us still carry with us the mistaken image of a "vast northern desert."

It is extremely important for Canadians to recognize the increased strategic significance of Canada's Arctic region because this change has a direct bearing on Canadian security and sovereignty. Yet it is equally important, in fashioning a response to the increased strategic significance of the region, to realize that the government of Canada and individual Canadians have little influence and almost no control over the factors which are responsible for the increase. Changes in superpower strategic doctrine are largely the province of the superpowers and a product of their interaction over time. Changes in international law are the product of complex multilateral negotiations and the practices of many individual states. Changes in military technology occur through the research and development efforts of other states—and Canada may then decide to buy, copy, modify, or react to them in some manner. Changes in the economic value of resources such as petroleum hydrocarbons are the result of the forces of world supply and demand as well as the decisions of private or state enterprises based on their investment and profitability requirements: the government of Canada has only a minor influence, through national policies and programmes, on the economic value of non-renewable resources. Of the five factors, only political change arises from forces indigenous to Canada: political development occurs because of the interplay of the interests of northerners and those of other Canadians in the arenas of territorial, provincial, and national politics.

Most of the factors responsible for the increase in the strategic significance of Canada's Arctic region are beyond Canada's control, yet these very same factors will continue to be instrumental in causing future changes in that region's strategic significance. Therefore any response that a Canadian government fashions to such changes will of necessity be a reactive one: a reaction to changes which it cannot control, or significantly influence, and cannot predict with any certainty over the medium term.

The difficulties inherent in creating policies that respond to largely unpredictable changes are augmented because the changes in the strategic value of the Arctic that have a direct bearing on Canadian security and sovereignty also affect the security of the United States and the northern and central European members of NATO. Thus, for maximum effectiveness, a Canadian response should be fashioned in consultation—if not concert—with the nation's continental and European allies. Such consultation may introduce constraints on the choice of policy options available to Canada, but it may also point the way to relieving cost constraints on the choice of other options. Canada's policy response to the increased strategic significance of the Arctic can be a unilateral one, a bilateral one in agreement with the United States, or a multilateral one in agreement with several or all of its NATO allies. Or it can be some combination of these three types which evolves over time. Given the unpredictability of change in the Arctic's strategic significance, which rests on the unpredictability of most of the individual factors determining it, and the importance of the Arctic region to Canadian security and sovereignty, an evolutionary approach might serve Canada's national interest best.

Looking at the 1987 white paper on defence from this perspective, some interesting conclusions emerge about the government's view of the strategic significance of the Arctic. The white paper is not a radical document either in terms of the shifts in emphasis that it proposes among Canada's main defence commitments or in terms of the equipment that will be acquired to meet those commitments. The increased strategic significance of the Arctic is acknowledged but not by altering the priority attached to the Atlantic and the Pacific. Instead, the white paper narrows the gap: the Arctic still ranks third, but a much closer third than it was before June 1987. The navy's main task of anti-submarine surveillance remains unchanged, but the equipment that will be acquired to perform that task represents a better balance for Atlantic and Pacific naval forces. In an era of Exocet-like missiles and even longer range SLCMS, it is only prudent to move away from nearly total dependence on surface vessels for Canada's relatively small navy. The acquisition of nuclear-powered attack submarines holds several other advantages for anti-submarine surveillance in the Atlantic and Pacific, but these are beyond the purview of this article. What is germane to this analysis is that this type of submarine is primarily designed for—and used in—open water. Its acquisition does provide an under-ice capability for Canada's navy, but this feature appears to have played a distinctly secondary role in the acquisition decision.

The white paper indicates certain other adjustments to equipment, force expansion, and tasking that will have the effect of improving the Canadian Forces' capabilities in the Arctic, such as constructing an under-ice detection grid and increasing the size of the Ranger reserve force. The adjustments that the government proposes to take in response to the increased strategic significance of the Arctic are, individually and collectively, incremental. At first glance, they also give the appearance of a strictly unilateral response. However, closer examination of these adjustments, particularly the length of the acquisition period and the modest rate planned for real increases in the defence budget, leads me to conclude that we are probably witnessing the first stage of an evolutionary response. The *intent* to acquire an improved Arctic capability is clearly established in the white paper. Only time will show whether that intent is converted into real capability. During that rather lengthy interval, time is also available for the government to get a better grasp of all of the effects of the Arctic's increased strategic significance and to examine various unilateral, bilateral, and multilateral responses with the view to selecting those that best serve Canada's interests. If this conjecture is correct, the 1987 white paper represents not a radical departure in Canada's defence policy, but a very carefully constructed minimal adjustment to a generally perceived significant increase in the strategic value of the Arctic region.

NOTES

1. L.B. Pearson, "Canada Looks 'down north,'" *Foreign Affairs*, 24 (July 1946), 638.

2. R.J. Sutherland, "The Strategic Significance of the Canadian Arctic," in *The Arctic Frontier*, ed. R. St. J. Macdonald (Toronto: University of Toronto Press, 1966), 256–78.

3. Ibid., 274–75, 278. Emphasis added.

4. This activity included passage of the Arctic Waters Pollution Prevention Act and the extension of Canada's territorial sea from three to twelve miles. See W.H. Critchley, "Canadian Naval Responsibilities in the Arctic," in *The RCN in Transition, 1910–1985*, ed. W.A.B. Douglas (Vancouver: University of British Columbia Press, 1988), 280–91 for a recent analysis of

the impact of these pieces of legislation on Canada's sovereignty claims in the Arctic.

5. Canada, Memorandum of understanding and exchange of notes on the modernization of the North American air defence system (March 1985).

6. Canada, House of Commons, *Debates*, 10 September 1985, 6463–64.

7. Ibid., 6462–63.

8. .One example of the symbolic nature of these activities is the flight by two CF-18 aircraft, unarmed on this occasion, to Iqualuit (formerly Frobisher Bay) and then from Iqualuit on a round trip to the geographic North Pole. The latter part of the flight apparently required three mid-air refuelling operations: "CF-18s Blaze Trail on 'Top of the World,'" *Globe and Mail* (Toronto), 17 June 1987, A20. While not wishing to detract from the real accomplishment that this exercise demonstrated—it was the farthest north that Canadian fighters had ever flown—it hardly represents a quick-reaction capability, given the dire shortage of Canadian Forces aircraft that can act as mid-flight refuelling carriers.

9. Canada, Department of National Defence, *Challenge and Commitment* (Ottawa: Supply and Services Canada, 1987), 17–22, 49.

10. Ibid., 23. Emphasis added.

11. Ibid., 17, 19, 23.

12. In this respect, it is useful to recall that the Canadian Patrol Frigate Project (the long-range plan to replace Maritime Command's aging fleet of destroyers) included the acquisition of eighteen new frigates in three batches, or stages, of six frigates each. Less than a decade after that project was announced, the 1987 white paper has changed it to include only two stages for a total of twelve new frigates when the planned acquisition is completed. Ibid., 54.

13. "The government has reviewed our defence effort. This review has confirmed that Canadian defence policy, as it has evolved since the Second World War, is essentially sound." Ibid., 89.

14. For a more detailed analysis of this development, see W.H. Critchley, "Polar Deployment of Soviet Submarines," *International Journal*, 39 (Autumn 1984), 828–65.

15. Geoffrey Till, "Strategy in the Far North," in *Northern Waters: Security and Resource Issues*, ed. C. Archer and D. Scrivener (London: Croom Helm, 1986), 69–80.

16. See press reports on a May 1987 joint exercise by American and British submarines in the Arctic. For example, "Ottawa knew about exercises, Beatty says," *Globe and Mail* (Toronto), 8 June 1987, A5.

17. For a detailed analysis of these and other changes, see P. Birnie, "The Law of the Sea and Northern Waters," in *Northern Waters*, ed. Archer and Scrivener, 11–41.

18. Shorter range, less accurate cruise missiles have existed since the 1950s, and the Soviet Union has deployed such missiles since the 1960s.

19. A more detailed discussion of this matter can be found in Critchley, "Polar Deployment of Soviet Submarines," 859ff.

CONTROLLING THE DEFENCE POLICY PROCESS IN CANADA: WHITE PAPERS ON DEFENCE AND BUREAUCRATIC POLITICS IN THE DEPARTMENT OF NATIONAL DEFENCE ✧

o

INTRODUCTION: DEFENCE WHITE PAPERS IN CANADA OVER THE LAST QUARTER CENTURY

On 5 June 1987 the Canadian government issued its White Paper on defence, *Challenge and Commitment: a Defence Policy for Canada*.[1] Typically, the paper has been reviewed, analyzed, and commented upon from several strategic studies points of view. . . .

The place of White Papers in the Canadian defence policy process is, at best, ambiguous. In fact, Canada's experience of operating without a White Paper since 1975 suggests to some officials that they are not needed. Generally, however, White Papers have been viewed by the Department of National Defence, the Canadian Forces, and the public as fundamental government statements intended to direct the policy process towards its political and operational objectives.

Only three defence White Papers have been issued since 1964; each possessed its own policy characteristics, yet were similar in certain important respects. Moreover, each paper was prepared in unique ways and prompted different bureaucratic responses. It is the purpose of this paper to compare these similarities and thereby to suggest that how defence policy takes shape within the defence establishment has a significant impact on how faithfully that policy will be implemented over the long term. . . .

✧ *Defence Analysis* 5, 1 (1989), 3–17.

THE 1964 WHITE PAPER ON DEFENCE

Paul Hellyer was not so much interested in international aspects of Canadian defence policies as he was in reorganizing the Canadian Forces. Although his *White Paper on Defence* of March 1964[2] included a lengthy strategic discussion, it was for many, as Opposition critic Douglas Harkness remarked in the House of Commons, merely a restatement of "what one might call the basic facts of defence, as they have been recognized by everyone . . . for the last 15 years." But Hellyer did try to reshape the strategic picture, if only to support his plans for reorganization. He rejected the notion that "mobilization" would be useful in future conflicts. This declaration for the "short-war scenario," especially in a NATO context, was remarkably anachronistic if only because in 1964–1965 members of Hellyer's staff were involved in NATO discussions that in 1967 introduced the new strategy of "flexible response" to the Alliance. This strategy implied, at minimum, a growing dependence on the mobilization of conventional forces.

In place of mobilization, Hellyer sought to increase Canada's reliance on professional forces-in-being. These forces were for the most part to be held in Canada for use as a worldwide mobile force ready to fight "brushfire wars" at the call of the United Nations or to defend particular Canadian interests. Interestingly this is exactly the same concept proposed by the New Democratic Party of Canada in its 1987 statement of defence.[3] The earlier Hellyer strategic plan failed because it contradicted Canada's alliance policies, was too expensive, especially in transport aircraft, and because, except on a small scale, no role for such a designated force ever materialized. Most military leaders were unconvinced of the wisdom of Hellyer's strategy which they saw as a fanciful vision harmful to "real" defence needs.

As with many things in life, the one who defines the "Problem" is also the one who can control the selection of policies intended to solve it. In Canada, as in other countries, a central defence concern is always to balance defence needs and resources. Imbalances and inadequate outcomes, however, can be attributed to many causes. In Hellyer's estimation the increasing costs of Canadian defence and the obviously declining return for defence expenditures were ascribed to a debilitating organization whose faults were evidenced in so-called inter-Service rivalries, duplications in support services, old habits of command and decision-making, and a "committee system" of policymaking that was controlled by three Service Chiefs each of equal authority.

Hellyer declared that he was forced "either to greatly increase defence spending or to reorganize. The decision was to reorganize." This reorganization was accomplished with two reform bills, the first of which, Bill C90, eliminated the independent Service Chiefs and replaced them with a single Chief of the Defence Staff (CDS). This appointment resulted in a reorganization of Service defence staffs into a unified Canadian Forces Headquarters as well as in the development of an integrated staff structure. Bill C90 also provided the legal basis for the "unification" of the Canadian Forces that was finally accomplished in 1967 after bitter debate of Bill C243, the Canadian Forces Reorganization Act.[4]

Hellyer expected significant returns from his efforts, predicting, for example, that the reorganizations would lead to savings that would permit 25 percent of the budget to be devoted to capital equipment. This goal was never reached. Although there was qualified support for the establishment of a single CDS, what was clear is that no consensus emerged in the defence establishment over the Minister's unification policies. Nevertheless, Hellyer purposefully left officials to sort out how unification was to be achieved and how it was to function. Paul Hellyer left the Department in 1967 and without his single-minded insistence and with no consensus as to what unification really meant in terms of "who gets what," the defence establishment dissolved into intense bureaucratic bargaining. Added to this confusion of aims was a continued erosion of defence budgets, indicating that the problem was not simply organizational. Hellyer, in fact, had both to reorganize and spend more money to balance Canada's long-term defence needs and resources.

The 1964 White Paper generally supported the continuation of the NATO, NORAD and UN roles the Canadian Forces were then performing. Therefore, Hellyer's plan to form a highly mobile force to fight brushfire wars from already limited capabilities required that the extant roles of the Canadian Forces and their equipments be made compatible with this strategy.

His plan was to have sizable land, air and naval forces deployed in and around Canada limiting other forces to "special tasks" and UN peacekeeping duties. The major portion of the flexible and mobile force was to devolve upon the Army, held to be the "key organization." But the demands placed upon it by this strategy divided the Army; for the role in NATO required mechanized forces while the mission of worldwide mobile warfare demanded light airportable equipment. Thus groups within the Army, for example those who favored tanks, were pitted against others who were intrigued with the mobile concept. It soon became evident, moreover, that the mobile role would, in fact, advance the cause of those in the Air Force who wished to buy long-range transport aircraft. Many Army leaders soon realized that in a world of limited and declining budgets support by the Army for the mobile role would, ironically, inevitably reduce the size and effectiveness of the Army, as funds got spent on transportation equipment.

In the early 1960s the Navy had been experiencing difficulties because its equipment was aging and procurement plans were incomplete. An "Ad Hoc Committee on Naval Objectives" was established to look into the whole future of Naval roles and equipment. It produced a detailed appreciation that considered the political, military and technological aspects of naval warfare and set a course for the Navy into the mid 1980s. The Report proposed the continued development of the Navy's antisubmarine warfare expertise that was to be balanced with other classes of ships including the aircraft carrier *Bonaventure* (expected to remain serviceable to 1975) and submarines, "the capital ships of the future." It was a comprehensive plan for a "Three Ocean Navy," which included plans for an Arctic base, and, eventually, nuclear submarines to patrol the Arctic Ocean.[5]

No firm decisions about the future of the Naval program were taken by the embattled Diefenbaker government in 1962. In April 1963 when the Liberals won control of Parliament and Paul Hellyer was appointed

Minister of National Defence (MND) the Navy's program was thrown into doubt. Hellyer announced a "review" of the plans that continued through 1964. In his White Paper he would only state that "studies" were ongoing to determine "the most effective mix of weapons systems" to provide the "maximum defence potential for the least cost." This decision was an explicit rejection of the Navy's study. As Hellyer's reorganization plans unfolded the Navy's main priority then became defending seas already won, and particularly as reductions under unification became apparent, to beat off cuts in organizations and personnel. Hellyer's principal Naval opponent during these unification fights was Admiral Jeffery Brock, Chairman of the ignored Ad Hoc Committee on Naval Objectives.[6]

The Air Force was a particularly specialized force by the time the 1964 White Paper was issued. It was wed to a short nuclear war strategy and as a consequence had equipped itself with two nuclear/capable combat aircraft, the CF104 in NATO and the CF101 in North America, both of which had only limited capabilities for other roles. Hellyer's policy to emphasize smaller-scale operations but to continue the NATO and NORAD roles split the Air Force by taking monies from programs aimed at reequipping the nuclear squadrons and giving them to a new CF5 tactical support aircraft.

Unification was an especially traumatic experience for the Air Force. Without a functional or geographical center around which to rally their units, the Air Force was gradually divided between the other Services to provide specialized support or organized into unified Canadian Forces components such as air transport and air training.[7] The Air Force struggled against these reorganizations into the 1970s until a reunited air force emerged as Air Command in 1975.

The 1964 White Paper on defence in effect multiplied the roles of the Canadian Forces. It also introduced a strategic concept that threatened the Services' own view as to what they should be doing to defend Canada. In response the Army began to develop a dual character. It continued to train and equip itself for both a NATO and a mobile role. The Navy entered a period of retrenchment and began its long painful adjustment to unification, but there was little heart of conviction in either undertaking. The Air Force, while continuing its nuclear roles in NATO and NORAD, was saddled with the limited CF5 aircraft, intended primarily for an army support role, something the Air Force had not done since the 1950s. This new role, and the rending of the Air Force under unification, opened great cleavages within the air operations fraternity. In summary, the contradictions between Hellyer's strategic concepts, his organizational ideas, the proposed roles for the Canadian Forces, and the predictably declining defence budgets created phenomenal bureaucratic pressures within the defence establishment, and these in turn contributed directly to the failure of Hellyer's policies.

The budget did not feature prominently in Hellyer's White Paper even though it was the concern for defence expenditure that largely motivated the production of the policy paper in the first place. Hellyer's assumption that better organization and program control would yield significant savings was not founded on any pragmatic or empirical research. There was, however, evidence that this might not be the case. The thinness of the budgetary assumptions underlying the White Paper and the speed with which unification came to mean retrenchment quickly destroyed any hope that the White Paper would point the way to new sources of capital funding. Within a few months the pressures caused by the imbalances between defence needs and resources proved fatal to any consensus that might have been built around the assumed efficiencies of unification.[8]

There seems little need to examine in any depth consensus building during the preparation of the 1964 White Paper, especially when one recalls the circumstances that developed after its release, when many senior officers retired or were fired, and the MND spoke of an "Admirals' revolt." But this bitter situation was to surface some months *after* the White Paper was issued and developed mostly around Bill C243. The defence policy was not without support as it was being developed. Much of the strategic assessment was, in fact, based on the work of individuals within the defence establishment, particularly Dr R. J. Sutherland.[9] But there was a curious misreading by the military leaders of the determination and speed with which Hellyer would institute his ideas on organization and budgeting. Their surprise at Hellyer's determination and their resentment of his neglect of their advice shattered any desire they may have had to correct inefficient organization in the Department of National Defence or Canadian Forces.

Hellyer wished to be an active manager in his Department and when questions were raised about certain aspects of the draft defence policies he tended to take them as attacks upon himself. As the process of preparing the White Paper came closer to completion Hellyer became ever more secretive. Reports and studies sent to his office by the defence establishment were often either ignored or never acknowledged. The Chiefs of Staff, in what was to be one of their last Chiefs of Staff Committee meetings just prior to the release of the White Paper, complained to each other that they knew neither what would be included in the policy statement nor what was to become of their committee or even themselves. Only Hellyer's strong personality, if not his leadership, carried his policies forward. Unfortunately for those policies, when he left the defence portfolio in 1967 there was no satisfied defence establishment able or willing to carry them any further.

DEFENCE IN THE 1970s

Donald Macdonald was MND from 1970 until 1972. In that short period he produced a defence White Paper, *Defence in the 70s*, that was to serve as Canada's defence policy for the next 16 years.[10] In fact, the policy directions in the paper survived for barely two years, and were replaced in all but name by incremental policy decisions taken by the defence establishment.

One way to correct the defence needs–resources imbalance is to eliminate a portion of the needs by reducing or canceling some commitment at home or abroad. Macdonald began by reducing the "Threat" as a precursor to eliminating commitments. Accordingly, his White Paper rested on a strategic assessment that was optimistic by any standard, one that depended critically on the existence of a lasting era of "détente" between the United

States and the Soviet Union. The paper predicted steady economic growth in Europe—a prospect that in Macdonald's opinion would allow Europeans to take over responsibilities for their own defence. Growing prosperity and stability was also expected for Third World countries. There was amongst this general euphoric mood one pessimistic note, and it concerned "peacekeeping" and the related failures of the United Nations (UN) to be a force for peace in the world. As a consequence Macdonald believed that "the scope for useful and effective peacekeeping activities now appears more modest than it did earlier." The message of the assessment, therefore, was to predict a more stable world, at least as far as Canada's interests were concerned, and to support a reduction in Canadian commitments to NATO and the UN.

This assessment largely ignored those produced by the defence establishment which tended towards a more classic "realist" image of international affairs, with the military leadership in particular sharing little of the Minister's optimism. They tended to regard the existing capabilities of the Soviet Union as a continuing menace, irrespective of the ongoing arms control debates. They also tended to believe in a positive relationship between power and security, and to see Canada's interests and defence intimately tied to those of the United States and Europe. The White Paper ran counter to this realist paradigm and failed to convince most officers, and many officials, who saw it simply as an expedient to reduce the already starved defence budget.

The problem in the defence portfolio in Macdonald's view was principally one of management. He believed that the "program was a mess," and that the Department of National Defence was asking for too much while at the same time the Canadian Forces were not reacting properly to Prime Minister Trudeau's new policy directions. . . .

There is little doubt that there were program problems in the Department of National Defence. The growing demands of the defence budgets and some notable spending gaffes, such as refitting the aircraft carrier *HMCS Bonaventure* and then scrapping her, did not encourage the Cabinet's confidence in the Department of National Defence. For many in the defence establishment, however, management problems were the result, *inter alia*, of Government indecision on such questions as the status of Canada's commitments to NATO; unification hangovers and organizational confusions; the lack of priorities for purchasing new equipments; and the general lack of funds for ongoing operations. There was no consensus about the nature of the defence problem and how it might be addressed, and no consensus was constructed during the preparations of the White Paper.

In his Calgary speech of 3 April 1969, Trudeau seemed to clarify the roles for the Canadian Forces in the future.[11] Briefly, they were to survey Canadian territory, defend North America in co-operation with the United States, "fulfill" agreed NATO commitments, and participate in peacekeeping missions from time to time. Unfortunately, the statement and subsequent direction left unanswered questions of relative priorities and resulted in confusion and argument within the defence establishment, and between it and other government departments and agencies. For example, officers from the Canadian Forces, the Department of National Defence, the Royal Canadian Mounted Police, and other departments and agencies could not decide what role the Canadian Forces ought to play in the Northern Territories where other departments had already been mandated to do the tasks that the Prime Minister's statement implied the Canadian Forces should do in the future. A large "working group" was assembled in Ottawa to try to "interpret" the government's policy, but they only concluded that more precise policy direction from the government would be necessary if they were to unravel this presumed first priority.[12]

Similar confusion confronted planners who tried to implement the other roles. The government, for instance, had arbitrarily withdrawn half of Canada's forces deployed in NATO Europe in 1969 and had intended to withdraw the remainder in 1970–1971. But this second withdrawal was halted, mainly because of a strenuous European reaction and because Leo Cadieux (then MND and who had not been consulted about the impending decision until the last moment), threatened a noisy resignation. The result was that the Department of National Defence was left with a policy hang-fire and with no indication of whether to proceed with NATO-related planning or not. In the absence of a decision the staff did whatever the interests of their own programs suggested.

The few short paragraphs in the White Paper about the defence budget were the essence of Macdonald's defence policy. He stated that "there is no obvious level for defence expenditures in Canada" and that statements of defence requirements could not be taken as the defence budget. The budget would have to be determined "in relation to other government programs."

The 1971 White Paper imposed severe constraints on the Department of National Defence and the Canadian Forces. Cutbacks were announced in both manpower and equipment acquisitions. The budget was frozen at approximately one percent above its 1971–1972 ceiling and the clear implication was that it would continue to be constrained in future years as, indeed, it was.

The budget announcement was more than a little perplexing to the defence establishment. That there is no obvious level for defence spending in Canada was not a particularly difficult notion for officials to grasp. Some might point out that there is no "objectively" obvious level of expenditure for any government program; rather, the level of spending is almost always determined in each case by the political process. All that defence officials had expected was to be given a fair hearing for their needs after which the political process would prevail. By freezing the budget and announcing an arbitrary level of defence spending before the needs of the White Paper had been considered by the defence establishment, the Minister undercut the whole policy process in National Defence Headquarters. In the view of one official, the budget process subsequently became "nothing more than an elaborate management exercise with no real purpose or end."

Besides upsetting the management schemes of the Department of National Defence, the budget restrictions presented the Canadian Forces with difficult, and some would contend insurmountable, operational problems. First, for most planners the levels of defence expenditure were not abstractions as Macdonald implied but rather had concrete relationships to the defence objectives set by the government. These relationships were particularly evident once the government had accepted commitments to alliances that were important not only to Canada's defence but also to the defence of other countries. Second, in the White Paper the government had announced new roles or emphasized older tasks but it was not obvious how these would be met from a diminished store of resources. There is some evidence that the government had an expectation that it could completely withdraw from NATO Europe and use the "saved" manpower and resources from that commitment for other things. The problem with that proposal was (and is) that few savings would accrue to Canada from such a withdrawal because almost all the expenditures in Europe are made in the form of salaries and operations and maintenance costs. Little "new money" would be made available unless the withdrawn units were also disbanded, but that would negate the rationale for the withdrawal in the first place. In any event, the withdrawal did not take place and the Canadian Forces faced the prospect of trying to perform its standing commitments while taking on new and as yet undefined (in terms of expenditure) tasks with a steadily decreasing budget.

The Management Review Group announced in the White Paper, and brought into action later in 1971, criticized the government's defence policy because of the "lack of any Department of National Defence contributions to either the statement of 3 April 1969 or the 1971 White Paper."[13] Other officials who worked in the Department during the period confirm that there was little if any attempt on the part of the Minister to engage them in a serious discussion aimed at building a shared analysis of the defence policy. Indeed, the White Paper seems to have been written under Macdonald's close supervision and only circulated in outline to officials. The final text was passed to them just before its release but with the admonition by the Minister that not a word was to be changed without his permission. But most officials were not overly concerned because they believed, as one senior officer remarked to the Management Review Group, that "the White Paper did not provide meaningful policy declarations from a military viewpoint" and they simply tried to work their way around it.

Once *Defence in the 70s* was issued the staffs attempted to make some plans to meet its objectives. Few in the defence establishment, however, believed that the strategic assessment was valid or that the government really could ignore NATO. Their dilemma, therefore, was to conduct operations as best they could without at the same time compromising any possibility or returning to the *status quo ante* when the time came—as most believed it would—when the government would be forced to resume a more "traditionally Canadian" defence posture. In fact, they had to wait less than two years.

Between 1970 and 1982 seven successive politicians held the defence portfolio in Canada. . . . During this period also a major upheaval, fully on the scale of unification, was taking place in the Department of National Defence. As a result of the Management Review Group Report of 1972, the entire headquarters structure and management system of the Department of National Defence and the Canadian Forces was changed. These reorganizations, in conjunction with the budget cuts, changes in NATO Europe commitments, and the uncertainties of the White Paper itself rendered the Department of National Defence and Canadian defence policy wide open to bureaucrats (in or out of uniform) who could get into the middle of the policy process as the "squeakiest wheel." By the end of 1973 it was evident that the "managers" interested in the program management process had prevailed over "operators" interested in defence capabilities—the outcome of the policy process.

THE WHITE PAPER OF 1987

The most recent White Paper on defence is radically different, in terms of its development and concepts, from the two previous policy papers. *Challenge and Commitment: a Defence Policy for Canada* is a "classic" defence statement in that it is focused primarily on defence needs and ways to meet them. It is the product of a political attitude that does not accept on faith that a natural "harmony of interests" exists in the world and as a consequence it reflects a belief that the military defence of Canada is of primary concern to the government. In this regard, therefore, the 1987 White Paper reflects also the beliefs and experiences of those who will have to administer it.

For this analysis the details of the strategic assessment are not as important as is the fact that it was developed "in-house." That is to say, the appreciation of the world situation was developed by defence officials using for the most part information drawn from Canadian, NATO, American and other Allied sources. Nor surprisingly, this assessment supports the "realist" preferences of the defence establishment: the Soviet Union plays a leading role in the world's security problems; power, in terms of military assets, is the currency of security; and the military defence of Canada is the aim of national defence policy.

The problem facing the MND, Perrin Beatty, was, as always, the needs–resources imbalance and the Minister, on tabling his White Paper in the House of Commons on 5 June 1987, identified it precisely:[14] "the existence of a significant gap between the capabilities required to meet the military commitments accepted by successive governments on behalf of the Canadian people and the capabilities possessed by the Canadian Forces." Once the "commitments–capabilities gap" was defined as *the* problem of Canadian defence policy other aspects of defence policy changed also.

Canadian defence planners have learned that operational assets cannot be developed simply by reorganizing peacetime units and that capabilities cannot be conjured up by designing some managerially more efficient policy process. Rather, the only criterion for measuring operational capabilities is a military model of efficiency and effectiveness in which the salient questions is, "Will it work in conflict?" Furthermore, it was recognized that there is a military social aspect to national defence and that military concepts and values would have to be considered at the outset of defence policy planning. With this definition of the problem and the implied acknowledgement that the military leadership would have a major role to play in the developing of government policy, Beatty went a long way towards building a consensus for his policy program.

In the history of the Canadian defence policy there has been a continuous tension between two poles, the defence of Canada at home and the defence of Canada overseas. Since 1949 that tension has been evident in the struggles in Canadian defence policies that have in turn emphasized NATO and Canadian priorities. Liberal government under Trudeau reduced Canada's commitments to NATO and starved the Canadian Forces of resources as much as they dared. In doing so, however, they eventually produced a backlash first from NATO and then from the Canadian people who literally became ashamed of Canada's "free-rider" image. This response was highlighted in comments from many segments of Canadian society— but not always, it must be acknowledged, from similar motives. So politically charged did the issue become that even the Liberal Party reacted to it by increasing defence spending in 1982, but nevertheless it surfaced as a telling election issue two years later.

Once they assumed power there was some sense in the Progressive Conservative Party that treating the NATO policy pole would cure Canada's defence problem.[15] Unfortunately for this assessment, other unexpected events and issues arose that eventually made it clear that both the Canadian territorial and the NATO poles would have to be addressed at the same time. These new pressures included challenges from the United States over Arctic passages; the growing influence of the so-called peace movement; discontent in Canada with President Reagan's "Star Wars" concepts; and such things as the large and successful nationalist rally, "True North, Strong and Free?" held in Edmonton in 1986.

The challenge for the Progressive Conservative government, therefore, was to fashion a defence policy that would address both home defence and NATO commitments at the same time. Never before had the tension between the two poles been so inelastic and public awareness of its existence so acute. This dilemma, as much as anything else, likely accounts for the delay the new government experienced in producing its defence policy document.

The key to solving the dilemma, within acceptable expenditure limits, was to find a force model and capabilities spectrum that would be as compatible as possible with both demands. This is precisely what the White Paper sets out to do and it is most evident in the enunciation of roles for the Services.

Challenge and Commitment announced that the government would "pursue a vigorous naval modernization program" aimed at producing a "Three Ocean Navy" for Canada. This naval program, not by coincidence, was to have capabilities that would enable the Navy to operate both in support of Canadian domestic defence objectives and NATO strategies. Thus the Minister could in good conscience deem the Navy to be a contribution to both defence poles at once.

The Air Force, through the policies of both the former Liberal government and the present Conservative one, reduced its inventory of combat aircraft from three to one, the new CF18. This aircraft is being deployed both in Canada and overseas. Again, as with the Navy, it is easy to argue that these deployments are not contradictory but mutually supporting; for it can be pointed out that air squadrons deployed in Europe could be quickly returned to Canada in a crisis if it were necessary or vice versa.

The Army, however, presented planners with a more awkward problem. When critics and politicians speak of withdrawing the Canadian Forces from Europe, it is usually the Army that they have in mind. The Army is the one Service most often identified as having little relevance except in its NATO Europe role, a role that leaves Canada to be defended by only a few widely dispersed regular Army units.[16] The prescription, therefore, is to bring the Army home from Europe to redress this incapability for territorial defence. The White Paper attempts to thwart this idea by promoting the Reserve Forces as the principal means of home defence. If this policy is accepted then Canada would in essence have two Armies—one more or less professional to meet overseas commitments, and another, Militia Army, to handle domestic disputes. On the assumption that a satisfactory plan to defend Canada with the Militia can be developed, it is the expectation of officials that the rationale for bringing the Army "home" may be diminished.

All these notions and plans have a great deal of support from the defence establishment. Each Service believes that it is being asked to do things that it should do and that would contribute to solving Canada's "real" defence needs. The Navy, for instance, feels that it is finally beginning to recover from the stinginess of funding of the 1960s. The Army, on the other hand, has come to embrace the integration of the Regular Army and the Militia, something that it had resisted doing since 1950. Even without the insistence of Beatty, the Services would likely champion the roles he has spelt out for them.

It is noteworthy that as the White Paper was being prepared a strong consensus developed among the Services that each was demanding only its necessary share of the budget and not poaching on the needs of the others. In this regard the CDS, General Paul Manson, must be given full credit for managing the demands of the Services so as to produce from them a coherent military structure and acquisition program acceptable to the Cabinet. Here, at least, Hellyer's initiative to establish a single CDS seems to have fulfilled its promise.

Beatty was determined to develop what he called "an honest funding program." In pursuit of the 15-year defence rebuilding plan that he had constructed, the Minister announced that the government was "committed to real annual growth in defence spending which, except in fiscal emergencies, will not fall below two percent." This level of expenditure is not as high as the defence establishment would have liked. Indeed, many were convinced that any level below four percent would only allow the Canadian Forces to "rust out" as equipment became increasingly worn out and was not replaced quickly enough to maintain capabilities. In any event, a strong consensus on resources requirements developed among officials of the Department of National Defence and senior officers of the Canadian Forces. The Minister's task was to wring that higher level of spending from the

Cabinet or, failing that, to find a compromise that would sustain his pledge of honest funding and the confidence of the defence establishment.

It was some members of Cabinet and officials of other departments, and particularly the Minister of Finance, who raised the most serious challenges to the defence program. The arguments, generally, were not about the needs but about the ability of the government to meet the fiscal demands of the program. As the White Paper was in production, debates continued around the value of each element of the program and ways to finance it. In the end the government settled on a funding formula that did not commit it to the entire 15-year program but allowed Beatty to complete the White Paper on time.

The budget compromise was "two percent and bumps." That is, 2 percent would be the budgetary floor, with the Cabinet, in annual reviews, considering incremental increases above this floor in order to finance selected major programs. In effect the defence budget process will become a "rolling five-year funding plan." The compromise, however, leaves more than a little doubt in many minds about future funding and thus disturbs the consensus somewhat. It may yet inspire some competition among the Services and allow other departments' officials to return to arguments thought settled in the spring of 1987.

Consensus building in the defence establishment was a major activity during the preparation of the 1987 White Paper. The Minister was determined to present to Cabinet a document and policy proposals that had the support of those who would have to carry them out. The outline of the new White Paper had been under consideration for some time before Beatty arrived in the Department of National Defence and these discussions and false starts gave senior officers and officials some experience in drawing up the final document. When Beatty arrived with a clear direction to produce a new defence statement the policy gears were well oiled.

Even so the White Paper did not come easily from the establishment. Officials report that it went through some 20 drafts. There were discussions, some very strongly contested, in which the central themes and methods for the defence of Canada were worked out. It was not a case of the military producing a "wish list" for the Minister's signature but in some cases it was the Minister who challenged the conservative inclinations of the officials. In the end a consensus was reached as to the nature of the problem, the roles of the Canadian Forces, and the funding necessary to implement the policy. For many officers and officials the feeling of teamwork and honesty developed during this period was a new sensation. Within the Canadian Forces especially, there was confidence that officers were making a significant contribution to the policy process in ways that had been impossible under Hellyer or Macdonald. After the White Paper was issued this consensus was to pay dividends to the Minister and the Department of National Defence as the public and the members of the Canadian Forces came to understand and, in most respects, to trust the results of the defence policy review, even if some doubt that the political will can be sustained to provide the funds to support it all.

To what extent were the policies set out in Hellyer's and Macdonald's White Papers actually carried through to completion? Some policies were completed as announced but most were not. In both cases as soon as the Minister left office the preferences of the bureaucracy began to refashion the aims and priorities of the defence policy.

At the surface, Hellyer's White Paper produced the unified Canadian Forces that he desired. But over the long term the latent resentment in the Canadian Forces and in some elements of the public towards unification worked to restrict its complete implementation. Only the persistence of successive Liberal defence ministers held the policy in place. Once governments changed, the policy changed, and the defence establishment returned to a form of "integration" it preferred. Hellyer's White Paper produced no "mobile army," no new resources for capital expenditures, no sealift, and certainly no operational improvements as he had predicted.[17]

Defence in the 70s lasted as policy for less than two years after Macdonald left the Department of National Defence. It produced no new home defence or even sovereignty-surveillance forces, and the lack of decision about Canada's NATO commitments left the Canadian Forces with an emasculated structure in Europe. By 1974 the preferences of those committed to the NATO roles began to reassert themselves as confirmed by the history of Canada's decision to buy new tanks for the European-based units.[18]

Only in one respect did Macdonald's White Paper make a lasting impact, albeit indirectly. The Management Review Group study initiated by Macdonald resulted in a significant shift in bureaucratic power from the military leadership of the Canadian Forces to public servants, a power alignment that survived until 1982 when the definition of the problem changed from program management to operational performance.

Will *Challenge and Commitment* fare better in the long term than previous White Papers? If the degree of consensus in the defence establishment is the controlling criterion for consistent implementation of policy, then this White Paper may well remain the principal guiding document for Canadian defence decisions for years to come. There is a strong consensus about what needs to be done to defend Canada, how it should be done, and what it will take to do it. So long as that consensus holds, the White Paper of 1987 should provide a reliable policy base.

A number of factors, however, could disturb this now solid consensus. For example, should there be a major change in senior officials or officers, new clashes of priorities could develop. Certainly if Beatty were to leave, the direction of the defence policy could be altered unless a new Minister were equally dedicated and politically influential in the Cabinet. Obviously, if the funding formula breaks down such that internal stress is placed on the Services and programs there will be a scrabble as groups try to protect their own interests.

Probably the most unpredictable change, but potentially the most disruptive, would occur if there was a change of government. A new government would be faced with the task of reconciling itself to the consensus in the defence establishment, which means accepting essentially the 1987

White Paper, or building a new one of its own. Given the strength of the support for the present policy and public acknowledgments of it, changing the collective bureaucratic mind might not be easy.

It is hard to conceive of the defence establishment readily abandoning the 1987 White Paper. In a situation in which it was forced to do so Canada could face turmoil in its defence policy and maybe irreparable delays in equipment acquisitions while a new consensus was being constructed. If building a new consensus required the release of officers and officials who could not be persuaded to change their advice there would be a significant loss of morale and confidence in the Canadian Forces. The predictable end of such a scenario would be a government served by a reluctant defence establishment most likely looking to fashion, incrementally, defence policies that it prefers to support.

This analysis should not be taken to suggest that ministers must fall into line with the opinions and policies of their officials or be damned. To the contrary, most officials would be pleased to have clearly articulated policies from their ministers. Defence officials and officers in 1970 were, if anything, keen to try to impart some stability to defence policies after the Hellyer era. The difficulty, however, in trying to administer policies that officials believe to be flawed or, in their experiences, not in the interest of the country, is apparent; consequently ministers must argue convincingly, if they can, to change these attitudes. If they are simply arbitrary in their policy statements, ministers will find themselves compelled to active supervision of their officials, or else they will risk losing control of their policies. Left to their own interests and opinions, officials can only be expected to resolve their inevitable doubts and what appear to them to be contradictory policies by bargaining within the establishment for things they believe to be important. This bargaining process, as Charles Lindblom and others have demonstrated, tends to produce random results based not necessarily on government policies but on the strength of the factions within the establishment in question.

NOTES

1. Government of Canada, *Challenge and Commitment: A Defence Policy for Canada* (Department of National Defence [hereafter DND], Ottawa, 1987). Hereafter cited as 1987 White Paper.

2. Government of Canada, *White Paper on Defence* (DND, Ottawa, 1964). Hereafter cited as 1964 White Paper.

3. D. Blackburn, *Canadian Sovereignty, Security, and Defence: A New Democratic Response to the Defence White Paper* (Ottawa, 1987), 21–24.

4. Government of Canada, P. Hellyer, *Address on the Canadian Forces Reorganization Act* (House of Commons, Ottawa, December 1966), 13. Hereafter cited as Bill C243.

5. Canada, Ad Hoc Committee on Naval Objectives, *Final Report*, July 1961, 73/250 (DND Directorate of History, Ottawa). For a description of the 1960s discussion about submarine acquisitions, see S.M. Davis, "It Has Happened Before: The RCN, Nuclear Propulsion and Submarines,"

Canadian Defence Quarterly 17 (Autumn 1987), 34–40.

6. For a lively description of Brock's bureaucratic battles with Paul Hellyer see J. Brock, *The Thunder and the Sunshine* (Toronto: McClelland and Stewart, 1983).

7. For a complete description of Service reaction to unification, see V. Kronenberg, *All Together Now: Canadian Defence Organization*, Wellesley Papers 3/73 (Toronto: Canadian Institute for International Affairs, 1973).

8. For example, see D. Bland, *The Administration of Defence Policy in Canada: 1947–1985* (Kingston: R.P. Frye, 1987), 25–31.

9. R.J. Sutherland, *Canada's Strategic Situation and the Long Term Basis of Canadian Defence* (Ottawa: DND, 1962).

10. Government of Canada, *Defence in the 70s* (Ottawa: DND, 1971).

11. Government of Canada, Department of External Affairs (hereafter DEA), *A Defence Policy for Canada*. Statement to the Press by Prime Minister Pierre Trudeau, Calgary, 3 April 1969. DEA Statements and Speeches No. 69/7.

12. J.D. Anderson and J.C. Arnell, "Program Management in the Department of National Defence," *Canadian Defence Quarterly* (Autumn 1971), 31–33 .

13. Government of Canada, Department of National Defence, *Report to the Minister of National Defence on the Management of Defence in Canada*, Management Review Group (hereafter MRG) Report, (Ottawa, 1972), 25, 28. For an analysis of the MRG Report see Bland, *Defence Administration*, 59–85.

14. Government of Canada, House of Commons, Perrin Beatty Speech on tabling, 5 June 1987, *Challenge and Commitment*.

15. A major initiative at the "Shamrock Summit" held in Quebec City between Prime Minister Mulroney and President Reagan in March 1985 was Canada's announcement to increase its deployments in Europe and to increase defence activities generally.

16. NDP Defence Statement, 1987, p. 22.

17. For a critique of the era of unification see Government of Canada, *Task Force on Review of Unification of the Canadian Forces: Final Report* (DND, Ottawa 15 March 1980). The Report concluded that it was "dubious whether unification has achieved the intended goals" (p. 60).

18. On 27 November 1975, James Richardson, MND, reported to the House of Commons the results of a Department of National Defence study entitled, *The Defence Structure Review*. It was begun in the Department in late 1973 and substantially reversed *Defence in the 70s* by confirming the NATO role and announcing equipment purchases, such as tanks, to support it.

CIVILIANIZATION AND THE CANADIAN MILITARY

W. HARRIET CRITCHLEY

o

... Canada is both unique and radical in its organization of defence. Every other country in the world has organized its defence around the concept and tradition of at least three separate armed services—army, navy, and air force. Canada is unique not only in thoroughly integrating but also in *unifying* its armed forces into a single entity: the Canadian Forces. While some operational elements of the CF were discernible as army, navy, or air force units even before the recent decision to employ distinctively colored uniforms for all personnel, the majority of uniformed personnel in the CF have not been so easily identifiable.[1] In addition to the retention of a unified CF, the higher reaches of defence decision making in the Department of National Defence (DND) remain both integrated and unified across the different CF operational and support types *and* between military and civilian personnel.

The decisions, first, to unify Canada's armed forces and, later, to integrate the military and civilian personnel remain controversial to this day. In this author's opinion, the controversy persists largely because of the lack of any convincingly established rationales for these radical steps. Two recently published works, *Not Much Glory* and *The Administration of Defence Policy in Canada*, have raised these fundamental questions again.[2] In some respects these two works parallel, but do not mirror, two earlier thorough investigations of the same questions—one by the Task Force on Review of Unification of the Canadian Forces and the other by the Review Group on the Report of the Task Force on Unification of the Canadian Forces.[3] These latter reports remain the only comprehensive analyses of the unification policy available in the public domain.

⬦ *Armed Forces and Society* 16, 1 (Fall 1989), 117–36.

This article will focus on one set of problems identified in three of the above works and alluded to in the fourth: the so-called "civilianization of defence" that has resulted from the implementation of the unification policy over the past two decades.[4] Discussion of the problem of the "civilianization of defence" in Canada frequently consists of an unwittingly confused mixture of at least three distinct and different phenomena. Firstly, "civilianization" can be used to refer to the principle, established in democratic forms of government, of civilian control (by elected members in the executive and legislature of the national government) over the nation's military organizations. Secondly, "civilianization" is used to characterize the development of a so-called nine-to-five mentality among certain individuals and groups within the military profession. This mentality is disparaged by other military professionals in Canada, who feel strongly that, once having accepted "the Queen's shilling," they are bound to serve willingly at any time or place and that their duties are not confined to weekday civilian business hours. Finally, "civilianization" can also refer to the perceived marked increase in the number of civil servants employed by DND who are performing "strictly military jobs," or making "strictly military decisions" in National Defence Headquarters (NDHQ). . . .

This article will focus on the third phenomenon: the perception that civil servants are increasingly performing "strictly military jobs" and making "strictly military decisions" in NDHQ, and the further perception that this "problem" is a direct result of implementation of the unification policy.[5] The use of the term *civilianization* in the remainder of this article will refer only to these two perceptions. It will be argued here that, as of 1989 at least, the Canadian military has not been increasingly civilianized. To the contrary, the present organization of NDHQ gives the military a much greater influence—on a broader range and at a higher level—over defence decision making than in the past.

UNIFICATION AND CIVILIANIZATION

Because civilianization is so frequently linked to unification, discussions of "the problem" and potential solutions almost automatically involve the very important issues of command and control within and of the military organization as well as the broader issue of defence management in Canada. The following complaint is an example of how such issues are viewed by Canadian critics of civilianization.[6]

> There has been a significant increase in the numbers of civil servants filling NDHQ positions that require them to make decisions which have a direct impact on the lives and performance of the officers and other ranks in the field. These decisions should only be made by military officers and the positions, which were filled by military officers in the past, should revert to military officers. A direct result of this increase in civil servants in "military" positions in NDHQ is the corresponding decrease in military influence over defence decision making. This trend should be revised by restructuring NDHQ

to resemble the Canadian Chiefs of Staff system of the early 1960's, or the current U.S. Joint Chiefs of Staff system, or the current system in the United Kingdom.

The perception of increasing civilianization is a relative rather than absolute evaluation. That is, a comparison is being made between the number and role of civil servants in the unified NDHQ and the number and role of civil servants in defence headquarters at some point prior to unification.

This focus on "pre-unification" and "post-unification" comparisons is the most common approach. Closer examination, however, reveals a number of difficulties with this characterization. Pre-unification, in the correct use of the term, would refer to the situation immediately prior to 1 February 1968 when Bill C-243 (the Canadian Forces Reorganization Act) came into effect. However, many commentators use the term pre-unification to refer to the situation immediately prior to 1 August 1964 when Bill C-90 (Act to Amend the National Defence Act) came into effect. Since there were major structural changes in defence headquarters and the Canadian military forces after each of these acts was promulgated, clarity on the date for pre-unification is essential to any serious analysis of the problem of civilianization. Similarly, clarity on the date being used to mark post-unification is also essential. Although the unification policy came into effect in February 1968, since then a series of major changes has occurred in the structure of the CF and the NDHQ. Some of the changes were related directly to the implementation of the unification policy, but others were, at most, indirectly related.

In short, there has been an *evolution* in the structure of the unified CF and NDHQ from 1968 through to the present. Since each of the pre-unification and post-unification changes has a bearing on any evaluation of civilianization and since, by all appearances, this is not sufficiently appreciated by many commentators, it is useful to analyze what the *Task Force Report* calls the "evolution of unification."[7]

EVOLUTION OF THE POLICY

PRE-UNIFICATION PERIOD

Until 1964, Canada had three separate armed forces—the Canadian Army, the Royal Canadian Navy, and the Royal Canadian Air Force—with three separate headquarters in Ottawa.[8] Each headquarters was responsible for military plans and the consequent operational requirements in terms of manpower, training, equipment, supply, and the general welfare of military personnel. Co-ordination of military activities was to be achieved through a Chiefs of Staff Committee, which was headed by a chairman. In addition to these armed forces organizations, there was a departmental organization headed by the deputy minister. The latter organization was staffed predominantly by civil servants and had administrative, legal, financial, procurement, construction, and civilian personnel responsibilities.

Although the three service headquarters and the departmental organization were virtually four separate structures—with information flowing up and down within each structure—co-ordination was attempted by the creation of some 200 interservice committees and by meetings of the Defence Council.[9] The Defence Council included the minister, parliamentary secretary, deputy minister, associate deputy minister, and the Chiefs of Staff Committee. Depending on the issues being discussed, civil servants from other departments of the federal government and the Cabinet Secretariat would attend both Defence Council and Chiefs of Staff Committee meetings—indeed, they attended interservice committee meetings as well.[10] As many "purely military" policy decisions had an impact on Canada's foreign relations, domestic economic and fiscal policies, other domestic priorities for overall government action, and the federal budget, the attendance of officials from departments and agencies outside the DND and the armed services was entirely understandable. This important fact deserves emphasis because much of the Canadian commentary today on civilianization leaves the impression that under the chiefs of staff system of the early 1960s, "the military" was in charge of defence and defence policymaking. Such an impression is mistaken and seriously misleads the debate on civilianization: the "good old days," when the military was in charge of defence policymaking, never existed.

The organizational structure described above reflected a national priority accorded to defence that stemmed from Canada's military requirements, experiences, and activities during World War II and the Korean War. By 1961, however, governmental decisions on defence were beset by quite a different constellation of concerns: other domestic political priorities were competing for governmental attention and expenditure; new strategic concepts and developments in military technology were being discussed in the Western Alliance; the armed forces had suffered a series of embarrassing and expensive weapons procurement fiascoes; and personnel and administrative costs within the armed forces were rising dramatically to the point at which it was "predicted that, without significant budget increases, there would be no money for capital equipment acquisition by 1969."[11]

In 1961 the federal government appointed the Royal Commission on Government Organization—known as the Glassco Commission—"to examine the whole of the government service, to eliminate duplication and uneconomic operations, and to recommend improvements in decentralization and more efficient management practises."[12] The fact that defence was a major focus of the commission's inquiry is highlighted in its subsequent report in 1963:

> The dramatic change that has taken place in the level of federal expenditures since 1939, a twelve-fold increase, has two main sources—the broad enlargement of social services and the development of a peacetime defence organization on an unprecedented scale. . . . From a total of $385 million in the fiscal year 1950, defence expenditures increased to $1652 million in 1961–62. For the

first time in the history of Canada in peacetime, there are *over 125 000 serving in the Armed Forces, with almost 50 000 civilians also employed*. . . .

In this inquiry, the Department of National Defence has been singled out for a number of reasons. The most obvious are its size, the range and cost of its activities, and the impact of Western defence alliances. *Moreover, the composition of the department is unique, consisting as it does of two elements, military and civilian, differing in status, rank structure and terms of employment, although they function as an entity*. Also of significance is the character of the Armed Forces, whose members, organization and skills are predicated on wartime tasks, with the consequences that utilization in peacetime is a problem.

The $1652 million spent in 1961–62 by the Department of National Defence . . . represented 25 percent of total federal expenditures, but defence activities in terms of employment, equipment and other operating costs accounted for an even larger proportion of federal government operations . . . expenditures . . . more than 40 percent.[13]

The Glassco Commission made a series of recommendations with respect to the DND. The most important of these were the replacement of the chairman of the Chiefs of Staff Committee by an appointed chief of the Canadian Defence Staffs to direct the armed forces and control their common elements; the assertion of civilian control by giving the deputy minister greater responsibilities for reviewing the department's organization and administration on behalf of the minister; and the requirement to review civilian and military manpower needs in each of the three services.[14]

These and other recommendations also addressed the need to "promote efficiency . . . in light of their [the recommendations'] possible effect on operational effectiveness."[15] In short, the commission's recommendations were aimed at promoting greater co-ordination among the separate services and between the armed forces' headquarters and the department. It was felt that with such co-ordination and some further integration of support functions (including greater utilization of civilian personnel), the true tasks of providing national defence could be achieved more efficiently and at less cost than had been previously predicted.[16]

The Glassco Commission did not investigate the concept and practice of civilianization as defined in this article, but its report did make some observations that are germane to the topic. In one part of the report the commission noted that the vast majority of the 50 000 civilians in the DND were employed as tradesmen or junior administrators, and that even in those non-combatant support areas in which there was extensive civilian employment, almost all senior positions were held by military officers. The commission observed that this situation had two effects: it limited opportunities for specialization and advancement among civilian employees, while the frequent rotation and early retirement of military officers undermined management continuity. In short, the system was inefficient in its use of personnel, and organization effectiveness suffered as a result.

In recommending a policy of greater employment of civilians in the (senior) administration of the armed services, the commission noted that

> experience elsewhere has shown that many senior administrative tasks of the Services can be efficiently performed by civilians, even in such fundamentally military staff functions as those dealing with plans and operations. . . . [Such civilians'] sole function should be to assist the Services and provide continuity in administering programs. . . . *One object of [this] policy is, in fact, to prevent the segregation of civilian and military elements into two separate organizations between which antagonisms can too easily develop.*[17]

The commission concluded that part of their report by making two recommendations:

1. Career opportunities be improved for civilian technicians and administrative personnel employed in the Armed Forces.

2. There be a greater interchange of Service and civilian officers, especially of the intermediate and senior rank, between the headquarters staffs of the Armed Forces and the organization of the Deputy Minister.[18]

It should be noted that these observations and recommendations were being made in the interests of managerial efficiency; although integration and even unification of the armed forces are mentioned in passing,[19] the whole tenor of the commission's report clearly assumes the continued existence of three separate services. In other words, the commission was concerned with improvements in the management of the existing defence organization and armed services.

The commission's report also contains some interesting data on the degree to which the services' headquarters were "military organizations" in the sense of being staffed predominantly, if not exclusively, by military personnel. These data were presented in the form of questions:

> Why does the Royal Canadian Air Force require 58 percent of its headquarters establishment in uniform and the Army 52 percent, while the Navy gets along with 20 percent? . . . What circumstances dictate that 54 percent of the Air Materiel Command should be in uniform, when the Navy can manage the similar function with approximately 5 percent?[20]

There can be little doubt that some of the variation highlighted in these questions was due to different rank structure requirements of each of the services and to the different functions performed by headquarters in each of the services. In addition, as quoted earlier in this article, a substantial portion of the civilians in the headquarters occupied clerical and junior administrative and technical positions. Nevertheless, given the high proportion of civilians in the headquarters organizations (42% to 80%) and the large numbers of interservice committees to be staffed, it is difficult to envisage how the military was in charge of *all* aspects of defence decision making, or how civil servants were *not* making decisions with direct impact on the lives and performance of officers and other ranks in the field.

The Glassco Commission report was published in 1963, and the analysis and recommendations it contained were eventually to exert considerable influence on the management of defence in Canada. Indeed, even before the report was written or published, internal Defence Department studies were undertaken on some of what were seen to be emerging as the major commission recommendations.[21] Before those recommendations could be implemented in a straightforward manner, however, a separate series of events intervened to change the whole context for seeking better co-ordination and efficiency.

IMPLEMENTATION OF UNIFICATION

In 1963, a new government took office and, among other things, decided to conduct a wide-ranging review of defence policy and organization. The result of this review was the publication of the *White Paper on Defence* in March 1964 and, shortly thereafter, the introduction of legislation to reorganize the headquarters of the DND and the field command structure. The legislation—Bill C-90—is commonly referred to as providing for the "integration" of Canada's three armed services.[22] The purpose of the reorganization was to allow Canada's armed forces to meet the objectives of Canadian defence policy, as enunciated in the *White Paper on Defence*, more efficiently and effectively.[23] However, the continued lack of intradepartmental co-ordination despite the existence of the many interservice committees, the slow progress being made in integrating common services and functions, and the escalating costs of providing for defence under the existing structure were equally important considerations in drawing up the legislation.[24]

Co-ordination of military activities at headquarters was to be achieved by appointing a single chief of defence staff and creating a headquarters organized along functional lines. In the past there were functional subdivisions within headquarters—personnel, comptroller, and technical services, for example; however, each of these was separated by service—that is, there were three different subdivisions for each function. Thus, there was a comptroller subdivision in naval headquarters, another in army headquarters, and a third in air force headquarters. The same was true for the personnel and technical services subdivisions. In addition, each service's chief of staff was the commander of the service headquarters and of the service's operational units in the field.[25]

The reorganization of 1964 reversed such organization priorities: functional divisions were created and each of the services contributed personnel to each division. The chief of defence staff commanded all of these functional divisions, as well as all of the operational units in the field. Over the next few years, near chaos reigned in headquarters as this reorganization was put into effect, but the overwhelming majority of personnel—including military personnel—worked with determination to make integration and functional organization succeed.

Two other changes in the headquarters organization are of direct relevance to the topic of this article: the new Defence Council and a change within the deputy minister's departmental organization.

Prior to the 1964 reorganization, the Defence Council consisted of four senior military officers (the Chiefs of Staff Committee) and five civilians, including the minister. Depending on the issue being discussed, other military officers and civil servants attended council sessions. After the 1964 reorganization, the Defence Council consisted of two senior military officers (the chief and the vice chief of defence staff) and three civilians, including the minister. Again, depending on the issue being discussed, other military officers and civil servants attended the meetings. Although the new Defence Council was somewhat smaller and the intention was to make this the forum in which final decisions were taken,[26] the relative proportion of military to civilian representatives had hardly changed.

Secondly, by comparison to the organizational revolution occurring in the chief of defence staff's part of the department, the deputy minister's part remained almost unchanged. The Comptroller General Inspection Services division was eliminated and the Information Services–Parliamentary Returns division was added. Neither change was significant, in view of the deputy minister's overall responsibility for financial supervision and the fact that the deputy minister's part of the department had traditionally handled parliamentary returns—that is, the preparation of answers for the questions put to the minister during Question Period in the House of Commons.

To summarize, the reorganization of 1964 had a dramatic effect on the services headquarters, but it left virtually intact the division between the "military" and "civilian" aspects within the overall departmental organization.

A second major revolution in the organization of defence was signaled less than three years later, in November 1966, with the introduction of Bill C-243 (the Canadian Forces Reorganization Act) in the House of Commons. This legislation is commonly, and correctly, referred to as providing for unification of Canada's armed forces: it abolished the Canadian Army, the Royal Canadian Navy, and the Royal Canadian Air Force and in their place created the Canadian Forces—a single service with a common uniform and common rank designations. After considerable public controversy, House of Commons committee hearings, and debate on the floor of the House, Bill C-243 was passed in April 1967 without major changes and came into effect on 1 February 1968. In 1964, the goal was—to put it bluntly—to integrate as much as possible, as rapidly as possible. In 1968, the goal was to unify everything, period. Headquarters personnel carried out this task in an unenviable environment—continued public controversy and the resignations of a number of senior and middle-ranking officers, for example—and without the benefit of any plan or even detailed appreciation of the meaning and scope of unification.[27]

For the military personnel at headquarters, as elsewhere in the CF, this demanded a personal reorientation of considerable proportions: they were now all in the one armed service; the new designations of the land, air, and maritime "environments" or "elements" superseded the terms army, air force, and navy—the latter having been abolished from speech and prose.

Admirals and air marshals became generals, while naval captains and lieutenant commanders became colonels and majors; certain procedures and modes of organization or administration as practiced in one or another of the former three services were chosen to be applied across-the-board in the new CF without prior consideration of the potential effects on operations in the different "environments."

Much of the commentary and scholarly attention to unification has focused on the above-mentioned types of changes. It is, therefore, rather surprising to see the degree to which the 1968 revolution in military organization did not produce basic change in the organization of the headquarters. Many of the major divisions within headquarters and the lines of authority continued as instituted by the 1964 reorganization. As before, headquarters continued to be divided into the predominantly military organization, including the operational units in the field, commanded by the chief of defence staff and the predominantly civilian organization headed by the deputy minister. The membership of the Defence Council remained unchanged, as did the basic organization of the whole headquarters along functional lines.

Within the military part of headquarters, the only change of any significance was the restriction of the vice chief of defence staff's responsibilities to supervising the Plans, Operations, and Reserves divisions, and the consequent "promotion" of the chiefs of Personnel and the Technical Services, and the Comptroller General to the vice chief's level in the organization. Each of the four officers now reported directly to the chief of defence staff. Other changes occurred within the divisions supervised by the vice chief. Among these were the replacement of the chief of operational readiness by a deputy chief of operations and the reorganization of that division along environmental rather than strictly functional lines. The Plans, Intelligence, and Program divisions were reorganized into one division headed by a deputy chief (rather than directors general). The deputy chief reported directly to the vice chief, and the position of assistant chief of defence staff was abolished. A new deputy chief, Reserves was created.

The only change of significance in the deputy minister's part of the department—by now it was commonly referred to as "DM's Alley"[28]—was the amalgamation of the divisions for Works and for Requirements (each headed by an assistant deputy minister, or ADM) into one Logistics division, headed by a single ADM.

To summarize, the implementation of unification did necessitate a revolution within the structure of the armed forces and a readjustment for many military personnel. However, it did not produce changes on anything approximating the same scope or scale in the basic organization of headquarters. That particular revolution came four years later.

In spite of consolidation into one department in 1946, creation of a chairman, Chiefs of Staff Committee, in 1953, integration of the armed services under the command of a chief of defence staff and reorganization of headquarters along functional lines in 1964, and the reorganization of the armed forces into a single service in 1968, "there were still problems in the Department of National Defence management."[29] Those problems continued to be the need for better co-ordination of planning and budgeting, better accountability and control of capital acquisitions, elimination of costly duplication of effort and expense, and more effective relationships with other government departments that had an input into defence decision making. The persistence of these basic problems led to the appointment of the Management Review Group (MRG)—known as the Pennyfather Commission—in 1971.[30]

The MRG was asked "to examine all aspects of the management and operation of the Department of National Defence."[31] The central finding of the MRG, as stated in the group's report, which was submitted to the minister of National Defence in July 1971, was as follows:

> We were favourably impressed with the leadership of the Canadian Armed Forces in the field, and *concluded that their principal difficulties in carrying out operational tasks were due to basic management problems within the civilian and military components of Headquarters.*
>
> We found that these problems were, in the main, a direct result of the Department not having adapted its basic management and organization to accommodate adequately the implications of unification, changing defence roles and priorities, and changing public attitudes.[32]

The report went on to itemize five basic inadequacies that were involved in this central finding. Of these, one is particularly germane to the topic of this article:

> lack of unity of purpose due to a high degree of parallelism and duplication of management responsibility among its three major divisions—the Deputy Minister's staff, the Canadian Armed Forces, and the Defence Research Board—and instead, *the development of adversary relationships and undue compartmentalization.*[33]

This MRG finding was exactly what the Glassco Commission had hoped to avoid in 1963 (before integration and unification) with its recommendations. For its part, the MRG recommended that the "first step toward a better, more effective and efficient defence establishment must be the restructuring of the Department as a single entity."[34] The subsequent implementation of many of the MRG recommendations, in effect, amounted to reorganization of the one area of the department that had remained constant in the post-World War II era: the separation of the "military" and "civilian" parts of the headquarters. The reorganization that followed can best be characterized as the *integration* of these two parts of the department to form a single organization.

Implementation of the MRG recommendations led to a comprehensive restructuring of the entire department, but particularly of the senior management structure. In the past, plans and projects priorities would often filter up through the military part of headquarters, becoming fully elaborated and approved within that part of the department, only to run into a virtual wall of rejection on "DM's Alley," which would then often require a complete reprocessing of the plan or project through the military part of headquarters.[35] This laborious, time-consuming, and basically wasteful procedure carried no guarantee of final approval on "DM's Alley." Critics traced these operational problems to basic flaws in headquarters structure: there was little meaningful interaction between the two parts of the department except at the very top. Integration from bottom to top of the two parts of the department into a single organization was seen as the optimal solution. This would require interaction from the first stages of any plan or project, and thus allow for difficulties or problems to be perceived—and resolved—earlier in the process. It would also provide for real co-ordination within the *whole* of the headquarters. The use of functional lines of organization, as set out in the 1964 reorganization and retained in the 1968 reorganization, was not only retained but reinforced in the 1972 reorganization.[36]

The following example indicates, to a very large degree, how integration within headquarters was accomplished. What had been the military part for personnel, headed by a chief of personnel, was amalgamated with the civilian subdivision for personnel, which had been headed by an ADM. The new amalgamated Personnel division was headed by an ADM who was aided by an associate deputy minister. If the ADM was a civilian, the associate DM would be a military officer and vice versa. Civil servants and military officers who had worked separately in their respective divisions were now also amalgamated. The same integration occurred for the division of Finance/Comptroller General, which became Finance, and for that of Technical Services/Logistics, which became Materiel.

Information and the Judge Advocate General's (JAG) divisions remained as distinct units, while Intelligence and Security was removed from the Plans division and became a separate division on its own. The remainder of the old Plans division was rearranged and expanded: one series of divisions, headed by an ADM, was responsible for overall planning and program control along with independent operational research and evaluation of plans and programs; another series of divisions, headed by the deputy chief of defence staff (a new position), was responsible for military planning, tactical doctrine development, and military operations, as well as reserve forces.

Since the early 1970s, then, the integrated headquarters has consisted of five major groups of divisions—Policy, Personnel, Finance, Materiel, and Military Plans, Doctrine, and Operations. Each group is headed by a group principal at the ADM or three-star rank, who can be a civil servant or military officer. In addition to the five major groups, there are the three independent divisions (Information, JAG, Intelligence and Security) mentioned above. The vice chief of defence staff acts as the general manager of the entire headquarters and is second in command to the chief of defence staff for the CF. The chief of defence staff and the deputy minister are seen as coequals who, while having distinct responsibilities, share the responsibility for the effective and efficient functioning of the integrated headquarters. They, in turn, are responsible to the minister of national defence.[37]

The Defence Council has undergone one change as a result of the institution of an integrated headquarters: the ADM (policy)—the senior of the five group principals—replaced the chairman of the Defence Research

Board (this position was abolished),[38] but the military/civilian balance remained unchanged at three to two.

The evolution of unification did not end with the reorganization of National Defence Headquarters that resulted when the MRG recommendations were implemented. Many intrasystemic problems—some recognized and others not officially acknowledged—remained and continue to create a variety of difficulties.

More importantly, in this author's opinion, three extrasystemic factors—that is, factors having no direct relationship to unification per se—had a significant and simultaneous impact on the functioning of the CF and the NDHQ during the late 1960s and the 1970s. These three factors were "decreases, in real terms, in the defence budget; decreases in authorized military personnel levels; and increases in the number and types of tasks set for the Canadian Forces."[39] While these factors had a broad impact on defence in general and on the CF in particular, they did not significantly affect the degree of civilianization and will, therefore, not be discussed here.[40] Nevertheless, they did continue to affect the evolution of unification.

Of the changes instituted since 1972, the most significant was the formation of Air Command in 1975 by drawing together the former Air Transport and Air Defence Commands as well as tactical units deployed in Europe and attached to Mobile and Maritime Commands. Again, it was the continuing necessity to cut costs and improve co-ordination in planning and operations that prompted this change. Although this change had virtually no impact on the organization of NDHQ at the time, it did mark the re-emergence of an air force (Air Command), a navy (Maritime Command), and an army (Mobile Command). But usage of the terms "army," "navy," and "air force" did not receive official public sanction until they appeared in the *Task Force Report* in 1980, and the official titles of the commands remain as they were in 1975.

CONCLUSION

Canada's defence headquarters, the armed forces as military organizations, and individual members of the armed forces have undergone a rather rapid series of major organizational upheavals over the past two and a half decades. During much of that same period, many of these same individuals have also been engaged in a sheer survival operation: that is, survival as an effective military organization able to fulfill government defence policy in the midst of apparent changes in that policy and of real decreases in the defence budget and in authorized military personnel levels. Given this complex and confusing context, it may not be too surprising that some military personnel perceived an influx—if not invasion—of civil servants into "military jobs" in headquarters. This is a misperception that unfortunately continues today.

This article has attempted to deal with the misperception by analyzing what exactly *did* happen to the headquarters organization from 1963 to the present. With respect to the issue of civilianization, it has been shown that the key changes occurred after 1972—when the military and civilian parts of headquarters were integrated—and not after 1968, when Canada's armed services were unified.

The Task Force on Review of Unification of the Canadian Forces did hear testimony on the issue of civilianization and felt bound to record a summary of these opinions.[41] Because of the complexity in the changes of 1964, 1968, 1972, and 1975, the data that *can* be obtained on numbers of civilian and military personnel in headquarters at various times during the period are noncomparable.[42] The review group also made an attempt to analyze senior-level personnel data for this period and, although they found no change in proportion of military officers to civilians at that level, they also recognized that "organizational changes have made valid comparisons virtually impossible."[43]

Both the task force and the review group encountered a perception of increasing civilianization at headquarters. The fact that both bodies regarded the issue as one of misperception is shown by the following excerpts from their respective reports:

> In part, this perception may be based on an incomplete understanding by those in the field about the nature and functioning of National Defence Headquarters, and in part, attributable to the composition of the Defence Management Committee which is perceived to be dominated by civilians. . . .[44]
>
> [P]roblems concerning civilianization in senior level decision making are ones of perception and education. . . . [T]here remains an educational task to ensure that the organization, division of responsibilities, and management system of NDHQ are clearly and simply understood throughout the Forces.[45]

As to the *reality* of civilianization, the *Task Force Report* made one terse statement: "The Task Force feels that the relationships between civilian and military personnel are of a high quality."[46] The *Review Group Report* made an almost identical statement.[47]

One aspect of the civilianization issue—the role of senior personnel in defence decision making in the current headquarters organization—has purposely been left for discussion in the conclusion of this article. Defence decision making within the DND is accomplished by a hierarchy of senior committees: the Defence Council, the Defence Management Committee, and the Program Control Board.

Changes in the composition of the Defence Council, in terms of the proportion of civilians to military members, has been mentioned earlier and traced through the various changes in headquarters organization. Although the size of the council declined slightly over time, the proportion of civilians to military officers has not: the civilians have always been in the majority.[48] The Defence Council is the most senior of the committees listed above and is the forum for the exchange of views between the minister (chairman of the Defence Council) and his senior advisers: the deputy minister, chief of defence staff, parliamentary secretary, vice chief of defence staff, assistant deputy minister (Policy), and the deputy chief of defence staff. This committee is also the ultimate intradepartmental vehicle for policy co-ordination.

The Defence Management Committee is the next major departmental co-ordinating committee. "It considers all significant matters of policy, plans, programs and administration that require the approval of the Minister, the Deputy Minister, or the Chief of Defence Staff."[49] It is chaired by the deputy minister, co-chaired by the chief of defence staff, and includes the vice chief of defence staff and the five group principals (the ADMs for Policy, Personnel, Finance, Materiel and the deputy and chief of defence staff). The proportion of military officers to civil servants on this committee depends on the occupants of the group principal positions, but recently it has been evenly divided.

The Program Control Board is the senior committee that controls the Defence Services Program, the comprehensive budgeting plan that matches expenditures to resources. This committee is chaired by the vice chief of defence staff (in his capacity as general manager of NDHQ) and includes the group principals, chief of program, and chief of review services. Recently, the membership of this committee has been evenly divided between civil servants and military officers.

In 1980, as a result of recommendations by the Task Force on Review of Unification of the Canadian Forces, the three senior environmental commanders (the commanders of Mobile Command, Maritime Command, and Air Command) were made members of the Defence Council and the Defence Management Committee. This change has increased the proportion of military officers on these two most senior committees, giving the military officers a clear majority on each for the first time. In addition, these three senior officers—along with the vice and deputy chiefs of defence staff and the ADM (Personnel)—form the Armed Forces Council, under the chairmanship of the chief of defence staff.

The membership of these various senior committees reflects the statutory responsibilities of the minister, the deputy minister, the chief of defence staff, and their respective staffs. Further, it reflects the organization of headquarters along functional lines and the subsequent integration of the military and civilian parts of the department.

This integration did not result in an influx of civil servants into the organization. It merely brought together the two sets of people—in two hierarchies, working largely separately from one another.

The aim of that integration and the organization of headquarters along functional lines was to provide for better co-ordination and management of defence in Canada. In the process, rather than allowing for increased civilianization, the military has—by virtue of its increased membership in each of the senior committees—*more influence*, over a broader range and at a higher level, on defence decision making than in the past. This is a fact that current commentators, particularly military personnel critical of the current organization of headquarters, would be wise to consider very carefully when entertaining ideas of returning to the Canadian headquarters system of 1963 or the adoption of the organizational systems of foreign headquarters.

NOTES

1. For example, in its entry under "Canada," *The Military Balance* (London: International Institute for Strategic Studies) traditionally adds a footnote to the effect that a particular number (usually at least half) of the total strength of the Canadian Forces (hereafter CF) is not identified by service.

2. Dan G. Loomis, *Not Much Glory: Quelling the F.L.Q.* (Toronto: Deneau, 1984), and D. Bland, *The Administration of Defence Policy in Canada: 1947 to 1985* (Kingston, ON: R.P. Frye and Co., 1987).

3. *Final Report, Task Force on Review of Unification of the Canadian Forces, March 15, 1980* (Ottawa, 1980) and *Report, Review Group on the Report of the Task Force on Unification of the Canadian Forces, August 31, 1980* (Ottawa, 1980). (Hereafter cited *Task Force Report* and *Review Group Report* respectively.)

4. Some of the material in this article was published previously and is used here as necessary supporting evidence for the thesis in this article. W.H. Critchley, "Changes in Canada's Organization for Defense, 1963–1983," in *Reorganizing America's Defense*, ed. R.J. Art, V. Davis, and S.P. Huntington (Washington, DC: Pergamon-Brassey's, 1985), 136–48.

5. The author of this article was a member of the Task Force on Review of Unification of the Canadian Forces (hereafter Task Force on Review) and co-author of the *Task Force Report*. The views expressed here are solely those of the author and should not be interpreted, in any way, as representing the views of other members of the task force, the Department of National Defence (hereafter DND), or the CF.

6. This is not a direct quotation from a particular individual or group. It is, rather, an amalgamation (in the author's own words) of a very large number of oral and written submissions to the Task Force on Review.

7. *Task Force Report*, 26.

8. Although some changes occurred in 1953—most notably the creation of the position of chair of the Chiefs of Staff Committee—the basic structure of the three separate headquarters and of the deputy minister's departmental organization remained unchanged from 1947 to August 1964.

9. *Task Force Report*, 27, and author's notes on testimony before the task force; see also *Royal Commission on Government Organization*, vol. 4 (Ottawa: Queen's Printer, 1963), 70.

10. Author's notes on testimony before the task force.

11. *Task Force Report*, 14, 27.

12. Ibid., 27.

13. *The Royal Commission on Government Organization*, 61–62 (emphasis added by this author).

14. *Task Force Report*, 27.

15. *The Royal Commission on Governmental Organization*, 64.

16. Ibid., especially 65–67, 72.

17. Ibid., 78–79 (emphasis added by this author).

18. Ibid., 79.

19. Ibid., 69.

20. Ibid., 83–84.

21. Ibid., 83, and *Task Force Report*, 28.

22. There remains considerable controversy as to whether Bill C-90 was a conscious preliminary step toward unification of Canada's armed forces, as set out in Bill C-243. Although the Task Force on Review looked at this question very thoroughly, it was concluded that the terms *integration* and *unification* were used so indiscriminately at the time that no clarification of the issue—one way or the other—was possible; *Task Force Report*, 5–6.

23. Canada, DND, *White Paper on Defence* (Ottawa: Queen's Printer, March 1964), sect. 5, pp. 21–26.

24. *Task Force Report*, 28.

25. In 1965 the field commands were reorganized into a new integrated structure that reduced the number of commands from 12 to 6: Mobile, Maritime, Air Defence, Air Transport, Material, and Training. It should be noted that this reorganization occurred *before* unification in 1968 and was retained until 1975, when Air Command was created. Therefore, the abolition of service designations for commands and their replacement by environmental/functional designations predated unification.

26. V.J. Kronenberg, *All Together Now: The Organization of the Department of National Defence in Canada, 1964–72*, Wellesley Paper no. 3 (Toronto: Canadian Institute of International Affairs, 1973), 108–109.

27. The designation "DM's Alley" was derived from the physical layout of the headquarters building, wherein all of the ADMs' offices were situated along one corridor.

28. Author's notes on testimony before the Task Force on Review.

29. *Task Force Report*, 31.

30. Ibid.

31. Management Review Group (hereafter MRG), *Report to the Minister of National Defence on the Management of Defence in Canada* (July 1972), i.

32. Ibid. (emphasis added by this author).

33. Ibid., ii (emphasis added by this author).

34. MRG, *Report*, iv.

35. Author's notes on testimony before the Task Force on Review.

36. One major set of MRG recommendations was not implemented. The set involves creating an "Office of the Minister," including the deputy minister in the "Office of the Minister," redesignating the chief of the defence staff to chief of the defence forces, creating an "Office of the Chief of Defence Forces," and making both the chief and the chief's office directly responsible to the "Minister's Office." If this set of rec-

ommendations had been implemented, the chief of the defence forces would have been responsible to the minister *through the deputy minister* and the deputy minister would therefore be senior to the chief. MRG, *Report*, iv–v, 59ff., 63ff., Exhibit II. Rather than implement this set of recommendations, the decision was taken to make the deputy minister and the chief co-equals.

37. The legal relationship of these three most senior officials are contained in *National Defence Act*, 1950, sect. 3, as amended in 1976–77, and sect. 18 (1), as amended in 1964. For further elaboration of these positions, see the *Task Force Report*, 10–13, and the relevant Acts of parliament cited in that report.

38. The Defence Research Board was disbanded as a result of the implementation of the MRG recommendations. Part of that division was amalgamated into the ADM Policy Group, and the remainder was amalgamated into the ADM Material Group.

39. *Task Force Report*, 10.

40. For further information on these factors, see ibid., sect. 3, especially 13–24.

41. *Task Force Report*, 40–41.

42. For further explanation of the problems, see ibid., 58.

43. *Review Group Report*, 16.

44. *Task Force Report*, 76.

45. *Review Group Report*, 19.

46. *Task Force Report*, 76.

47. *Review Group Report*, 20.

48. The term *membership* as applied to any committee in this section of the article should be understood to mean formal membership. As stated earlier in the text, a number of additional civil servants and military officers may attend specific meetings of these committees, depending on the issues being discussed.

49. For the formal membership of all three committees, see *Task Force Report*, 34.

CANADA AND DEFENCE
INDUSTRIAL PREPAREDNESS:
A RETURN TO BASICS? ⬦

DAN MIDDLEMISS

○

The 1987 white paper on defence has focussed attention on Canada's defence commitments and roles, the mission priorities of the Canadian Forces (CF), and new weapons procurement programmes. Somewhat lost in the public debate on these admittedly important issues has been the evidence of the government's growing awareness of the importance of defence preparedness. In the long run, renewed interest in preparedness could have far-reaching consequences for the Forces' military doctrine, for the structure and orientation of Canada's defence industry, for the level and allocation of defence expenditures, for the role of government in industrial and emergency planning, and for Canada's relations with its principal military allies in the North Atlantic Treaty Organization (NATO).

Defence preparedness is a catch-all phrase which may have different meanings, depending on how it is used. In common military parlance, however, it is usually associated with the concept of mobilization—"the art of preparing for war or other emergencies through assembling and organizing national resources."[1] Mobilization is itself a broad concept and encompasses several interrelated aspects of defence planning including manpower, industry, and emergency measures.

To the extent that mobilization has been seriously studied in Canada, the focus usually has been on manpower, often in the context of the appropriate functions and roles of the reserve forces in Canadian defence policy. Until comparatively recently, the industrial dimension of mobilization has received very little in the way of systematic analysis either inside or outside government circles. However, the Canadian government has now embarked

⬦ *International Journal* 42 (Autumn 1987), 707–30.

upon a serious reappraisal of its defence industrial preparedness requirements and policies.[2] This review is a long overdue, but welcome, step in the direction of bridging the widely acknowledged gap between the commitments and capabilities of the Canadian Forces. Indeed, the white paper places considerable emphasis on the importance of defence industrial preparedness in providing for both the initial readiness and the follow-on sustainment of the Forces.[3] This in itself is a marked departure from the neglect of this subject in previous defence policy statements.

This essay seeks to add to our understanding of the role of defence industrial preparedness in Canadian defence policy. It examines the broad rationales for defence industrial preparedness in Canada, provides a brief overview of the major developments in Canadian industrial preparedness planning both past and present, and concludes by noting those factors which may influence the future direction of Canadian policy in this area.

RATIONALE FOR DEFENCE INDUSTRIAL PREPAREDNESS

There are two main types of rationale for defence industrial preparedness in Canada: one is military, the other economic.

From a general military standpoint, the Canadian Forces require an indigenous industrial base to support their defence commitments in both peace and war for several reasons. First, the CF need to be assured of adequate sources of supply of both weapons and other equipment. In the event of hostilities, experience has shown that a self-sufficient domestic industry is the most reliable source of supply. Second, a domestic defence industrial base is required to supply the CF with the specialized weapons and equipment developed specifically for its unique operational needs. Third, the CF require in-service support of existing and future equipment—that is, a capability for timely maintenance, repair, overhaul, and modification of sophisticated weapons systems under possibly adverse conditions. Given the likely preoccupation of one's allies with the support needs of their own services, this capability can be most reliably provided by a domestic industry familiar with, and responsible for, the equipment of the CF.

From a domestic economic standpoint, a defence industrial base serves several functions. First, it generates production and employment in Canada. This in turn reduces the drain on Canada's balance of payments from offshore defence procurement. Second, developing technological expertise concerning state-of-the-art weapons systems enables Canadian firms to expand into foreign markets. Third, by spreading research and development costs over the longer production runs generated by export sales, a healthy defence industrial base not only helps to maintain the productive capacity required by the Forces, but also acts to reduce the unit and life-cycle costs of weapons procurement for the CF in the longer term. Finally, it provides the federal government with an instrument to help enhance the technological competence of Canadian industry in non-defence areas, to foster regional economic development, and to increase Canadian ownership and control of an important sector of the economy.

However compelling these rationales are in the abstract, in practice there are serious obstacles to defence industrial preparedness planning in any country. The sophistication and complexity of modern weapons technology have meant rapid obsolescence and higher costs. In addition, the reduced warning times, faster pace, and higher attrition rates of modern warfare have placed enormous strains on the responsiveness of industry to both immediate and long-lead-time military requirements. Furthermore, there are peacetime structural asymmetries in the productive capacity and responsiveness of different levels of the defence industrial base.

Taken together, these factors have both raised the cost of military sustainability and complicated the task of industrial preparedness planning. Defence officials are thus forced to make difficult choices between satisfying immediate operational and force modernization requirements and pursuing future mobilization needs relating to the retention, expansion, and creation of new capacity in an industrial mobilization base. Because of budgetary limitations, the greater visibility and deterrent impact of combat-ready forces, doctrinal preferences, and the uncertainties concerning the feasibility of industrial mobilization, most NATO countries, including Canada, have opted until recently to bring their peacetime and war establishment forces to higher states of readiness at the expense of preparations to sustain and replace them during an intra-war mobilization period. An active, rather than potential, force posture has been the priority. In effect, sustainability has been sacrificed to readiness.

The relative neglect of mobilization planning in Canada has been reflected in the predominance of economic over military considerations in shaping the Canadian defence industrial base. To understand why this has been the case, a brief review of the major developments in Canadian defence policy and industrial mobilization planning is in order.

HISTORICAL CONTEXT

During the short period between the end of World War II and the onset of the Korean War, Ottawa's main efforts were directed to the de-mobilization of its forces and the re-conversion of its defence industry to fill peacetime, civilian consumption needs. On the assumption that any attack on North America would be diversionary, Canada's defence policy was geared to maintaining a small standing force for home defence needs. This peacetime cadre force was to be formed around an organizational and administrative nucleus which could be expanded rapidly in the event of an emergency.[4] Conceived "primarily in terms of mobilization potential" along the lines of World War II, this defence posture had the added virtue of using fewer scarce resources.[5]

The wartime experience had nevertheless impressed upon Department of National Defence (DND) officials the need for some form of defence industrial preparedness planning. In 1947 the defence minister, Brooke

Claxton, listed as one of Canada's long-term defence objectives: "Close integration of the armed forces, the defence purchasing agency, government arsenals and civilian industry, looking towards standardization and industrial organization to permit the speedy and complete utilization of our industrial resources."[6] To this end a small coterie of Canadians sought to ensure the survival of certain key sectors of Canada's defence industry and advance planning for industrial mobilization. For example, the chief of the air staff, Air Marshal W.A. Curtis, was an early exponent of Canadian self-sufficiency in fighter production.[7] The chairman of the Canadian section of the Permanent Joint Board on Defence, General A.G.L. McNaughton, was another tireless advocate of industrial mobilization planning, although he, more than Curtis, saw a need for close co-operation with the United States, albeit with certain safeguards for Canadian industry.[8]

Notwithstanding these efforts, progress in defence industrial preparedness remained desultory and somewhat tentative.[9] There was no concerted effort to maintain a large infrastructure for the defence industrial base in Canada. Funds were lacking, and most of the Forces' equipment consisted of surplus World War II stocks or was purchased abroad. Moreover, there was little sense of urgency to spur co-operation between Canada and the United States on industrial mobilization.

The outbreak of hostilities in Korea changed Canadian defence policy dramatically. Under the energetic leadership of C.D. Howe, the government moved quickly to mobilize both its military and industrial resources. A three-year, $5-billion, accelerated rearmament programme was instituted, and new legislation was enacted to facilitate the rapid mobilization of the country's industrial potential. In addition, arrangements were concluded with the United States for greater co-ordination and integration of the productive resources of the two countries along the lines of the highly successful principles of bilateral economic defence co-operation established in World War II.[10]

Canada's response to the Korean crisis followed traditional precepts of military and industrial mobilization. Nevertheless, some important changes had occurred which were to have a lasting effect on Canada's subsequent approach to defence preparedness. Recognizing that its defences had been caught unprepared by Korea, Canada sought to maintain a much larger peacetime military establishment to carry out its sizeable commitments for the defence of Western Europe and for North American air defence. The equipment and training methods of these forces were to be standardized according to United States practice because of what was now perceived to be a common defence effort. These commitments required regular "forces-in-being," rather than the mobilization potential of reserves or industry.

The Department of National Defence and other government departments and agencies thus came to view defence preparedness in the context of nuclear deterrence. The prevailing strategic conception was that of a short, rapidly escalating nuclear war in which conventional military forces, to the extent to which they would have any role at all, were to be drawn from adequate peacetime forces-in-being. Under this scenario, there would be no time for the extensive mobilization of troops or industry along the lines followed in the previous two global conflicts. For industry this meant that measures to ensure the production and supply of the weapons and equipment required by these forces would have to be provided for well in advance of any major future conflagration. Both the military and industry would have to rely on in-place capabilities;[11] mobiliz*ed* rather than mobiliz*able* military and industrial power was what mattered in the nuclear age. As the 1964 white paper on defence later noted, this was "in strategic terms, the equivalent of a transference from credit to cash."[12]

Nevertheless, the Korean build-up had provided Canada with a small, specialized defence industry which was capable of satisfying some of its own Forces' equipment requirements as well as competing selectively for United States weapons contracts.[13] Furthermore, this industrial base had been created and sustained through the active involvement of the federal government This federal role included the placing of procurement contracts with Canadian suppliers, financial support, and assistance to help Canadian firms penetrate defence markets abroad, especially in the United States. Finally, a panoply of legislative and administrative measures had been set in place to govern defence industrial preparedness activities not only within Canada but also between Canada and the United States.[14] The issue for Canada was, therefore, what was to be done with this nascent defence industrial base.

An active, bilateral approach to defence industrial preparedness proved to be short-lived. The recessionary contractions in their economies caused by post-Korean reductions in defence spending led both the Canadian and United States governments to revert to their former protectionist military procurement practices.[15] From 1953 on, therefore, the essentially military rationale of Canada's industrial preparedness programmes began to recede in importance as more pressing economic and political rationales for the support of a Canadian defence industry began to assert themselves. Whereas in 1951 the chief concern of the Department of Defence Production had been the rapid growth and expansion of Canada's defence industry to meet the heavy demands of war, by the mid-1950s the priority became the "maintenance and stabilization" of this industry to provide a sustained rate of growth and stable employment for this sector of the economy.[16]

Having nurtured a specialized defence industry through its birth and infancy, the government now had to sustain it through adolescence and adulthood. Because access to the lucrative United States defence market was becoming increasingly restricted, the government turned to the requirements of its own military as a means of protecting its investment in this domestic defence industrial base. A policy of sectoral specialization—in aircraft, electronics, shipbuilding, and munitions—and government financial assistance was instituted, of which the chief beneficiary was to be the fledgling Canadian aircraft industry. This policy of limited defence production self-sufficiency focussed on the CF-105 Arrow jet interceptor programme. Unfortunately, this proved to be too rich for the government's blood. In 1959 the Arrow programme was cancelled amidst recriminations on all sides, leaving the aircraft and avionics industries to face an uncertain and unpromising economic future.[17]

The Arrow's demise marked the collapse of Canada's brief experiment in creating a national defence industrial base geared to the unique requirements of the Canadian military. However, it is important to remember that even before this the Department of National Defence had embraced the concept of forces-in-being and that the government was actively exploring ways to facilitate the access of Canadian defence firms to the United States defence market. Thus, while there remained a concern to preserve the nucleus of a domestic defence industry in Canada, the military rationale for an industrial mobilization capacity was being subordinated to, if not entirely supplanted by, the more pressing economic and political considerations relating to balance of payments, employment, and the maintenance of an important sector of the Canadian manufacturing industry. As one authoritative source noted, by 1959, "no serious attempt was being made to maintain a mobilization base."[18]

As has been recounted elsewhere, these considerations led to the adoption of a series of co-operative bilateral measures between 1959 and 1963 which have come to be known as the Canada–United States Defence Production and Development Sharing Arrangements (DPDSA).[19] They proved to be a major watershed in Canada's approach to the notion of defence industrial preparedness.

While the short-term goal of the arrangements was to ensure the economic survival of Canada's defence industry through increased participation in the production and support of weapons and equipment programmes for the defence of North America, the longer term objectives encompassed the more traditional military rationale for defence industrial preparedness. These arrangements included: greater integration of Canadian and United States military development and production; greater standardization of equipment; wider dispersal of production facilities; establishment of supplemental sources of supply; removal of obstacles impeding the flow of defence supplies between the two countries; and the most economical use of defence resources. Significantly, a major goal was "to seek the best possible co-ordination of the materiel programs . . . including actual integration so far as practicable of the industrial mobilization efforts of the two countries." This would involve: "the determination of Canadian production facilities available for the supply of United States current and mobilization requirements, and the furnishing of planned mobilization follow-up schedules to Canadian contractors producing for the United States as guidance in the event of full mobilization."[20]

Notwithstanding the ostensible emphasis accorded defence industrial mobilization in the DPDSA, in reality very little was done in Canada to promote defence industrial preparedness per se. The bilateral arrangements did help to preserve Canada's defence industry, but did so at the expense of the military rationale for maintaining an indigenous defence industrial base. By the mid-1960s, the Defence Department and Canada's defence industry were following divergent paths.[21]

On the one hand, the department remained firmly committed to the concept of forces-in-being as the cornerstone of its defence posture.[22] Weapons and equipment for these forces would continue to be acquired from foreign (mainly United States) suppliers, thus recognizing that "only the largest powers can economically design, develop and produce all their weapons."[23] On the other hand, Canada's defence industry became increasingly export-oriented and concentrated on producing sub-assemblies, components, and parts primarily for United States defence contractors.

To the extent that defence industrial mobilization was considered at all, it was in the context of United States, not Canadian, military requirements and planning. For example, the DPDSA made provision for Canadian manufacturers to be included in the mobilization production planning arrangements of United States military departments under the United States Industrial Readiness Planning Program.[24] A 1970 Canada–United States memorandum of understanding formalized the procedural arrangements whereby Canadian defence firms could be registered as "planned producers" under the United States Industrial Mobilization Production Planning Program.[25] Canadian defence firms have been granted a unique status under this programme inasmuch as the United States has not extended it to its other military allies. Canadian firms registered as planned producers have acquired several advantages: they are always entitled to bid on restricted United States mobilization contracts; they are guaranteed some share of the ensuing contracts; they acquire greater visibility to United States prime contractors; and they obtain exemptions to the United States small business set-asides which hamper other Canadian firms under the DPDSA.[26]

Unfortunately, until fairly recently, the Department of Defense, the military services, and the defence industry in the United States have not given much substance to the concept of defence industrial preparedness.[27] As a consequence, the fate of industrial mobilization in Canada has largely been held hostage to the gradual recrudescence of interest in this policy area in the United States.

From this admittedly simplified overview, several important conclusions emerge concerning the Canadian approach to defence industrial preparedness. In the first place, from the mid-1950s on, DND has adhered firmly to a short-war, forces-in-being military doctrine. This approach, which corresponded to United States military doctrinal preferences, emphasized the need for operationally ready forces rather than manpower or industrial mobilization potential. For Canada, this meant that the Regular Forces were to have preference over the Reserves in both manning and equipment; indeed, the 1971 white paper noted that the Reserves had been "designated as part of the 'forces in being.'"[28] Although Canada's policy coincided with that of its alliance partners, other factors, most notably the department's funding constraints and the limited capabilities of Canadian industry, played an important role in reinforcing this orientation. As one study noted, "'Forces in being' fitted rather neatly with the previous Canadian policy decisions concerning the requirement for small operational units, the variations in types and levels of equipment, and the lack of a mobilization base."[29]

Secondly, in the postwar period Canada had developed a defence industrial base of sorts. However, save for the Korean War years, this industry had not been primarily oriented to serving the requirements of the

Canadian armed forces either for immediate weapons and equipment or for longer term mobilization needs.

Thirdly, as time passed, it became clear that the survival of Canada's defence industry had become increasingly dependent upon continued exports abroad, especially to the lucrative United States defence market.

Finally, the federal government had played a key role in establishing and maintaining this defence industry, both through financial assistance and marketing support. Through programmes such as the DPDSA and the Defence Industry Productivity Program (DIPP),[30] Ottawa sought to retain specialized sectors of a defence industrial base in Canada.

RECENT DEVELOPMENTS

Of late the government has acknowledged the importance of defence industrial preparedness to Canadian defence planning. In part this is a lagged response to developments in both the United States and NATO military doctrines which now emphasize a longer, conventional phase in any future conflict with the Soviet Union and its Warsaw pact allies. This doctrinal shift—itself a belated recognition of the requirements of NATO's "flexible response" strategy in response to apparent changes in Soviet military doctrine—has led to greater scrutiny of the ability of the allies, and especially of the United States, to support and sustain conventional military operations for a much longer time.[31] This in turn has focussed allied attention on the potential contributions a robust, well-planned, and organized defence industrial base can make to both deterrence and defence.

In part, too, this renewed interest in Canada has been a response to the growing recognition of the failure of previous government policies to serve Canadian defence industrial preparedness needs. The essential point of departure in this regard was the public acknowledgement by Canadian political leaders that the capabilities of the Canadian Forces had reached the point of no return by the mid-1970s. Years of deliberate budgetary starvation had left the CF undermanned and ill-equipped to carry out their numerous commitments.

In the autumn of 1974, in response to the warnings of the dire consequences of this ever-widening commitment-capability gap,[32] the Trudeau cabinet authorized a Defence Structure Review to assess the tasks, effectiveness, and appropriate funding levels for the Forces. As a result of this review, a long-term re-equipment programme, based on a more stable and steadily expanding capital funding formula, was instituted. However, while the review initiated the process of rectifying the force structure deficiencies of the CF, it did so primarily in the context of the Forces' peacetime, rather than wartime, requirements. As a consequence, it did not adequately address the mobilization and sustainability needs of the Canadian military.[33]

Nevertheless, since 1975, the commitment-capability gap has been the priority concern for the CF, one which politicians could no longer afford to ignore. Public awareness and concern was intensified and focussed by a welter of academic critiques, parliamentary reports, and other external studies. While the majority of these accounts tended to emphasize the operational shortcomings of the Forces, some viewed the re-equipment programmes of the CF as an opportunity to re-structure Canada's defence industry the better to serve both military and economic objectives.

But active political leadership was required to translate general expressions of concern into concrete policy directives. The turning point in this regard was the speech of the defence minister, Gilles Lamontagne, to the Conference of Defence Associations on 13 January 1983. In this seminal speech, Mr Lamontagne announced that the government had agreed in principle that "the Canadian Forces should be able to meet and fully sustain their commitments in an emergency and, if so directed, further expand their capabilities." He emphasized that this was a significant policy shift inasmuch as it acknowledged that the previous "almost exclusive reliance on 'forces-in-being' . . . will be insufficient in the future." He added that the long-term ability to sustain the operations of the CF during a crisis was "dependent on the state of our national industrial readiness." Because fulfilment of this objective went beyond the mandate of his department, he announced his intention to pursue this issue with the responsible departments and agencies. Finally, the minister indicated that the government would back up this decision by providing additional funds "specifically for increased readiness and sustainability."[34]

However, by the time Canadian politicians and senior military planners had become converted to the idea of defence preparedness and the related notions of enhanced readiness, sustainability, and mobilization, the efficacy of the government's post-1959 approach to ensuring the economic survival of Canada's defence industry was being called into question. From the early 1960s on the Canadian industry had become increasingly divorced from the operational requirements of the Canadian Forces. In the words of one analyst, "the arms industry in Canada is . . . not so much a 'defence' industry concerned with Canadian military needs as it is a military industry with economic objectives."[35] The industry's survival was predicated upon an economic, not a military, rationale. Until the mid-1970s, Ottawa had relied upon the market access afforded by the DPDSA coupled with government financial assistance and marketing support to sustain an export-oriented defence industry. This approach was never entirely satisfactory owing to the advantages enjoyed by larger and better established competitors in the United States and because of growing legislative and administrative protectionism in United States defence contracting.

As a result, Ottawa turned to the leverage of DND's large-scale weapons procurement programmes to provide a wide range of socio-economic benefits to the country in general, and to the defence industrial sector in particular. Beginning with the CP-140 long-range patrol aircraft programme of 1976, this approach emphasized contractually stipulated "industrial benefits" which included the traditional "Canadian content" provisions of earlier programmes, but went beyond this to specify non-military procurement offsets that were often unrelated to the particular programme.[36]

This industrial benefits procurement strategy has proved to be a mixed blessing for Canada. It has been criticized by spokesmen for the Canadian defence industry for providing only low-technology "scraps" of weapons systems developed abroad in return for "the political illusion of jobs."[37] For DND, the strategy has undoubtedly helped to "sell" its expensive procurement programmes to cabinet, but at the same time it has added complications in the form of partisan tampering by federal, provincial and even municipal politicians and bureaucrats seeking employment and industrial spin-offs from these large military purchases. The result has been delivery delays and extra procurement costs for DND, as well as minimal technological enhancement of direct benefit to the Forces' long-term operational needs. Finally, the strategy has come under increasing fire from the United States administration and Congress as a violation of the spirit of defence industrial co-operation embodied in the DPDSA.[38]

The federal government has responded to these concerns in several ways. In 1982 it authorized the Department of Regional Industrial Expansion (DRIE) to undertake an interdepartmental review of its industrial benefits policies and practices and directed the Department of External Affairs to review the trade and industrial activity implications of the DPDSA. In addition, following Mr Lamontagne's 1983 announcement, DND policy directives were instituted covering a wide range of topics relating to defence preparedness (procurement, research and development, interim mobilization planning, and readiness and sustainment). The Ministry of State for Science and Technology set up a Task Force on Federal Policies and Programs for Technology Development which reported in 1984. A ministerial Task Force on Program Review, established in September 1984 and chaired by the deputy prime minister, Erik Nielsen, published a study team report, *Government Procurement: "Spending Smarter,"* in June 1985 which dealt with aspects of DND's procurement policies.

These and related policy reviews and initiatives have yielded tangible results. During the March 1985 Quebec summit, Prime Minister Mulroney and President Reagan reaffirmed their support for the principles of the DPDSA and agreed to take joint steps to reduce barriers to the flow of defence goods between the two countries and to strengthen the North American defence industrial base. This agreement paved the way for the establishment of DND's Defence Industrial Preparedness Task Force. The group's final report, which has been approved by the high-level Defence Management Committee, recommended the development of closer bilateral ties in establishing a North American defence industrial base and the incorporation of defence industrial preparedness considerations into *all* aspects of defence capital acquisition planning on a routine basis.[39] Since "routine consideration" of these matters has been virtually non-existent until now, this is an important new departure for DND.

The Department of Supply and Services has established a Defence Industries and Emergency Planning Branch and a Supply Administration Emergency Planning Working Group to assess and plan procedures to enable that department to respond quickly to DND's preparedness needs in the event of an emergency. The United States–Canada Precision-Guided Munitions Surge Joint Task Force (September 1986–March 1987) provided a detailed assessment of the capability of Canadian defence firms to increase production rapidly to satisfy the operational requirements of both defence departments in this area of critical materials. Canadian government officials and industry representatives had also begun to participate regularly in workshops held in the United States on the North American defence industrial base. Finally, in May 1987, the defence minister created the Defence Industrial Preparedness Advisory Committee to advise him on preparedness matters of mutual concern to DND and the Canadian business community.[40]

FUTURE PROSPECTS

This analysis suggests that, for a variety of reasons, the concept of defence industrial preparedness has now regained military, economic, and, most important, political respectability in Canada. What, then, might be the future direction of Canadian policy in this area? What problems lie ahead and what factors are likely to influence the government's approach to possible solutions?

Analysts have identified two fundamental causes of the weakened state of Canada's defence industrial base: the fragmentation of political authority within the federal government and inadequate co-ordination with the United States government on industrial preparedness issues. To restore the severed link between defence policy and the defence base, some observers have proposed that the Department of National Defence be designated the lead department responsible for developing, in co-ordination with other departments and in consultation with industry itself, a clear, comprehensive, and coherent defence industrial preparedness strategy.[41] The rationale for this is simple: hitherto the department, as the ultimate user of the defence industrial base, has been both the most affected by the overlapping and often contradictory responsibilities of other departments relating to defence industrial policy in Canada and yet the least involved in the formulation of this policy.

Such suggestions, however sensible from a purely military standpoint, are likely to spark organizational turf battles within the federal government. Departments which have gradually assumed responsibility for maintaining Canada's defence industrial base in the wake of DND's earlier neglect and indifference, cannot be expected to relinquish their mandates—and, in the case of the DRIE, the funds (DIPP) that go with it—willingly. Signs of just such a political boundary dispute surfaced publicly in the spring of 1987. The minister of supply and services, Monique Vézina, criticized the minister of national defence, Perrin Beatty, for his neglect of the "national interest" in his department's equipment procurement policies. Mme Vézina believes that defence spending should be used to promote regional economic development, even if it means that the Defence Department would have to pay a premium for Canadian-made products, as part of an extended "strategic source" purchasing policy (that is, exclusive long-term contracts) being considered by her department.[42] Following so soon after the publication of the defence white paper, this interdepartmental squabble reveals clearly that the old tension between the military and economic rationales for Canada's defence industrial base has by no means disappeared.

But even if this dispute is resolved in favour of DND, it is not self-evident that the result would be a coherent defence industrial policy for Canada. The military is itself divided on the exact meaning and implications of defence industrial preparedness.[43] Within the CF, the army holds the traditional view that the defence industry should be directed towards the peacetime stockpiling of war matériel and to measures to ensure the mobilization and sustainment of field operations once hostilities have begun. The navy and the air force, however, both rely on sophisticated, long-lead-time technology and see the fundamental role of a defence industrial base to be the provision of the essential tools of their trade *before* the outbreak of war. A similar dichotomy exists between the operations and the matériel branches of the department. The former supports a "preparedness" orientation geared to long-run mobilization and sustainment objectives, while the latter prefers an orientation tied to immediate production, procurement, and logistics goals. It would be unrealistic to assume that such differences in fundamental inter- and intra-service operational concepts will disappear. In fact, the opportunity presented by the new political interest in defence industrial preparedness can be expected to sharpen, rather than mute, such pre-existing organizational and service prejudices.

These internal managerial concerns, while important, should not detract attention from the larger issue of the government's overall approach to defence industrial preparedness. It is clear from the new defence white paper and other government statements that the revitalization of Canada's defence industrial base is being conceived in a North American, rather than a purely national, context. As in 1959, the "go-it-alone" option has been rejected on economic and technological grounds in favour of a continental approach. A joint Department of National Defence/Department of Defense watchdog committee is envisioned which will be responsible for overseeing the gradual elimination of the many barriers to a North American defence industrial base in which the military and industrial planning requirements of both countries would be more fully integrated and in which joint funding of improvements to industrial preparedness responsiveness might be provided.[44]

However laudable such an approach may be, it suffers from one serious weakness: it has been tried before and found wanting. Many of these same objectives formed the basis of the long-standing co-operative economic defence relationship which has evolved between Canada and the United States. Nevertheless, even the comparatively brief crises of World War II and the Korean War proved insufficient to overcome the enormously disparate military interests and requirements and defence industrial capacities of these two allies. Despite the recent emergence in both countries of the political and military will conducive to an improved North American defence industrial base, these military and economic realities could ultimately render bilateral institutional tinkering ineffectual.

The time element could prove critical in this regard. Expressions of greater political will to co-operate may dissipate quickly as a result of changes in administration on either or both sides of the border or in the course of implementing a bilateral free trade agreement. The concept of integrated military requirements may prove to be of limited utility in the face of differing national strategic and operational demands, and it may be politically unpalatable in Canada where nationalist tendencies find expression in calls for a "home-grown" defence policy.

Free trade could also pose difficult problems. The DRIE introduced a new industrial benefits policy in June 1986 in response to domestic criticisms. The policy now aims at long-term industrial and regional development by focussing on high-quality benefits (investments, technology transfer, joint ventures, and licencing arrangements), greater life-cycle support in Canada, and improvements to Canada's defence industrial base. While it de-emphasizes offset maximization and project-oriented balancing of regional benefits, the new approach, along with the DIPP grants which support it, may run afoul of the "equal national treatment" provisions in a Canada–United States free trade agreement. American national security restrictions could also prove a serious impediment to establishing a more closely integrated North American defence industrial base.

Interdepartmental infighting can be expected to intensify as politicians and bureaucrats seek to protect hard-won mandates pertaining to various aspects of Canada's defence industrial base. Within DND and the CF, parochial service and bureaucratic interests may lead to longer term defence industrial preparedness objectives being sacrificed to immediate procurement priorities in the competition for scarce funds. Given such uncertainties, Canada's defence industry itself may resist inducements to invest in what may prove to be an illusory and short-lived government flirtation with the concept of defence industrial preparedness.

This is not to suggest that defence industrial preparedness is not worth pursuing in Canada. Far from it. The idea makes eminent sense from both military and economic standpoints. The real challenge is in devising practical—and acceptable—means to achieve it. This problem is not new. Indeed, it is but another manifestation of Canada's perennial difficulty in reconciling its own military, economic, and political interests, needs, and goals with the inescapable realities of sharing a continent with a powerful superpower ally.

In re-emphasizing such venerable concepts as military and industrial preparedness, Canadian defence policy has returned to its basic post-World War II foundations: old problems have evoked old responses. Whether or not these responses are appropriate to the challenges ahead depends largely on how well Canada's defence planners have learned from the lessons of the past.

NOTES

1. Ralph Sanders and Joseph E. Muckerman II, "A Strategic Rationale for Mobilization," in *Mobilization and the National Defense*, ed. Hardy L. Merritt and Luther F. Carter (Washington: National Defense University Press, 1985), 8.

2. In mid-1985 (NDHQ Action Directive D/35, 30 May 1985), a ten-person Defence Industrial Preparedness Task Force (hereafter DIPTF) was established in the Department of National Defence, under the direction of the assistant deputy minister (matériel),

to focus and co-ordinate efforts to develop plans, systems, and procedures upon which defence industrial preparedness could be based. The task force completed its report in June 1987.

3. Canada, Department of National Defence (hereafter DND), *Challenge and Commitment: A Defence Policy for Canada* (Ottawa: Supply and Services Canada, 1987), 74–76.

4. See, Brooke Claxton in Canada, House of Commons, *Debates*, 24 June 1948, 5779–810, and 11 November 1949, 1662–66.

5. *White Paper on Defence* (Ottawa: Queen's Printer, 1964), 8.

6. *Debates*, 9 July 1947, 5272.

7. James Eayrs, *In Defence of Canada*, vol. 3, *Peacemaking and Deterrence* (Toronto: University of Toronto Press, 1972), 60, 101–05.

8. John Swettenham, *McNaughton*, vol. 3, 1944–1946 (Toronto: Ryerson, 1969), 196–98.

9. For details on some of the measures taken, see Danford W. Middlemiss, "Economic Defence co-operation with the United States 1940–63," in *An Acceptance of Paradox: Essays on Canadian Diplomacy in Honour of John W. Holmes*, ed. Kim Richard Nossal (Toronto: Canadian Institute of International Affairs, 1982), 88–91. Also, Lawrence R. Aronsen, "Planning Canada's Economic Mobilization for War: The Origins and Operation of the Industrial Defence Board, 1945–1951," *American Review of Canadian Studies* 15 (Spring 1985), 38–58.

10. For example, the Exchange of Notes Between Canada and the United States of America Giving Formal Effect to the "Statement of Principles for Economic Co-operation" (26 October 1950) echoed the broad objectives of the Hyde Park Declaration (20 April 1941). *Canada Treaty Series*, 1950, no. 15.

11. John J. Kirton, "The Consequences of Integration: The Case of the Defence

Production Sharing Agreements," in *Continental Community? Independence and Integration in North America*, ed. W. Andrew Axline et al. (Toronto: McClelland and Stewart, 1974), 4–5.

12. *White Paper on Defence* (1964), 9.

13. Jon B. McLin, *Canada's Changing Defense Policy, 1957–1963: The Problems of a Middle Power in Alliance* (Baltimore: The Johns Hopkins Press, 1967), 176.

14. The most enduring of these was the Defence Production Act of 1951 which provided the minister of the new Department of Defence Production with sweeping powers to handle all aspects of defence industrial mobilization in war or peace. Under a departmental reorganization in 1969, this became the Department of Supply and Services.

15. As C.D. Howe noted: "When defence contracts are let these days, both of our Governments are inclined to turn first to their domestic production facilities and only because they have to, or for some very special reason, do they look across the border. This may be regrettable but it is one of the facts of life and must be accepted as such." Address to the Canadian Club of Boston, 4 October 1954, in Howe Papers, 89-2, folder 48, Public Archives of Canada (hereafter PAC).

16. C.D. Howe, in *Debates*, 27 June 1956, 5456–57, and 29 June 1956, 5520.

17. For accounts of the rise and fall of the Arrow, see James Dow, *The Arrow* (Toronto: Lorimer, 1979), and McLin, *Canada's Changing Defense Policy*, 61–84.

18. *White Paper on Defence* (1964), 9.

19. See Middlemiss, "Economic Defence Co-operation with the United States," especially 95–109.

20. Canada, Department of Defence Production, *Production Sharing Handbook: Canada–United States Defence Production Sharing*, 4th ed. (Ottawa: Queen's Printer, 1967), appendix A, US Department of Defense Directive 2035.1, "Defense

Economic Cooperation with Canada," 28 July 1960.

21. William Yost, "Sustaining the Force: Industrial Mobilization in Canada," in *Guns and Butter: Defence and the Canadian Economy*, ed. Brian MacDonald (Toronto: Canadian Institute of Strategic Studies, 1984), 78.

22. *White Paper on Defence* (1964), 24.

23. Ibid., 28.

24. *Production Sharing Handbook*, 15.

25. Memorandum of Understanding: United States/Canada Mobilization Production Planning Arrangements, 21 October 1970, signed by J.S. Glassford, President, Canadian Commercial Corporation, and Barry J. Shillito, United States assistant secretary of defense for installations and logistics.

26. This program has been re-named the United States Industrial Preparedness Production Planning Program. For details, see *Production Sharing Guidebook: Canada–United States Defence Production Sharing Program* (Ottawa: Defence Programs Bureau, Department of External Affairs, ca 1985), 9; and the remarks of Thomas Chell, Department of Industry, Trade and Commerce, in *Industrial Preparedness and National Security* (Toronto: Canadian Institute of Strategic Studies, 1980), 45.

27. For critical studies, see Roderick L. Vawter, *Industrial Mobilization: The Relevant History*, rev. ed. (Washington: National Defense University Press, 1983), and Timothy D. Gill, *Industrial Preparedness*, National Security Affairs Monograph Series 84-6 (Washington: National Defense University Press, 1984).

28. Canada, Department of National Defence, *Defence in the 70s: White Paper on Defence* (Ottawa: Information Canada, 1971), 45.

29. Task Force on Review of Unification of the Canadian Armed Forces, *Final Report* (Ottawa, 15 March 1980), 14.

30. For details of this program, see: William Yost, *Industrial Mobilization in Canada* (Ottawa: Conference of Defence Associations, 1983), annex D, 95–99; and the testimony of Robert Gillespie, acting director general, Office of Industrial and Regional Benefits, Department of Regional Industrial Expansion, in House of Commons, Standing Committee on External Affairs and National Defence, *Minutes of Proceedings and Evidence*, 33rd Parl., 1st sess., no. 34 (3 October 1985), 18–21.

31. Two United States military exercises—Nifty Nugget (1978) and Proud Spirit (1980)—had revealed the severe inadequacies of United States mobilization capabilities. These findings prompted a further congressional investigation by the House Armed Services Committee's Defense Industrial Base Panel. The panel's report, *The Ailing Industrial Base: Unready for Crisis* (31 December 1980), found evidence of substantial and continuing deterioration of the United States defence industrial base. Other studies confirmed these conclusions.

32. For example, General Dextraze, chief of the defence staff, noted, in response to DND's funding allocation for fiscal year 1975–76, that he had advised the government "that we could not reduce below that assigned level without running a grave risk of being unable to carry out all of our assigned tasks as well as of denuding the Canadian military profession to an unacceptable point. I also stressed that if we were forced to accept any further loss of combat capability that loss should be in response to basic defence policy objectives and not just current budgetary objectives." Address to the Conference of Defence Associations, 16 January 1976.

33. These shortcomings were noted in the "Report of the Auditor General of Canada to the House of Commons for the Fiscal Year ending 31 March 1984", 12–3 and 12–4.

34. Address to the Conference of Defence Associations, 13 January 1983. Mr Lamontagne expanded on this theme

of readiness and sustainability in his *Statement on Defence Estimates, 1983/84* before the Standing Committee on External Affairs and National Defence, 15 March 1983. He outlined in general terms to the committee his department's new emphasis on a "total force" concept as well as mobilization planning and indicated that the specific allocation for increased readiness and sustainability would increase at a rate of .25 percent annually to a maximum of two percent of the total defence budget.

35. Ernie Regehr, *Arms Canada: The Deadly Business of Military Exports* (Toronto: Lorimer, 1987), 68.

36. See R.B. Byers, "Canadian Defence and Defence Procurement: Implications for Economic Policy," in *Selected Problems in Formulating Foreign Economic Policy*, research coordinators Denis Stairs and Gilbert R. Winham, vol. 30 of the research program of the Royal Commission on the Economic Union and Development Prospects for Canada (Toronto: University of Toronto Press, 1985), 184–90.

37. See the critical testimony of industry representatives in Canada, Senate, Special Committee on National Defence, *Proceedings*, 32nd Parl., 2nd sess., no. 4, 14 March 1984, 13, 19.

38. A good summary of United States objections to the offsets component of Canada's industrial benefits strategy can be found in Colonel Robert Van Steenburg, "An Analysis of Canadian–American Defense Economic Cooperation: The History and Current Issues," a paper presented at the Conference on the Canadian Defence Industrial Base: Domestic and International Issues and Interests, held by the Centre for International Relations, Queen's University, Kingston, Ontario, 18–19 June 1987, 30–35.

39. "Defence Industrial Preparedness: A Foundation for Defence," executive version of the final report of the Defence Industrial Preparedness Task Force, November 1987.

40. Department of National Defence, news release, 29 May 1987.

41. See, for example, Beth L. Thomas of the DIPTF. "The Environment for Expanding the North American Defence Industrial Base," 26, and Colonel Douglas L. Bland, "The Canadian Defence Policy Process and the Emergence of a Defence Industrial Preparedness Policy," 21–23, both papers presented to the Conference on the Canadian Defence Industrial Base: Domestic and International Issues and Interests, Kingston, 18–19 June 1987. Also, Byers, "Canadian Defence and Defence Procurement," 175–78.

42. For Vézina's views, see the *Chronicle-Herald* (Halifax), 12 May 1987, 5, and the *Globe and Mail* (Toronto), 12 June 1987, A1.

43. For an informed analysis of this issue, see Bland, "The Canadian Defence Policy Process," 7–11, 22–23.

44. For suggestions along these lines, see Thomas, "The Environment for Expanding the North American Defence Industrial Base," passim.

THE ECONOMICS OF DEFENCE◇

ROD BYERS

o

Normally, security commitments and defence responsibilities—within the context of threat assessments—should guide the development of force structures and military capabilities. . . . There *should* be a reasonably close relationship between military-strategic considerations and the allocation of resources for defence, with the former guiding the latter. In the Canadian case, the opposite has been the norm: fiscal considerations have had a more important impact on the evolution of military capabilities than have commitments and responsibilities.[1] In the early 1960s, Canadian governments allocated a much higher fiscal priority to "social policy" than to "defence policy." Consequently, "defence as economics," more than "defence as strategy," determined the size, structure and equipment of the Canadian Forces.

Since then, defence has been seriously underfunded and the cumulative effect has been four-fold: first, the commitment-capability gap has emerged as a major defence problem; second, Canada's military reliability within the Western Alliance has been called into question; third, in the mid-1980s the Canadian Forces lack the capabilities to make a positive and credible contribution to conventional deterrence (and, more importantly, should deterrence fail, would be unable to perform their assigned roles and missions adequately); and, fourth, defence procurement has been adversely affected by non-defence considerations related to industrial benefits and regional economic development.

RESOURCES FOR DEFENCE

All resources allocation decisions by government are influenced by a wide range of issues and competing demands. Decisions are made within the

◇ Excerpted from R.B. Byers, *Canadian Security and Defence: The Legacy and the Challenges*, International Institute of Strategic Studies, Adelphi Papers, 214 (Winter 1986), 31–44.

context of overall governmental and societal priorities. Defence dollars should purchase, at least in theory, a unique commodity: national security. However, given the strategic situation and the nature of the threat, the commodity purchased is impossible to quantify in dollar terms. For Canada in particular it is difficult to establish any direct correlation between the level of defence spending and the degree of security purchased. This point was made in 1983 by the then Minister of National Defence, Gilles Lamontagne: "it is virtually impossible convincingly to demonstrate that Canada's national security against threats of military attack is diminished by reducing the size and capability of the Canadian Forces or is increased by increasing them." Nevertheless, it can be stated with some certainty that, over time, the Canadian Forces have not received sufficient funds to meet the commitments entered into by the government.

From 1960–61 to 1984–85, defence spending exhibited a number of characteristics (see table 1). First, overall federal spending in Canada, as in most NATO countries rose, more substantially than defence spending. Overall government spending went up from $C6 billion to $C109 bn—an average annual increase of over 12.0 percent, but defence spending increased from $C1.5 bn to $C8.9 bn—an average annual increase of only 5.7 percent. Defence as a percentage of total government spending thus fell from 25.5 percent to 8.1 percent, and as a percentage of Gross National Product (GNP) from 4 percent to approximately 2 percent. Second, the growth rate, in constant dollars, was uneven. Spending varied considerably and on some occasions fell below that of the previous year. Third, with budget freezes and inflation, the defence budget experienced negative real growth in 11 fiscal years between 1960–61 and 1984–85. For the entire 25-year period the average annual real growth was nil.

The rationales for defence funding decisions have been diverse and, for the most part, unresearched.[2] Immediately after World War II, it seemed that the pre-war situation of financial neglect would prevail. Even with the formation of NATO, some members of the Canadian government initially expected that the military requirements of the Alliance would be less significant than the political. However, during the Korean War defence expenditures increased substantially and ranked high as a government financial priority in the 1950s.

Major financial difficulties for DND developed during the early 1960s. Funding levels were inadequate to meet increased personnel costs as a result of the equation of military salaries with those of the unionized civil service. This led to reductions in manpower levels without any appreciable change in commitments, the first of many decisions which contributed to the commitment-capability gap. In an attempt to rationalize the allocation process, the government adopted a formula-funding approach to defence expenditure at the time of the 1964 White Paper. In one form or another, formula-funding, whereby the Cabinet approves annually resources for defence based on explicit percentage changes in either overall expenditures and/or specific components of the defence budget, has been in place ever since.

The Pearson Government claimed that formula-funding would stabilize the defence budget and provide for a better balance between personnel and

TABLE 1   *TOTAL FEDERAL & DEFENCE*
EXPENDITURES

	Federal expenditures		Defence expenditures		
Year	$Cm	current % change	$Cm	current % change	real % change
1987–88◇	119,965	+2.8	10,485.0	+6.3	+2.0✦
1986–87◇	116,740	+4.0	9,860.0	+6.1	+2.7✦
1985–86◇	112,250	+2.9	9,290.0	+4.5	n/a
1984–85✦	109,115	+21.9	8,892.0	+12.0	+3.1
1983–84	89,540	+12.0	7,972.2	+14.0	+6.9
1982–83	79,872	+17.5	6,991.1	+16.0	+3.3
1981–82	67,959	+8.9	6,027.7	+18.7	+2.8
1980–81	62,377	+19.1	5,077.1	+15.7	+1.5
1979–80	52,364	+11.6	4,389.3	+6.8	–1.9
1978–79	46,923	+9.4	4,108.0	+8.9	+0.5
1977–78	42,882	+10.1	3,771.0	+11.9	+2.4
1976–77	38,930	+17.3	3,371.2	+13.4	+0.9
1975–76	33,181	+27.3	2,973.7	+18.4	+3.2
1974–75	26,055	+29.9	2,511.9	+12.5	–3.6
1973–74	20,055	+24.4	2,232.0	+15.5	+6.4
1972–73	16,121	+8.6	1,932.2	+2.0	+4.6
1971–72	14,841	+12.6	1,895.0	+4.2	–1.6
1970–71	13,182	+10.6	1,817.9	+1.6	–3.8
1969–70	11,921	+11.0	1,789.5	+1.6	–5.5
1968–69	10,738	+9.6	1,760.8	+0.4	–3.9
1967–68	9,789	+12.4	1,753.5	+5.8	+1.4
1966–67	8,718	+17.7	1,640.4	+5.9	–2.8
1965–66	7,735	+7.2	1,548.4	+0.8	–1.9
1964–65	7,218	+5.0	1,535.6	–8.8	–11.6
1963–64	6,872	+4.6	1,683.5	+7.2	+3.1
1962–63	6,570	+0.8	1,571.0	–3.4	–5.7
1961–62	6,521	+9.4	1,626.1	+7.2	+5.4
1960–61	5,958	+4.5	1,517.5	+0.2	–2.7

◇ Figures are government estimates.
✦ DND estimates.

Price indices obtained from *Canadian Statistical Review*, Statistics Canada, May 1984.
GNE index for total government spending used for this series.
Source: Public Accounts (Canada), 1950–1984; Department of Finance, *The Fiscal Plan*, 1986.

operating costs on the one hand and capital costs on the other. Defence
Minister Paul Hellyer also argued that integration and unification of the
three services would reduce personnel requirements and produce substantial savings. As a result, the first formula-limited defence increases to 2 percent per annum for five years, commencing with the 1964–65 base budget of
$C1.5 bn. Hellyer stated that the formula would allow for capital expendi-
tures to reach 25 percent of the budget in order to re-equip the forces. For
the most part, budgetary ceilings were maintained and a capital programme
was completed; but capital expenditures continued to decrease and personnel costs rose despite manpower reductions of some 20 000. While the
defence budget increased to $C1.79 bn in current dollars during the five-
year period (see Table 1), inflation and equipment cost increased rendered

the formula ineffective. In real terms, the defence budget actually declined during the mid- to late 1960s. The negative impact of the Hellyer formula-funding for defence should be clearly understood, particularly since some critics attribute Canada's defence problems solely to the Trudeau era.[3] This view is unwarranted and overlooks the major problems which arose while Hellyer was responsible for the defence portfolio.

The end of the first five-year formula-funding cycle coincided with the 1968–69 Foreign and Defence Policy Review. By this time, it was clear to senior DND officials that substantial increases in the defence budget would be required to maintain the existing defence effort. However, defence was not a high priority for the first Trudeau Government. Defence spending rose from $C1.8 bn in 1970–71 to $C1.9 bn in 1972–73 while social pro-grammes had increased priority. For DND, insufficient funding was offset by force reductions in Europe, by cutbacks in naval capabilities, and by the downgrading of continental aid defence. The government was unconcerned by the problems caused by increases in the price of equipment and the increasing obsolescence of equipment. During this period manpower levels had dropped by a further 18 089 and capital expenditures fell to 8 percent of the defence budget—an all-time low.

Trudeau's first serious attempt to come to grips with defence deficien-cies was the 1973–74 Modernization and Renewal Program, which included 7 percent annual increases for five years. This was the first formula to stipu-late priorities for the various components of the budget. Personnel cost increases were limited to 6 percent, and operation and maintenance increases to 4 percent. Priority was allocated to the capital component of the budget which was to increase to 20 percent by the end of 1977–78. This for-mula allowed for only a 5 percent inflation rate for capital equipment, which was seriously undermined once the inflationary shocks of 1974 took effect. In an attempt to cope even partially with inflationary pressures, the 1975–76 Defence Budget was increased by 18.5 percent to $C2.97 bn. Overall government spending increased more (see Table 1).

The Defence Structure Review (DSR) was initiated in December 1974. In late 1975, Phase II of the Review led to a more realistic and stable funding formula for defence. All components of the Defence Budget were indexed in full to the rate of inflation; the capital component was to increase by an additional 12 percent. Despite a substantial increase in capital investment, the capital component did not reach the target level of 12 percent real growth between 1977 and 1979. In 1978 the government agreed to imple-ment NATO's proposal for 3 percent real growth in defence spending levels through to 1983–84. The DSR formula and the 3 percent real growth target were almost identical in terms of overall effect. Thus, the defence budget became more realistic. Commencing with the 1980–81 fiscal year (FY), it experienced real growth of 1.5 percent, 2.8 percent, 3.3 percent and 6.9 per-cent through to the end of FY 1983–84.

By mid-1984, defence spending stood at $C8.77 bn, with the formula based on the following considerations: to continue to meet the NATO goal of 3 percent real growth; to maintain capital acquisitions at approximately

25 percent of the total defence budget; to increase military strength to authorized levels of 82 740; to maintain current levels of operations and maintenance; and to allocate funds for readiness and sustainability.[4] This was a significant departure from the formula of early 1970s. Liberal defence spending was projected to increase from $C9.5 bn in 1985–86 to $11.1 billion by 1987–88. Had the Liberals funded defence within this framework during the 1960s and early 1970s, current force capabilities would probably be sufficient to meet commitments.

DEFENCE PROCUREMENT

All governments which maintain modern armed forces face problems with defence procurement. The cost, complexity and technological requirements of modern weapons platforms and systems have made the weapons acquisition process lengthier, costlier and more complicated. The economic problems of the Canadian Forces have been aggravated by inadequate action by successive governments in defence procurement. Procurement has tended to be *ad hoc*, and decisions by government have lacked a clearly defined and consistent policy on defence procurement and Canada's defence industries. Since the late 1950s, *all* Canadian governments have lacked the political will and interest to develop a long-term defence procurement policy. Immediately after World War II, no extensive procurement was required as large stocks of relatively modern equipment were still available after demobilization. With the formation of NATO and the outbreak of the Korean War, the Canadian Forces were re-equipped, mainly with American equipment, and their World War II stocks were transferred to the European allies. The 1958 Diefenbaker Conservative Government was unwilling to carry through the procurement policies of the previous Liberal government, and its own policy for defence procurement contributed to the politically disastrous handling of defence issues in the early 1960s. The inability of the Conservatives to appreciate that nuclear weapons platforms required nuclear systems, coupled with the cancellation of the Avro *Arrow* fighter/interceptor and the acquisition of US aircraft and missiles, contributed to their defeat in the 1963 General Election.[5]

The Liberals did produce the 1964 White Paper which established equipment priorities, and a capital programme, including destroyers, rotary and fixed-wing aircraft, armoured personnel carriers, and self-propelled howitzers, was carried out. However, given the lack of resources, procurement was either insufficient or inappropriate to declared roles and missions (as in the case of the CF-5 fighter). With the defence freeze of the early 1970s, the Trudeau Government saw no need to articulate equipment requirements and thus it had no need for a procurement policy. Between 1976 and the election of the Mulroney Government in September 1984, DND had a 15-year re-equipment programme based on the DSR, but no overall Liberal procurement policy and strategy emerged. This can be attributed to at least five factors.

First, Trudeau governments made no attempt to link procurement policy with the defence sector. There was a general procurement strategy in the form of the Procurement Review Mechanism (PRM) of 1980 but it made no reference to the defence sector and was linked primarily to regional development and industrial benefits. The omission of the defence sector was particularly surprising, since defence capital expenditures accounted for nearly 50 percent of total federal capital expenditures.

Second, the Trudeau cabinets were unwilling to make a long-term commitment to continued real growth in the defence sector. Thus, it was difficult to give substance to DND's 15-year equipment programme in a manner that allowed for a procurement strategy to be developed which could benefit both the Canadian Forces and Canadian industry. Cabinet defence acquisition decisions were dealt with on an *ad hoc* basis.

Third, DND's approach to equipment acquisition was as much a function of force structure survival for the various components of the Canadian Forces as it was of long-term planning. Priorities were assigned almost entirely by the advanced obsolescence of equipment which required replacement simply in order to retain some semblance of operational effectiveness. Equipment was acquired sequentially with a view to maintaining a balance of capabilities across the land, sea and air environments in a situation where all components of the Canadian Forces lacked sufficient capabilities to meet commitments.

Fourth, as discussed later, there has been undue emphasis on linkages between defence procurement and industrial benefits and offsets.

Lastly, the primary objective of DND's procurement policy clashed with the primary objective of the government's PRM. According to the PRM, the primary purpose of capital acquisitions is "to obtain lasting benefits . . . beyond the immediate impact of the procurement expenditure itself, toward the economic or social development of Canada."[6] The DND Directive states that "the primary objective of DND procurement policy is to provide the materiel and facilities required for the maintenance of fighting capability in Canada's Sea, Land and Air Forces. As a first priority the Canadian Forces must be suitable and adequately equipped and trained for combat."[7] In the absence of a clearly-defined policy and strategy for defence procurement, the PRM and DND policy directives operate at cross-purposes. DND is not required to link procurement to the Canadian industrial base. In fact, since "preference normally will be given to the procurement of equipment which is proven and in production," national manufacturing capabilities for the defence sector have been discouraged.

The lack of a coherent government policy has led to sweeping criticism from some sectors of the defence industries. For example, the Aerospace Industries Association of Canada (AIAC) has argued that "the linkage between defence policy and the defence base has almost been lost." The AIAC cites the overlapping jurisdiction of various government departments with different responsibilities for defence procurement: "What we have then, appears to be an incoherent and unmanageable situation with respect to defence preparedness and its relationship to defence policy.... [Departmental] reorganizations were made for good reasons but the side effect has been a complete disintegration of any compatibility between defence and industrial policy...."[8]

DND'S RE-EQUIPMENT PROGRAMME

A consistent long-term and planned approach to equipment acquisition should be a major component of defence policy. In the absence of a structured re-equipment programme, military capabilities are bound to be affected adversely and, as the Canadian case indicates, there is a point beyond which essential roles and missions can no longer be performed adequately. Canada's commitment-capability gap is, in large part, a result of the unwillingness of governments to address seriously the re-equipment requirements of the Forces on a long-term and sustained basis. This problem can be traced to the Conservative governments of 1957–63 but was exacerbated during the Pearson and early Trudeau years.

Canada's current re-equipment programme stems primarily, but not exclusively, from Phase II of the DSR. The basic DND equipment analysis had been largely completed prior to the DSR and the 1972 decision by the government to acquire a long-range maritime patrol aircraft—between May 1980 and July 1981, the Canadian Forces took delivery of 18 CP-140 *Aurora* (modified P-3C *Orion*) aircraft—was one factor which led to the Review. Following the DSR, the government approved a number of important equipment purchases:

  i)   1976 – the purchase of 128 C-1 *Leopard* main battle tanks;

 ii)   1976/7 – the general-purpose armoured vehicle programme;

iii)   1977 – the New Fighter Aircraft programme;

 iv)   1977 – the Canadian Patrol Frigate (CPF), Phase I;

  v)   1979 – the military truck programme.

During its last year in office, the Trudeau Government approved the acquisition of new small arms, new jeeps and trucks, additional self-propelled artillery, and opened contract bids for a low-level air defence system for Canadian Forces Europe.

According to the 1986–87 defence estimates, capital expenditures of $C2.59 bn are projected (26 percent of defence expenditures). The current Defence Services Program contains some 134 major capital projects—86 equipment, 18 development and 30 construction—with total approved funding of $C16 bn. More than 95 percent of this amount has been allocated to equipment purchases. According to current estimates, expenditures to April 1987 have been forecast at $C10.2 bn, which means that $C5.8 bn have been approved against future expenditures commencing in FY 1987–88.

In terms of contracted programmes, the most costly are the New Fighter Aircraft (NFA) and Canadian Patrol Frigate (CPF). In April 1980 the government signed a contract with McDonnell Douglas to supply 138 F-18A.

According to the Estimates, the cost of this programme was $C4.9 bn, with $C3.7 bn to be spent by the end of 1985–86. This constituted approximately one-third of the capital programme. Aircraft deliveries are scheduled over a nine-year period to replace the CF-104 in Europe and the CF-101 for North American air defence. A contract for CPF Phase I of six City-class patrol frigates at an estimated cost of $C5.25 bn was signed in July 1983. The lead frigate is scheduled for delivery in early 1989 and the last in 1992. This particular expenditure represents over one-third of the current equipment programme. These programmes indicate the extent to which two major acquisitions can consume the majority of the capital budget over extended periods of time.

Despite substantial improvements in the defence budget since the mid-1970s, capital funds are insufficient to meet the 15-year re-equipment programme. The magnitude of the problem can be seen by reference to the list of unapproved and unfunded equipment projects which constitute DND's 15-year programme through to the end of the century. In 1983 DND estimated overall equipment requirements over 15 years at approximately $C55 bn (in 1983 dollars) for some 315 projects. These funds were deemed the minimum required to update current capabilities. Before its defeat, the Liberal government had approved $C28 bn in capital funding—in other words, some 51 percent by value remained in limbo.

DND'S PROCUREMENT PROCESS

The DND procurement process has its origins in the Defence Services Program (DSP) which "defines and reflects the planned and costed activities of DND for the current year and the following fifteen or so years, and the allocation of resources to these activities."[9] The process is to include a military assessment of the strategic environment and military threats, technology and other military considerations related to procurement. The Defence Program Management System (DPMS) provides the framework for decision-making and implementation of decisions arising from the DSP, including the framework for the equipment acquisition process. For each major procurement, a DND project office is established to oversee the project from start to finish.

The approval process within government includes, in addition to DND: the Department of Supply and Services for contract purposes; the Department of Regional Economic Expansion for industrial and economic benefits; the Department of External Affairs; the Treasury Board for financial approval; and Cabinet. The process, while complex, is structured in a manner which should allow for relatively rational decision-making. However a number of concerns have emerged.

First, the length of time to complete the procurement cycle has been excessive. The following time periods are illustrative of the problem: CP-140 – 12 years; CF-18 – 13 years; CPF – 17 years; truck replacement – 8 years; armoured vehicle general purpose (AVGP) – 10 years; Leopard tank – 4 years. In principle, Cabinet approval for major projects has generally required two or three years. This length of time has often been a function of the inability or unwillingness of Cabinet to reach a decision. However, there have been instances, such as the March 1985 approval of eight *Challenger* aircraft, when Cabinet has proceeded quickly, but in this instance the main concern was political—to assist Canadair. After Cabinet approval in principle, it has taken, on average, three years to contract finalization. This reflects the need to assess and negotiate industrial benefits and offset requirements more than it does the comparative assessment of military systems. It has been argued that the time taken could be shortened by 25–30 percent. This would produce financial savings, make the process more cost-effective and bring new equipment into service more quickly.

A second major area of concern has been DND's management of the procurement process. Unlike the US,[10] there has been insufficient information on how effectively DND has managed the procurement cycle. Government/DND secrecy has generally prevailed and thus criticisms have been muted or ill-informed or both. The veil of secrecy was parted partially by the 1984 Report of the Auditor General to Parliament. The Report noted that "it was difficult for DND to explain how the numbers and types of equipment selected demonstrated due regard for economy and efficiency in relation to a long-term force structure plan, or the extent to which they represented an optimum strategy for maximizing overall defence effectiveness within available resources."[11] The Report acknowledged, however, that part of the problem has been the escalation of development costs and fixed budget ceilings on capital projects.

In reply to these criticisms, C.R. Nixon, Deputy Minister of DND from 1975 to 1982, claimed that:

[I]n all of the major capital procurement items . . . there was . . . no instance where the capital budget had been adequate to procure the numbers of equipment which could have been and records show, were justified. Because of the limited budget, the numbers of equipment bought have been in all cases the minimum to retain an effective capability in the area where the equipment is applicable. Given the restrained defence budget, it's difficult to make a poor decision on resource allocation for major equipments.[12]

The cost of the NFA reflects the problems involved. The Canadian House of Commons Standing Committee on Public Accounts held hearings on the NFA programme in the spring of 1985. DND officials defended the management of the project, but the parliamentarians were concerned that the government-approved component of the CF-18 project did not contain the entire life-cycle costs. Cabinet had approved the original contract cost of $C3.1 bn in 1979–80 dollars ($C5.2 bn through to 1988) for 138 aircraft and associated equipment, including three years of logistic support, but two other components of the project were less clear in terms of approval authority. Most importantly, DND had drawn up a list of long-term associated projects estimated to cost some $C3.6 bn through to late 1990, including attrition replacement aircraft, enhanced weapons and other equipment.

Second, DND had allocated $C770 million through to FY 1987–88 for the project from the operations and maintenance (O&M) component of the defence budget.[13] The Public Accounts Committee recommended that DND "establish life cycle costs at the start of each major capital project, including the Canadian Patrol Frigate project" and "include these life cycle costs in the initial project brief to the Treasury Board."[14] DND officials concurred with the recommendation even though some felt the Auditor-General and the Public Accounts Committee were demanding a degree of perfection not found elsewhere in government.

INDUSTRIAL BENEFITS: A PROCUREMENT MILLSTONE?

Was the primary purpose of defence spending during the Trudeau years to purchase security? The evidence suggests that from the perspective of capital expenditures, defence was often *not* the primary objective. For political reasons issues related to regional economic development and, more particularly, industrial benefits all too often became as, or more, important than defence considerations. The re-equipment programme for the Canadian Forces has been closely and explicitly linked to the government's requirement that federal capital acquisitions produce significant industrial benefits.

Since 1975, all major DND equipment acquisitions have had an industrial benefits component, either in terms of offsets or Canadian content or both. The most complicated industrial benefits package involved arrangements for the CF-18 aircraft, but the contending bidders were not given specific guidance within the framework of an industrial strategy. The frigate (CPF) programme constituted a different approach as major construction and direct employment would occur in Canada, and Canadian technology for ship and system design would be utilized. In this case, both offsets and Canadian content became important factors. On the surface, the industrial benefits approach has produced the desired economic results. For example, the *Aurora* CP-140 contract price of $C950 m (in 1976 dollars) produced guaranteed offsets with Lockheed worth 96.2 percent of the contract price and a further 54.9 percent on a best-effort basis. The *Leopard* tank offset requirements constituted 40 percent of the contract price of $C187 m (1976) and a further 20 percent on a best-effort basis. For the CF-18 contract, McDonnell Douglas agreed to firm offsets totalling $C2.9 bn based on a 1977 contract price of $C2.3 bn which was valued at $C3.3 bn in terms of projected real economic activity. The benefits were expected to generate between 60–70 000 person-years of employment. For the CPF, Saint John Shipbuilding agreed to 67 percent Canadian content and firm offsets of $C763 m (31.8 percent of the $C2.4 bn 1983 contract price) which is expected to create 30 000 person-years of employment.

The 1984 Auditor General's Report and the House of Commons Public Accounts Committee expressed concern with the definition and management of benefits, including the lack of clearly-defined objectives, policies and plans. Concerns were also raised regarding the complexity of the inter-departmental arrangements and the regional distribution of the benefits since the Province of Ontario has been the main beneficiary.[15] However, neither the Auditor General nor the Parliamentary Committee questioned the wisdom and long-term effects of the industrial benefits approach.

The limited impact on regional economic development has been noted, but four other concerns are equally important. First, on balance there have only been limited transfers of technology to Canadian industry and it is unrealistic to expect that such transfers would be significant in the future. Second, defence procurement has become increasingly politicized. Due to the action of some cabinet ministers, some potential prime contractors and some provincial authorities the Federal Government has, on occasion, made non-defence compensatory resource allocations to those provinces which lose out on major defence contracts.[16] Third, the long-term considerations of countertrade must be taken into account. A 1983 report for the DEA noted that many of Canada's trading partners have demanded countertrade arrangements.[17] Canadian-based defence industries are primarily export-oriented and are not structured to meet the equipment requirements of the Canadian Forces; thus the long-term effects of Canadian offset policy may undermine Canada's defence export trade markets. Countertrade has become an irritant to Canada's trading partners, and countertrade propos-als are being increasingly linked to the sale of Canadian products abroad.

The fourth aspect of this problem has specific implications for Canadian–US economic relations and possible adverse effects on the Defence Production Sharing Agreements (DPSA) which were initially agreed upon during the Diefenbaker Government. The DPSA was envisaged primarily as a means to facilitate Canadian defence purchases by allowing Canadian industry fuller access to the US defence procurement market. In effect, it serves as the major mechanism to conduct and regulate trade in defence products between Canada and the United States. Since its inception in 1959, two-way trade in defence products has totalled some $C22.5 bn (June 1985) with a $C1.6 bn surplus in favour of the US. During the Vietnam War the trade balance favoured Canada, but with the current defence re-equipment programme a Canadian deficit has been increasing rather rapidly—for example, by $C251 m in 1983 and $C377 m in 1984. In terms of overall defence trade, the United States is by far Canada's largest trading partner; since 1959 overseas two-way defence trade has totalled only $C5.1 bn but with a Canadian surplus of $C2.7 bn.

THE CONSERVATIVE RESPONSE

To date, the response by the Mulroney Government to the wide range of economic issues which affect the military viability of the Canadian Forces has been disappointing. The government has shown no inclination to develop a clearly-stated defence procurement policy and strategy. This could (and should) be addressed in the Defence White Paper, but the gov-ernment's record when dealing with complex interdepartmental policy problems and structures leaves little room for optimism. However, the Conservative government could depart from past Liberal practice. As a first step, DND could be assigned responsibility as the leading department for the enhancement of Canada's defence industrial base. In addition, the PRM could be modified to reflect the uniqueness of defence procurement—that is, that the major purpose of defence equipment is to enhance security. Under these circumstances, it would be incumbent upon DND to alter its objectives and procedures to offer greater scope for the participation of Canadian-based industry and to structure military R&D in a manner which enhances Canadian industry. These steps would only be effective if the government were to adopt a longer-term and more substantial commitment to real growth in the defence sector than it has to date.

With respect to capital acquisitions, the Mulroney Government has continued to implement the Liberal re-equipment programme. The capital projects it has approved, such as the joint Canadian–US NWS for NORAD modernization, the Low Level Air Defence (LLAD) system for Canadian Forces Europe and the *Tribal*-class destroyer Update and Modernization Program (TRUMP), were all initiated under the Trudeau Government. The Mulroney Government has given no clear indication of its long-term approach to capital expenditures for defence. In theory, the Defence White Paper should address this issue in detail, but to date the signals received also have not been reassuring. Some Liberal-projected expenditures have been reduced. However, a number of Conservative decisions have budgetary implications—for example, up to $C80 m for new uniforms; $C50 m for the first year and then $100 m thereafter for a 1220 personnel increase for Canadian Forces Europe; and possibly up to $500 m for a Type 1500 icebreaker (Polar 8 Project). If capabilities are to be brought into line with existing commitments, the *ad hoc* approach of the Mulroney Government will have to be replaced by a long-term procurement programme, along with additional resources beyond those projected by the Liberals.

The DPSA trade trends suggest that at the political level the Canadian government should pay greater attention to the agreement reached in 1963 that defence trade between the US and Canada would be roughly in balance. This principle was reaffirmed at the March 1985 meeting between Prime Minister Mulroney and President Reagan, but officials in the DEA have not been particularly concerned with the pattern which has developed during the 1980s. Unless the trend reverts to several years of Canadian surpluses— which is unlikely without more active political intervention—Canada's overall defence trade surplus will disappear. It can be argued, of course, that the offset approach to defence procurement has helped to arrest the DPSA trend, and this appears to have been the case. However, concerns have been expressed in America that the industrial benefits objectives are primarily protectionist in nature (which they are), while the underlying principles of the DPSA are those of managed trade.

The Canadian government may be forced to reconcile the philosophical differences between the PRM and the DPSA as a result of free-trade negotiations with the US. The Mulroney Government has placed considerable emphasis on the desirability of free trade with the United States, and agreement has been reached to proceed with negotiations. Canada must be prepared to address the issue of defence trade, including American objections to the industrial benefits approach. Should the Conservatives formulate a defence procurement policy and strategy, it could remain intact in free-trade negotiations. Canadian–US defence co-operation and US interest in having Canada make a more viable contribution to the defence of the West could influence the outcome.

Irrespective of the implications for Canadian–US trade relations, the Mulroney Government could alter the industrial benefits approach adopted by the Liberals. It is an approach which has obviously added to the overall costs of defence procurement. DND pays a premium when industrial benefits become central to the acquisition process. At a minimum, non-defence premiums should be allocated from regional development funds. The major issue is the extent to which defence procurement is a function of defence as defence, as distinct from defence for industrial benefits. Since the Conservatives have declared defence to be a priority, one test of their resolve will be whether they attempt to re-formulate defence procurement objectives. To date there has been no sign that this is likely to occur.

As previously noted, the Conservatives had pledged that defence spending would be increased to "allow for replacement of obsolete equipment in sufficient quantities so that the Armed Forces can meet their obligations" and to "raise the ceiling on [regular force] manpower levels to 90 000. . . ." The first Minister of Defence in the Mulroney Government, Robert Coates, publicly proclaimed that defence spending increases could reach 6 percent real growth per annum.[18] However, the reality of the government's fiscal situation, which was inherited from the Liberals, became apparent in November 1984 when the Minister of Finance, Michael Wilson, announced budget cuts of some $C4 bn, including a $C154 m reduction in the defence budget. Thus, while the Liberal defence expenditure plan for FY 1985–86 was $C9.53 bn, the estimates tabled by the Conservatives in March 1985 only allocated $C9.37 bn for defence.

The government pointed out that the rate of inflation had declined and that this justified a modified defence expenditure plan. Nevertheless, the initial financial signals indicated that to move beyond projected Liberal funding levels would be more difficult than anticipated. According to the Prime Minister, the Conservative government was faced, with a "bare cupboard." Overall government spending had to be brought under control and the national debt had to be reduced. By the spring of 1986, the Mulroney Cabinet decided that, given the adverse fiscal situation, the Liberal defence spending projections were unrealistic. Hence the budget of 26 February 1986 estimated defence expenditures of $C9.86 bn for FY 1986–87 and $C10.48 bn for FY 1987–88. According to the government, this constituted a reduction of 1 percent real growth from their earlier projections, but would still allow for 2.75 percent and 2.0 percent real growth after inflation. For planning purposes, the Conservatives set defence increases at 2 percent real growth from 1987–88 and subsequent years.[19] Obviously, the commitment-capability gap is not about to be breached by higher defence spending.

The existing financial situation has been compounded further by the Conservative government's decision to abide by the 1984 Liberal decision to employ the Gross National Expenditure (GNE) deflator (which reflects price increases in the economy as a whole) rather than the DND deflator (which reflects price increases in the defence sector) for purposes of calculating the rate of inflation. Before the 1982–83 budget year, the DND deflator was used for this purpose. Since inflation has been higher in the defence sector, the shift to the GNE deflator conveyed the impression that real growth was higher than was actually the case.[20] In addition, devaluation of the Canadian dollar against foreign currencies has adversely affected real growth since nearly one-fifth of the defence budget is allocated for off-shore purchases. Consequently, the Conservative real growth projections, in all probability, will be lower than announced. Over time, these factors could have a substantial adverse affect on defence spending.

To date, the Conservative approach to defence has been remarkably similar to that of the previous Liberal government: to resolve resource allocation issues first and then to address the substance of defence. Yet, if the Mulroney Government intends to upgrade Canada's military capabilities and to close the commitment-capability gap, alternative approaches for the allocation of resources must be considered. Formula-funding is too easily modified on grounds of political or economic expediency. It has also contributed to the lack of long-range planning, especially for capital procurement, and allowed for commitment-capability issues to be ignored or played down by elected politicians, including cabinet ministers.

Given the historical record of formula-funding, the Mulroney Government might serve Canada's defence interests best if it were to base the defence budget on a commitment-capability funding model based on the following considerations: that defence objectives and defence commitments be identified and articulated in specific terms; that force structure requirements be clearly identified in terms of declared commitments; that equipment requirements be established in relation to declared commitments; that personnel and infrastructure requirements be identified; and that funding allocations be based on a detailed assessment of the above factors and authorized on a multi-year basis. This approach would allow Cabinet to assess whether defence resources are appropriate within a military-strategic framework. Should funding levels be excessive on either political or economic grounds, the government could modify commitments without adversely affecting the relationship between commitments and capabilities. . . .

NOTES

1. This Chapter draws on the author's study "Canadian Defence and Defence Procurement: Implications for Economic Policy" for the Royal Commission on the Economic Union and Development Prospects for Canada (Macdonald Commission), published in Denis Stairs and Gilbert R. Winham, *Selected Problems in Formulating Foreign Economic Policy* (Toronto: University of Toronto Press, 1985), 131–95.

2. For an analysis of defence spending during the Trudeau years, see Danford W. Middlemiss, "Department of National Defence," in *Spending Tax Dollars: Federal Expenditures, 1980–81*, ed. G. Bruce Doern (Ottawa: Carleton University, 1980), 75–98. Also see Brian Macdonald, ed., *Guns and Butter: Defence and the Canadian Economy* (Toronto: CISS, 1984).

3. Gerald Porter, *In Retreat: The Canadian Forces in the Trudeau Years* (Ottawa: Deneau and Greenberg, 1978).

4. Department of National Defence (hereafter DND), *1984–85 Estimates*, Part 3, Expenditure Plan (Ottawa: MSSC, 1984).

5. John B. McLin, *Canada's Changing Defence Policy, 1957–1963* (Baltimore: Johns Hopkins University Press, 1967) and Peyton V. Lyon, *Canada in World Affairs, 1961–1963*, vol. 12, Canadian Institute for International Affairs (hereafter CIIA) (Toronto: Oxford University Press, 1968).

6. Treasury Board Canada, *Administrative Policy Manual*, Procurement Review, Chapter 305 (Ottawa: MSSC, March 1980), 5.

7. DND Procurement Policy, NDHQ Policy Directive P. 16 (Issue 2), September 1981.

8. Senate Special Committee on National Defence, Minutes of Proceedings and Evidence, 14 March 1984, 4A:13.

9. DND, *An Introduction to the Defence Services Program* (Ottawa, 1981), 1–2.

10. For example, Franklin C. Spinney, *Defense Facts of Life: The Plans/Reality Mismatch* (Boulder, CO: Westview Press, 1985).

11. Auditor General of Canada, *Report of the Auditor General of Canada to the House of Commons: Fiscal Year Ended 31 March 1984* (Ottawa: MSSC, December 1984), 2–22.

12. C.R. Nixon to Kenneth M. Dye, 21 January 1985.

13. House of Commons Standing Committee on Public Accounts, Minutes of Proceedings and Evidence, Issues no. 10, 28 February 1985, no. 11, 5 March 1985, and no. 12, 19 March 1985.

14. Ibid., Third Report to the House (Comprehensive Audit of the Department of National Defence), Issue no. 15, 30 April 1985, 15:5.

15. Committee on Public Accounts, *Third Report to the House*, 15:3.

16. See note 1.

17. DEA, *A Review of Canadian Trade Policy* (Ottawa: MSSC, 1983), 57–58.

18. *Ottawa Citizen*, 18 October 1984, A3.

19. DND, News Release, AFN: 7/86, 27 February 1986.

20. *Financial Post*, 7 December 1985, 45.

FURTHER READINGS

○

This listing supplements sources found in the chapter endnotes. Owen A. Cooke's reference guide, *The Canadian Military Experience: A Bibliography* (1984), includes literature on Canadian defence policy. The varied historiography of both defence and general military policy is treated in R.G. Haycock, "Mars and Clio in Canada: Historical and Contemporary Dimensions of Military History," in the Centenary Issue: "Military History Around the World," of the *Journal of Military History*, Royal Netherlands Army, 1991. For other works on defence policy, see J.M. Hitsman's chapter in Robin Higham, ed., *British Military History: A Guide to Sources* (1969), Haycock's chapter in the revised edition (1988), and the bibliography in Middlemiss and Sokolsky, *Decisions and Determinants*, cited below.

This list would be best used in conjunction with those found in two other readers in the New Canadian Readings Series: J.L. Granatstein, *Toward a New World: Readings in the History of Canadian Foreign Policy* and Norman Hillmer, *Partners Nevertheless: Canadian–American Relations in the Twentieth Century*.

SELECTED WORKS

Bland, Douglas. *The Administration of Defence Policy in Canada, 1947–85*. 1987.

Boutilier, James A., ed. *The RCN in Retrospect, 1910–1968*. 1982.

Cuthbertson, Brian. *Canadian Military Independence in the Age of the Superpowers*. 1977.

Douglas, W.A.B., ed. *The RCN in Transition, 1910–1985*. 1988.

———. *The Creation of the National Air Force: The Official History of the RCAF*, Vol. 2. 1986.

Eayrs, James. *In Defence of Canada*. 5 vols. 1964–83.

Granatstein, J.L., and Norman Hillmer. *For Better or for Worse: A History of Canadian–American Relations*. Toronto, 1991.

Granatstein, J.L., and J.M. Hitsman. *Broken Promises: A History of Conscription in Canada*. 1977.

Gray, Colin S. *Canadian Defence Priorities: A Question of Relevance*. 1972.

Haglund, David, ed. *Canada's Defence Industrial Base: The Political Economy of Preparedness and Procurement*. 1988.

Harris, Stephen. *Canadian Brass: The Making of a Professional Army, 1860–1939*. 1988.

Holmes, John W. *The Shaping of Peace: Canada and the Search for World Order, 1943–57*. 2 vols. 1979–82.

Middlemiss, D.W., and J.J. Sokolsky. *Canadian Defence: Decisions and Determinants*. 1989.

Morton, Desmond. *A Military History of Canada*. 1985.

——. *Ministers and Generals: Politics and the Canadian Militia, 1868–1904*. 1970.

Peden, Murray. *Fall of an Arrow*. 1987.

Preston, R.A. *The Defence of the Undefended Border: Planning for War in North America, 1867–1939*. 1977.

——. *Canada and Imperial Defence: A Study of the British Commonwealth's Defence Organization, 1867–1919*. 1967.

Ross, Douglas A. *In the Interests of Peace: Canada and Vietnam, 1954–1973*. 1984.

Sokolsky, J.J., and J.T. Jockel. *Fifty Years of Canada–United States Defence Cooperation: The Road from Ogdensburg*. 1992.

Stacey, C.P. *Canada and the Age of Conflict*. 2 vols. 1977; reprint, 1981.

——. *Arms, Men and Government: The War Policies of Canada, 1939–45*. 1970.

——. *The Military Problems of Canada: A Survey of Defence Policies and Strategic Conditions Past and Present*. 1940.

Taylor, Alastair, et al. *Peacekeeping: International Challenge and Canadian Response*. 1968.

Wise, S.F. *Canadian Airmen and the First World War: The Official History of the Royal Canadian Air Force*, Vol. 1. 1980.

Wiseman, Henry, ed. *Peacekeeping: Appraisals and Proposals*. 1983.

An honest attempt has been made to secure permission for all material used, and if there are errors or omissions, these are wholly unintentional and the Publisher will be grateful to learn of them.

Carman Miller, *"Sir Frederick William Borden and Military Reform, 1896–1911,"* Canadian Historical Review 50, 3 (September 1969), 265–84. *Reprinted by permission of University of Toronto Press.*

Barry Morton Gough, *"The End of Pax Britannica and the Origins of the Royal Canadian Navy: Shifting Strategic Demands of an Empire at Sea,"* is taken from The RCN in Transition 1910–1985, edited by W.A.B. Douglas (Vancouver: UBC Press, 1988). *Copyright University of British Columbia Press. All rights reserved. Reprinted with the permission of UBC Press.*

Desmond Morton, *"'Junior But Sovereign Allies': The Transformation of the Canadian Expeditionary Force, 1914–1918." Reprinted by permission from the eighth issue number one of* Imperial and Commonwealth History *published by Frank Cass & Company Limited, 11 Gainsborough Road, London E11, England. Copyright Frank Cass & Co. Ltd.*

Stephen Harris, *"The Canadian General Staff and the Higher Organization of Defence, 1919–1939,"* War and Society 3, 1 (May 1985), 83–98. *Reprinted with the permission of the journal.*

Norman Hillmer, *"Defence and Ideology: The Anglo-Canadian Military 'Alliance' in the 1930s,"* International Journal 33, 3 (Summer 1978), 588–612. *Reprinted with the permission of the journal.*

Adrian W. Preston, *"Canada and the Higher Direction of the Second World War, 1939–1945,"* RUSI Journal 110, 637 (February 1965), 28–44. *Reprinted with the permission of the Royal United Services Institute for Defence Studies.*

Robert Bothwell, *"'Who's Paying For Anything These Days?' War Production in Canada, 1939–1945,"* Mobilization for Total War: The Canadian, American and British Experience 1914–1918, 1939–1945, ed. N.F. Dreisziger (Waterloo: Wilfrid Laurier University Press, 1981), 59–69. *Reprinted with the permission of the publisher.*

J.L. Granatstein, *"The American Influence on the Canadian Military, 1939–1963,"* Canadian Military History 2 (1993). *Reprinted with the permission of the author.*

Joel J. Sokolsky, *"A Seat at the Table: Canada and its Alliances,"* Armed Forces and Society 16, 1 (Fall 1989), 11–35. *Reprinted with the permission of Transaction Publishers.*

Joseph T. Jockel, *"The Military Establishments and the Creation of Norad,"* American Review for Canadian Studies 12, 3 (Fall 1982), 1–16. *Reprinted with the permission of The American Review for Canadian Studies.*

Rod B. Byers, *"Peacekeeping and Canadian Defence Policy: Ambivalence and Uncertainty,"* Peacekeeping: Appraisals and Proposals, ed. H. Wiseman (Toronto: Pergamon Press, 1983), 130–56.

W. Harriet Critchley, *"The Arctic,"* International Journal 42 (Autumn 1987), 769–88. *Reprinted with the permission of the journal.*

Douglas L. Bland, *"Controlling the Defence Policy Process in Canada: White Papers on Defence and Bureaucratic Politics in the Department of National Defence,"* Defence Analysis 5, 1 (1989), 3–17.

W. Harriet Critchley, *"Civilianization and the Canadian Military,"* Armed Forces and Society 16, 1 (Fall 1989), 117–36. *Reprinted with the permission of Transaction Publishers.*

Dan Middlemiss, *"Canada and Defence Industrial Preparedness: A Return to Basics?"* International Journal 42 (Autumn 1987), 707–30. *Reprinted with the permission of the journal.*

Rod Byers, *"The Economics of Defence,"* from Rod Byers, Canadian Security and Defence: The Legacy and the Challenges, International Institute of Strategic Studies, Adelphi Papers, 214 (Winter 1986), 31–44. *Reprinted with the permission of The International Institute for Strategic Studies.*